新工科建设之路·软件工程系列教材

Android Studio 应用开发
——基础入门与应用实战

方　欣　杨　勃　主　编

胡青萍　徐剑波　副主编

電子工業出版社

Publishing House of Electronics Industry

北京·BEIJING

内 容 简 介

本书从初学者的角度出发，对 Android Studio 开发环境搭建及 Android 应用程序开发进行了介绍。全书共 10 章，主要内容包括：Android 概述、Android 中的项目、Android 常用基本组件、Android 中的事件处理、Android 常用高级组件、Android 组件之前的通信、Android 多媒体技术、Android 数据存储技术、Android 网络通信技术，最后介绍一个具体的投票系统 APP 端设计，将前面的知识贯穿。由浅入深、循序渐进，将理论知识和实例紧密结合进行介绍、剖析和实现，加深学生对 Android 基础知识和基本应用的理解，帮助学生系统全面地掌握 Android 程序设计的基本思想和基本应用技术，快速提高开发技能，为进一步深入学习 Android 应用开发打下坚实的基础。

本书的配套资源包括教学课件和程序源代码等，读者可以通过华信教育资源网（http://www.hxedu.com.cn）注册免费下载。

本书可作为本科计算机科学与技术、计算机网络、信息工程、电子信息等专业的程序设计课程的教材，也可以作为 Android 程序设计技术的培训教材，同时可供自学者及从事计算机应用工程技术人员参考。

图书在版编目（CIP）数据

Android Studio 应用开发：基础入门与应用实战 / 方欣，杨勃主编. —北京：电子工业出版社，2017.8

ISBN 978-7-121-32220-4

Ⅰ. ①A… Ⅱ. ①方… ②杨… Ⅲ. ①移动终端—应用程序—程序设计—高等学校—教材 Ⅳ. ①TN929.53

中国版本图书馆 CIP 数据核字（2017）第 167681 号

策划编辑：戴晨辰
责任编辑：裴　杰
印　　刷：北京盛通数码印刷有限公司
装　　订：北京盛通数码印刷有限公司
出版发行：电子工业出版社
　　　　　北京市海淀区万寿路 173 信箱　邮编　100036
开　　本：787×1 092　1/16　印张：19.75　字数：550 千字
版　　次：2017 年 8 月第 1 版
印　　次：2024 年 1 月第 11 次印刷
定　　价：59.80 元

凡所购买电子工业出版社图书有缺损问题，请向购买书店调换。若书店售缺，请与本社发行部联系，联系及邮购电话：（010）88254888，88258888。

质量投诉请发邮件至 zlts@phei.com.cn，盗版侵权举报请发邮件至 dbqq@phei.com.cn。

本书咨询联系方式：dcc@phei.com.cn。

前言

随着移动通信与 Internet 向移动终端的普及，网络和用户对移动终端的要求越来越高，Google 为此于 2007 年 11 月推出了一个专为移动终端设备设计的软件平台——Android。由于它开源以及使用 Java 作为开发语言的特点，受到越来越多程序设计人员的青睐，支持的厂商也在不断增加。目前，在市面上的几大手机操作系统中，Android 的市场占有率最高，上升速度最快，具有很大的市场发展潜力。

2016 年 9 月，Google 发布了 Android Studio 2.2 的开发环境，同年 11 月宣布正式终止了对 Eclipse+ADT 开发工具的支持，因此目前开发环境以 Android Studio 为主。

本书于 2016 年 12 月份开始筹备，至 2017 年 6 月份编写完成，在此期间不断与外界公司、企业沟通，了解它们的需求，整个教材的编写充分结合软件企业的用人需求，经过了充分的调研和论证，具有系统性、实用性等特点。目的是让尽量多的开发者少走弯路，从而尽快掌握基础知识，创造出更多、更好的 Android 应用程序，满足用人单位的需要。

本书从初学者的角度出发，通过通俗易懂的语言、丰富多彩的实例、关键代码的分析，详细介绍了 Android 基础知识以及进行 Android 项目开发应该掌握的基本应用技术，全书共 10 章，主要内容包括：Android 概述、Android 中的项目、Android 常用基本组件、Android 中的事件处理、Android 常用高级组件、Android 组件之间的通信、Android 多媒体技术、Android 数据存储技术、Android 网络通信技术，最后介绍一个具体的投票系统 APP 端设计，将前面的知识贯穿。由浅入深、循序渐进，将理论知识和实例紧密结合进行介绍、剖析和实现，加深学生对 Android 基础知识和基本应用的理解，帮助学生系统全面地掌握 Android 程序设计的基本思想和基本应用技术，快速提高开发技能，为进一步深入学习 Android 应用开发打下坚实的基础。

本书是编者多年来教学和软件开发经验的总结，编者对书中的内容进行了精心设计和安排，力求达到内容丰富、结构清晰；书中给出的实例简单实用，易于教学和读者自学。通过阅读本书，并结合上机实验就能在较短的时间内基本掌握 Android 项目开发的基本技能。本书除了纸质内容之外，还为教师配备了教学课件，附带了书中给出的 118 个实例源代码，所有源代码都经过反复调试，在 Android 开发平台能直接导入运行。

本书适用面广，可作为本科计算机科学与技术、计算机网络、信息工程、电子信息等专业的程序设计课程的教材，也可作为 Android 程序设计技术的培训教材，还可供自学者及从事计算机应用的工程技术人员参考使用。

本书要求读者最好具有一定的 Java 语言基础，具有面向对象基础和其他 GUI 设计经验的人员也可以学习本书。

本书的配套资源包括教学课件和程序源代码等，读者可以通过华信教育资源网（http://www.hxedu.com.cn）注册后进行免费下载。

全书由方欣、杨勃、胡青萍、徐剑波老师编写，其中第4、7、8、9章由方欣编写，第3、5、6章由杨勃编写，第1、2章由胡青萍编写，第10章由徐剑波编写。廖艳等人对本书做了一些图表的绘制、校对和纠错等工作，李煌峰等对本书中的代码进行了调试。

本书的编写得到了"受复杂系统优化与控制湖南省普通高等学校重点实验室"的资助。

本书的编者大都来自教学一线，在工作之余完成本书。虽然编者的目标是编写一本优秀的教材，但是由于水平有限，为了使本书尽早和读者见面，编写过程难免略显仓促，虽然经过审校，书中可能依然存在一些不足之处，敬请读者和同行专家批评指正。

编　者

目　录

第 1 章　Android 概述

学习目标：

- 了解智能手机的发展史及常见的手机操作系统。
- 了解 Android 操作系统的发展及其特点。
- 搭建 Android 系统开发环境。
- 利用 Android Studio 开发第一个 Android 程序。
- 了解 Android APK 封装过程。

随着人们生活水平的提高，手机已经逐渐从奢侈品发展成为十分普及的电子产品。经过多次技术变革，手机不再仅仅是一个语音通信工具，它已经成为具有独立操作系统的智能设备。

1.1　智能手机的发展

1．智能手机的定义

智能手机（Smartphone）是指像个人电脑一样，具有独立的操作系统，可以由用户自行安装软件、游戏等第三方服务商提供的程序，通过此类程序对手机的功能进行扩充，并可以通过移动通信网络实现无线网络接入的手机的总称。

2．智能手机的发展

1973 年 4 月 3 日，摩托罗拉公司前高管马蒂·库珀在曼哈顿的实验网络上测试了他的一部电话，他把电话打给了贝尔实验室的一名科学家，这是世界上公认的第一台手机，马蒂·库珀也被称为“现代手机之父”。

随着时间的推移，手机功能也在不断扩充，除了打电话之外，还具备了 PC 机的功能，例如，玩游戏、收发电子邮件及网页浏览等，这就是所说的智能手机。

全球首款智能手机是美国 IBM 公司在 1994 年投放市场的“IBM Simon”，这款手机配备了使用手写笔的触摸屏，除了通话功能之外，还具备 PDA 及游戏功能，操作系统采用的是夏普 PDA 的“Zaurus OS”。

1996 年，芬兰诺基亚公司推出了名为“Nokia 9000 Communicator”的折叠式智能手机。Nokia 9000 Communicator 受到了商务人士的青睐，后来逐步演变为 1998 年上市的“诺基亚 9110”和“诺基亚 9110i”，继而又推出了采用 Symbian 系统的机型。1997 年，瑞典爱立信公司推出了与 Nokia 9000 Communicator 相似的“GS88”手机，该手机的说明书中首次出现了“智能手机”一词。

进入 2000 年以后，市场上出现了很多采用面向 PDA 及嵌入设备的通用操作系统的智能手机。这些手机使用 Symbian、Palm OS 或 Windows CE 等操作系统。

首次采用 Symbian 操作系统的智能手机是爱立信“Ericsson R380 Smartphone”。之后，诺基亚公司也于 2000 年投放了采用 Symbian 操作系统的智能手机（后来诺基亚的智能手机便一直使用 Symbian 操作系统），Symbian 操作系统一度成为占据市场主导地位的手机操作系统。

2001 年 2 月配备 Palm 操作系统的手机“Kyocera 6035”上市。

美国微软公司于 2002 年发布了“Microsoft Windows Powered Smartphone 2002”，该手机配备的是 Windows CE 智能手机系统，后来更名为“Windows Mobile”，韩国三星电子及夏普等公司向市场投放了多款采用这种操作系统的智能手机。

加拿大 RIM（Research In Motion）公司于 2003 年推出了首款“黑莓”（BlackBerry）手机。该手机融合了电子邮件、SMS 及 Web 浏览等功能。

以上这些手机均以企业用户为目标，以嵌入商务软件的形式提供，基本未向普通消费者推广。掀起让普通消费者购买并使用智能手机潮流的是美国苹果公司于 2007 年 6 月投放市场的 iPhone。这款手机配备有以触摸屏完成的用户界面（UI）、基本与个人电脑等同的 Web 浏览器和电子邮件功能，以及与 iTunes 软件联动的音乐播放软件等，从而将智能手机提高到了任何人都能使用的水平。

随后，美国谷歌公司于 2007 年 11 月发布了智能手机软件平台 Android 系统。2008 年，美国 T-Mobile USA 公司推出了首款配备 Android 系统的智能手机——T-Mobile G1。此后，美国摩托罗拉移动公司、三星电子，以及日本与瑞典的合资公司索尼爱立信移动通信等公司都相继推出了基于 Android 系统的智能手机。

微软公司在 iPhone 与 Android 成功之后也转变了市场方针，于 2009 年 2 月宣布开发面向普通消费者的“Windows Mobile 6.5”及“Windows Phone 7”。采用 Windows Mobile 6.5 系统的手机于 2009 年 10 月投放市场，Windows Phone 7 手机则于 2010 年 10 月问世。

2011 年后，“双核”智能手机推出。摩托罗拉公司、LG 公司以及三星公司发布了采用双核处理器的智能手机产品，而 HTC 公司发布的双核处理器智能手机主频更是已经高达 1.2GHz。智能手机的硬件发展进入了一个新的阶段。

未来的手机将偏重于安全和数据通信：一方面加强个人隐私的保护，另一方面加强数据业务的研发，各种多媒体功能将被引入进来，手机将会具有更加强劲的运算能力，成为个人的信息终端，而不是仅仅具有通话和文字消息的功能。

3．智能手机与 4G

4G（Fourth Generation）指的是第四代移动通信技术，也是 3G 的延伸。

相对于第一代模拟制式手机（1G）、第二代 GSM 和 TDMA 等数字手机（2G）、第三代手机（3G 指支持高速数据传输的蜂窝移动通信技术，速率一般在几百千位每秒以上），4G 集 3G 与 WLAN 于一体，并能够传输高质量视频图像，它的图像传输质量与高清晰度电视不相上下。4G 系统能够以 10Mb/s 的速度下载，比拨号上网快 200 倍，上传的速度也能达到 5Mb/s，并能够满足几乎所有用户对于无线服务的要求。此外，4G 可以在 DSL 和有线电视调制解调器没有覆盖的地方部署，然后扩展到整个地区。

5G 也称第五代移动通信技术，是 4G 之后的延伸，速度有望提升 100 倍，正在研究部署中。

1.2 智能手机操作系统

智能手机就是安装了某个操作系统的手机，能够安装在手机上的操作系统有：Android、iOS、Windows Mobile、Symbian、BlackBerry、Palm 等。

1．Android

Android（中文名：安卓）系统是由 Google 公司推出的基于 Linux 平台的开源手机操作系统，由于其开源以及使用 Java 作为开发语言的特点，越来越受到青睐，支持的硬件厂商也越来越多。目前，在市面上的手机操作系统中，Android 系统的市场占有率最高，上升速度最快。

2. iOS

iOS（iPhone OS 的简称）是由苹果公司为 iPhone 开发的基于 Mac 环境的操作系统，采用 Objective-C 为主要开发语言，主要用于 iPhone、iPod Touch 以及 iPad 等终端设备。iOS 支持多点触控，能给用户提供全新的体验，目前只能应用于苹果公司的设备上。

3. Windows Phone 7

Windows Phone 7（前身为 Windows Mobile）是 Microsoft 公司为移动设备推出的 Windows 操作系统，该系统有很多先天的优势，有庞大的用户群，但是由于硬件要求极高，导致终端设备价格也高，在一定程度上限制了它的发展。

4. Symbian

Symbian 是一个实时、多任务的 32 位操作系统，具有功耗低、内存占用少等特点，非常适合手机等移动设备使用。Symbian 操作系统曾经是市场占有率最高的手机操作系统，随着越来越多手机操作系统的出现，尤其是 Android 系统的出现，Symbian 系统的发展遇到了瓶颈，被迫于 2010 年 2 月进行开源。

5. BlackBerry

BlackBerry 是 RIM 公司开发的手机操作系统，此系统曾经显赫一时，现在由于面临着 Android 和 iOS 两大阵营的冲击，其用户群在逐渐减少。

6. Palm

Palm 操作系统是 Palm 公司推出的 32 位嵌入式操作系统，早期主要应用于掌上电脑，该公司于 2010 年被惠普收购，惠普公司在 Palm 系统的基础上推出了 Web OS，现在成为惠普平板电脑上的操作系统。

7. Bada

Bada 是韩国三星公司自主研发的智能手机平台，支持 Flash 界面，对于 SNS 应用有着很好的支持，于 2009 年 11 月 10 日发布。

1.3 Android 操作系统

1.3.1 Android 操作系统的发展

Android 一词最早出现于法国作家利尔亚当在 1886 年发表的科幻小说《未来的夏娃》中，他将聪明美丽的机器人女孩起名为 Android。

美国 Google 公司早在 2002 年就进入了移动通信领域，可是由于手机操作系统企业和手机企业相对封闭，提高了行业的进入门槛，谷歌的目标是将传统互联网和移动互联网进行融合，但没有合适的手机系统作为合作伙伴。

Android 公司由安迪·鲁宾创办，谷歌公司于 2005 年收购了这个公司，安迪·鲁宾继续负责 Android 项目的研发工作。

2007 年 11 月 5 日，谷歌公司正式向外展示了 Android 1.0 操作系统，提供了基础的智能手机功能：闹钟、API 示例、浏览器、计算器、摄像头、联系人、开发工具包、拨号应用、电子邮件、地图（包含街景）、信息服务、音乐、图片、设置等。

该系统发布之后不久就有一款装有 Android 1.0 系统的手机 T-Mobile G1 问世，手机由运营商

T-Mobile 定制，中国台湾 HTC 公司代工制造。T-Mobile G1 是世界上第一款使用 Android 操作系统的手机，手机的全名为“HTC Dream”。

2009 年 4 月，谷歌正式推出了基于 Android 1.5 系统的手机，加入了输入法框架支持、视频录像等功能。同年 9 月，谷歌发布了 Android 1.6 系统，并且推出了装载 Android 1.6 正式版的手机——HTC Hero G3，凭借出色的外观设计以及全新的 Android 1.6 操作系统，HTC Hero G3 成为当时全球最受欢迎的手机。

2009 年 10 月，谷歌发布了 Android 2.0 操作系统，改进了桌面主题、联系人管理，完善了蓝牙通信以及对 OpenGL ES 2.0 的支持，新增了多点触控的支持。Android 2.0 版本的代表机型为 NEXUS One，这款手机为谷歌旗下第一款自主品牌手机，由 HTC 代工生产，NEXUS One 手机于 2010 年 1 月正式发售。

2010 年 5 月，谷歌正式发布了 Android 2.2 操作系统，支持应用安装到 SD 卡上，运行效率有了大幅的提升，支持更大内存，开始支持 Flash 播放器和 FLV 视频媒体解码。采用 Android 2.2 操作系统的手机比较出众的有 HTC Desire HD、三星的 GALAXY S。

2010 年 12 月，谷歌正式发布了 Android 2.3 操作系统，在多媒体库方面有了大幅的改变，同时引入了近距离数据通信协议的支持。Android 2.3 代表机型有 GALAXY SⅡ、HTC Sensation 等。

2011 年 2 月 3 日，谷歌发布了专用于平板电脑的 Android 3.0 系统，对大屏幕高分辨率的平板电脑进行了界面的优化，同时支持多核 CPU、高性能 2D 和 3D 图形性能，在娱乐方面有了大幅的增强，同时全新的开发附件协议，使其在 USB 外设上有了大幅的支持，这是首个基于 Android 的平板电脑专用操作系统。

2011 年 5 月 11 日，Google 发布了 Android 3.1 操作系统，部分功能做了小幅改进，在虚拟键盘等方面有了小幅的变化。新版本最大的改变是将 Android 手机系统和平板系统再次合并，方便了开发者。

2011 年 7 月 13 日 Google 发布了 Android 3.2 操作系统，对 7 英寸的屏幕在 1024×600 分辨率的设备进行了界面的优化，解决了早期系统仅支持 10.1 英寸大平板的问题。

2011 年 10 月 19 日在中国香港发布了 Android 4.0 操作系统，最明显的是 Android 4.0 界面 UI 做了重新设计。在系统性能方面也做了大幅改进，同时适用于手机和平板系统。

2012 年 6 月 28 日发布了 Android 4.1 操作系统，它使系统变得更快、更流畅，优化了系统操作体验，增加了包括 Google Now 和更丰富的通知中心在内的很多新功能。

2012 年 10 月 30 日发布了 Android 4.2 操作系统，增强了 Google Now 功能，增加了对航班信息查询、酒店和餐厅预订、电影和音乐推荐的支持，并且平板用户还能自由切换账户。

2013 年 7 月 25 日谷歌发布了 Android 4.3 操作系统，新增了用户账户配置，可以在拨号盘中输入号码和人名时自动搜索联系人，WiFi 关闭后保持定位功能。

2013 年 9 月 4 日谷歌发布了 Android 4.4 操作系统，支持两种编译模式，针对 RAM 占用进行了优化，配色更加简约，图标风格进一步扁平化，整体来说页面更漂亮，占用资源更少。

2014 年 10 月 15 日谷歌发布了 Android 5.0 操作系统，使用全新的 Material Design 设计风格，更好地应用了语音搜索功能，优化了面部解锁，改进了快速设置界面，以及改善了多任务视窗。

2015 年 9 月 30 日谷歌发布了 Android 6.0 操作系统，整体设计风格依然保持扁平化的 Material Design 风格，在对软件体验与运行性能上进行了大幅度的优化。

2016 年 8 月 22 日谷歌发布了 Android 7.0 操作系统，为提供统一的用户体验，加入了 3D Touch 功能，原创支持应用分屏，更改了下拉通知栏中控制中心的样式使其更合理，可调节字体和图标大小，在系统层面对移动 VR 做出了配合。2017 年即将推出的版本是 Android 8.0 操作系统。

现在，Android 系统不但应用于智能手机，而且延伸到其他便携式和嵌入式设备（平板电脑、电子书、上网本、高清电视等）。支持 Android 系统的主要厂商包括 HTC、三星、摩托罗拉、华为、中兴、联想、小米、LG、戴尔、宏碁、华硕、海信等。

开放手机联盟（Open Handset Alliance）是 Google 公司于 2007 年 11 月 5 日宣布组建的一个全球性的联盟组织。这一联盟支持 Google 发布的 Android 手机操作系统或者应用软件，共同开发名为 Android 的开放源代码的移动系统。开放手机联盟包括手机制造商、手机芯片厂商和移动运营商等。目前，联盟成员数量众多，这也是 Android 迅猛发展的一个原因。

1.3.2 Android 操作系统的特点

Android 系统是基于 Linux 开放性内核的操作系统，具有如下特点。

（1）开放性，Android 平台允许任何移动终端厂商加入 Android 联盟。开放性可以使其拥有更多的开发者，专业人士可以利用开放的源代码进行二次开发，打造出个性化的 Android 系统。开放性可以缩短开发周期，降低开发成本，也有利于 Android 的发展。

（2）应用程序无界限，Android 系统上的应用程序可以通过标准 API 访问核心移动设备功能。

（3）应用程序是在平等条件下创建的，移动设备上的应用程序可以被替换或扩展。

（4）应用程序可以轻松地嵌入网络。应用程序可以轻松地嵌入 HTML、JavaScript 和样式表，还可以通过 Web View 控件显示网络内容。

（5）应用程序可以并行运行。Android 系统是多任务环境，应用程序可以并行运行。

Android 操作系统的缺陷如下。

1）安全问题

由于 Android 系统的开源和快速发展以及应用程序审核机制的不完善等原因，导致 Android 程序应用方面出现了一些恶意软件。2009 年 11 月 10 日 Android 平台出现了第一个恶意间谍软件——Mobile Spy。2010 年 8 月 12 日，出现了第一个木马病毒——Trojan-SMS.Android OS.FakePlayer.a。在这些恶意软件的影响下，用户的隐私在不经意间就可能泄露，隐私不能得到充分的保障。因此，2011 年 11 月 20 日，Google 公司宣布启动 Android 应用审核、取缔、清扫行动，定期对电子市场中的不合格、低质量、违法、恶意程序进行清理。

2）稳定性问题

由于 Android 系统的开源，各个厂商都能对代码进行二次开发，由于开发水平的原因，一些厂商开发出来的应用程序可能会造成系统崩溃等后果。

3）必须用高配置弥补系统上的缺陷

Android 的 UI 渲染遵循传统电脑模式的主线程普通优先级，当触摸 Android 手机屏幕的时候，系统后台的程序并没有停止，仍然在继续运行之中，这就是 Android 系统不流畅的原因之一。Android 系统缺乏有效的硬件加速也是一个原因，在不同的 Android 手机上的硬件加速存在巨大差异。

1.3.3 Android 操作系统与 iOS 操作系统的对比

1）流畅性

从流畅性来讲，iOS 系统更具有优势。Android 系统采用了虚拟机的运行机制，这样的运行机制需要消耗更多的系统资源。Android 系统使用一段时间后就会变得卡顿，而 iOS 系统几乎不会出现卡顿的现象，并且 Android 系统的桌面滑动的灵敏性不如 iOS 系统。

2）性价比

从性价比的角度来讲，Android 系统要优于 iOS 系统。苹果公司对 iOS 系统拥有专利，如果其他手机生产厂商想使用 iOS 系统，就会收费。Android 系统是 Google 公司提供的免费、开源的系统，并且 Android 比 iOS 开放了更多的应用接口，可以很方便地实现各种功能。

3）系统稳定性

虽然 iOS 更稳定不易死机，但一旦出现死机情况不能通过拆电池来重启。一般的 Android 手机死机后可以直接通过拆电池来重启。一般而言，iOS 系统的界面比较单一，而 Android 的界面可以根据自己的喜好来设置，比较多样化。

4）安全角度

iOS 系统相对比较安全，因为苹果公司会为 iOS 系统添加功能之类的操作，会开发并测试很长时间。Android 系统要求较低，能运行起来即可，开发测试时间较短，漏洞也比较多，极端情况下只需简单发送一条彩信便能在用户毫不知情的情况下完全控制手机。

5）后台执行程序

iOS 系统根本不需要清理后台，当应用程序不在前台运行时，除了 GPS 服务、音频播放服务和 VoIP 服务以外，其他的应用在 10 分钟后将被系统自动挂起，从技术上来说，被挂起的意思等同于不执行，只是数据驻留在内存而已，iOS 会在后台维护这个服务以实现假的多任务，并且所有的应用程序都会共用这一服务通道。而 Android 系统的后台软件很难关掉，因为有的软件会以各种形式自启，占用系统资源。

6）省电

从省电的角度来讲，iOS 系统是更加省电的，这是系统机制决定的。Android 会占用更多的资源来支撑系统运行，导致它会比较耗电。

Android 系统和 iOS 系统各有优缺点，它们也在一些方面互相借鉴，各自自主创新，希望 Android 系统和 iOS 系统永远竞争下去，期待它们给我们带来更多的惊喜。

1.4 搭建 Android 系统开发环境

谷歌在 2013 年为开发者提供的 IDE 环境工具 Android Studio，在几次更新之后，Android Studio 已经成为非常强大的 IDE 开发环境。2016 年 9 月，Google 发布了 Android Studio 2.2 版本的开发环境，同年 11 月宣布正式终止了对 Eclipse+ADT 开发工具的支持，但开发者可以将自己的项目从 Eclipse 迁移至 Android Studio 环境。因此，目前开发环境以 Android Studio 为主。

Android 系统的开发环境可以搭建在 Windows XP 及以上的操作系统中，在 Windows 10 下的安装方法与 Windows XP 下的安装方式大致相同，需要注意的是，Windows 10 下需要 64 位的文件，而 Windows XP 下需要 32 位的文件，下面在 Windows 10 企业版环境下进行 Android Studio 的安装。

在搭建环境之前，需要准备下面的两个文件。

Java JDK，下载地址：http://www.oracle.com/technetwork/java/javase/downloads/。

Android Studio 安装文件，下载地址：http://www.android-studio.org/或 http://www.androiddevtools.cn，建议选择带有 Android SDK 的 Android Studio 下载。

1.4.1 安装 JDK

1）安装 JDK 程序

（1）双击“jdk-8u112-windows-x64”文件，运行该程序，进入如图 1.1 所示的安装界面。

（2）单击“下一步”按钮，进入如图 1.2 所示界面，单击“更改”按钮，可以更改 JDK 的安装路径，例如，更改为“D: \Java\jdk1.8.0_112\”，然后单击“下一步”按钮。

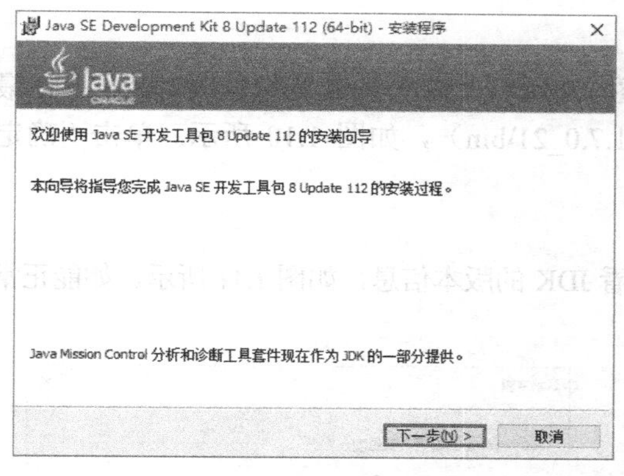

图 1.1　安装向导界面

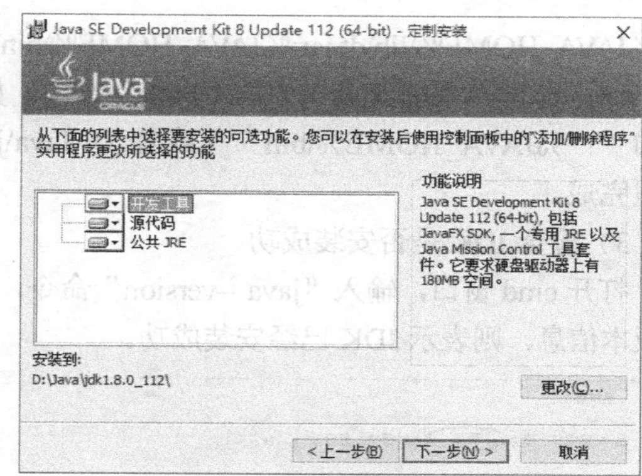

图 1.2　更改 JDK 安装目录

（3）进入进度界面，如图 1.3 所示。

（4）安装完成后，提示安装 JRE，建议和 JDK 安装在同一个盘符下，例如，目录为“D:\Java\jre8”，如图 1.4 所示。

（5）单击“下一步”按钮，开始安装 JRE，直至进入安装成功界面。

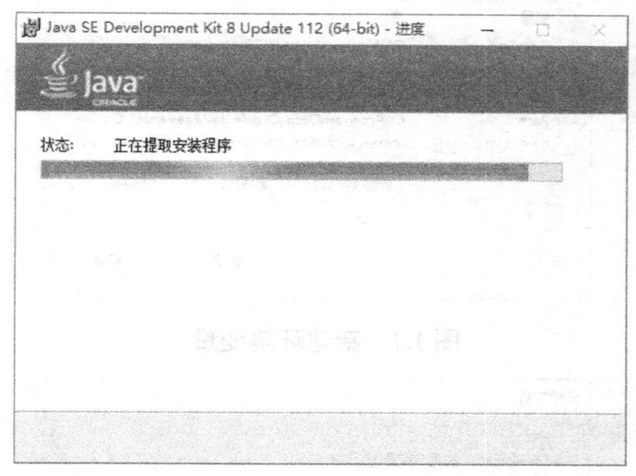

图 1.3　进度界面

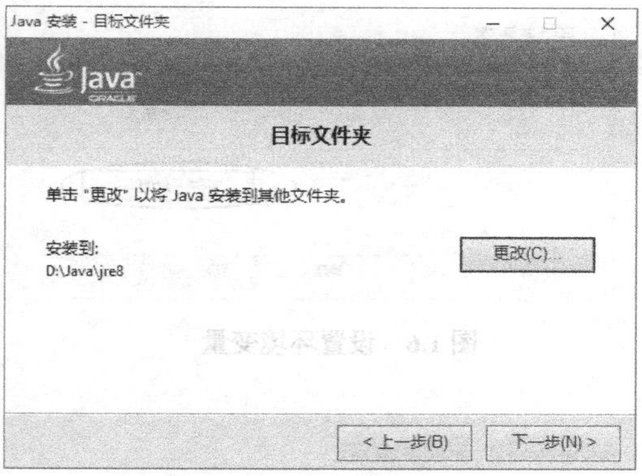

图 1.4　更改 JRE 安装目录

2）设置环境变量

（1）右击“我的电脑”图标，选择“属性”选项，在弹出的窗口的左边列表中选择“高级系统设置”选项，如图 1.5 所示。

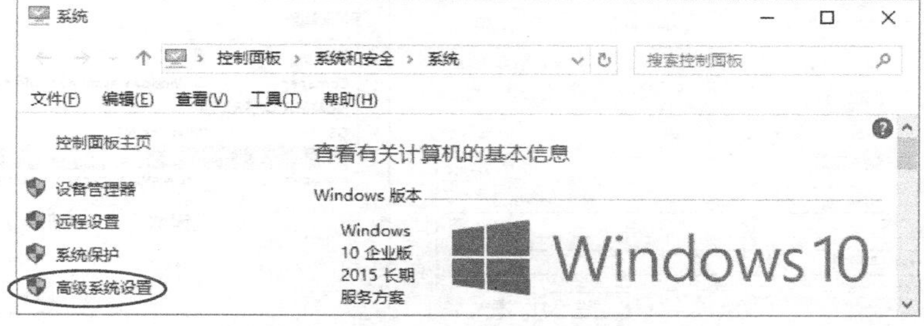

图 1.5　高级系统设置

（2）在弹出的对话框中选择“高级”选项卡，再单击“环境变量”按钮，如图 1.6 所示。

（3）单击对话框上半部分的“新建”按钮，如图 1.7 所示。设置 JAVA_HOME 变量的值为“D:\Java\jdk1.8.0_112”，如图 1.8 所示。类似的，新建 classpath 变量，其值为“.;%JAVA_HOME%\lib\tools.

jar;%JAVA_HOME%\lib\dt.jar;%JAVA_HOME%\bin;”。

（4）选择“系统变量”中的“Path”选项，如图 1.9 所示。弹出 Path 变量修改对话框，在最后添加“; %JAVA_HOME%\bin”（或者 D:\Java\jdk1.7.0_21\bin），如图 1.10 所示。单击“确定”按钮完成。

3）检查 JDK 是否安装成功

打开 cmd 窗口，输入“java –version”命令，查看 JDK 的版本信息，如图 1.11 所示。如能正常显示版本信息，则表示 JDK 已经安装成功。

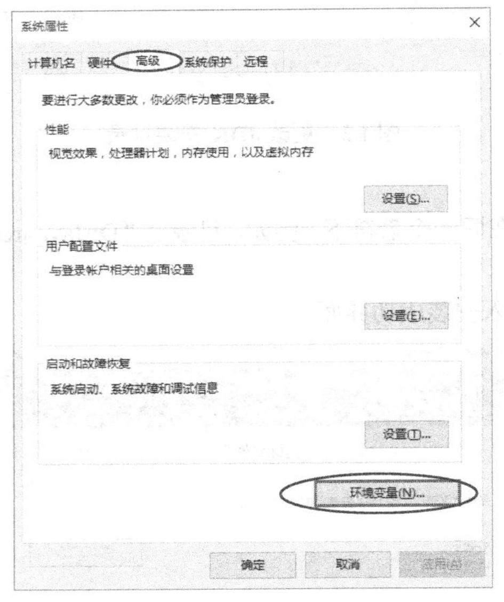

图 1.6 设置环境变量

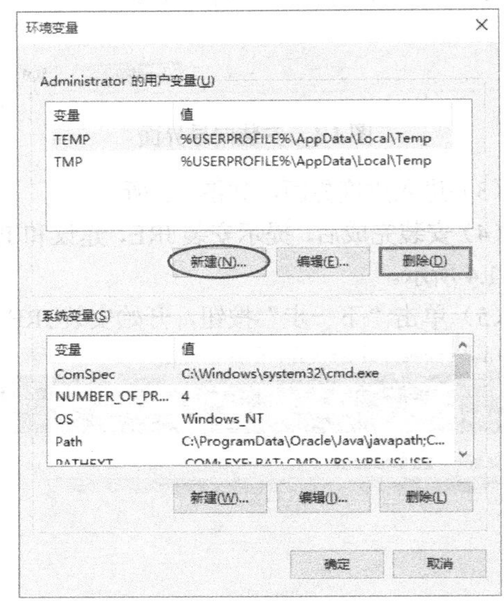

图 1.7 新建环境变量

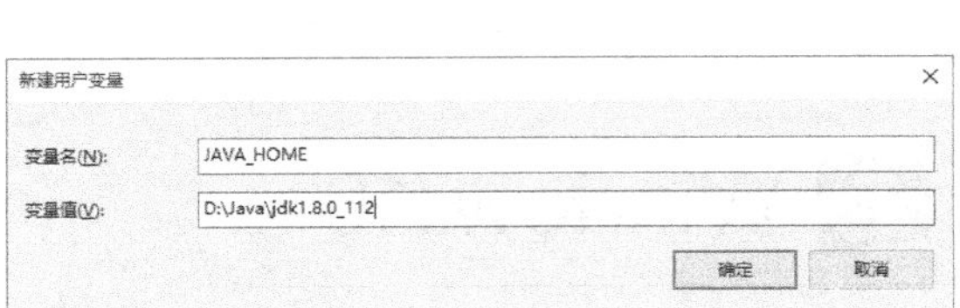

图 1.8 新建环境变量 JAVA_HOME

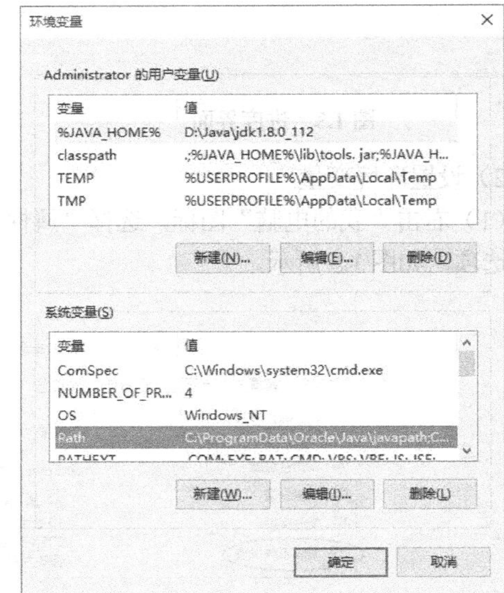

图 1.9 修改系统变量 Path

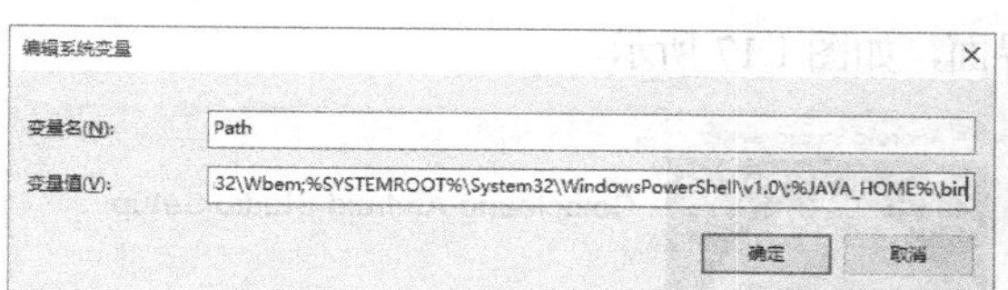

图 1.10 在最后添加值%JAVA_HOME%\bin

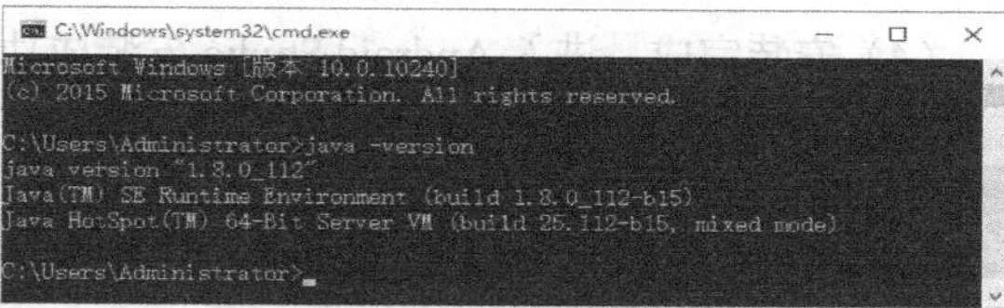

图 1.11 查看 JDK 的版本信息

1.4.2 安装 Android Studio

（1）双击下载的“android-studio-windows.exe”文件，出现 Android Studio 安装页面，如图 1.12 所示。

（2）单击“Next”按钮，在插件选择中，默认会选择“Android Studio”选项，同时选择“Android SDK”选项，表示同时安装 Android SDK。如果要在电脑上使用虚拟机调试程序，选择“Android Virtual Device”选项，如图 1.13 所示，然后单击“Next”按钮。

图 1.12 Android Studio 安装

Android Studio Setup
Choose Components
Choose which features of Android Studio you want to install.
Check the components you want to install and uncheck the components you don't want to install. Click Next to continue.
Select components to install:
Android Studio
Android SDK
Android Virtual Device
Description
Position your mouse over a component to see its description.
Space required: 4.2GB
< Back
Next >
Cancel

图 1.13 选择 Android Virtual Device

（3）在进入的许可证协议界面单击“I Agree”按钮，如图 1.14 所示。进入配置界面，选择 Android Studio 的安装目录和 SDK 的安装目录，可以更换路径，如图 1.15 所示。单击“Next”按钮，选择开始菜单中是否有 Android Studio 文件夹界面，若不需要，选中“Do not creat shortcuts”复选框，如图 1.16 所示。然后单击“Install”按钮，开始安装，可能需要较长时间。

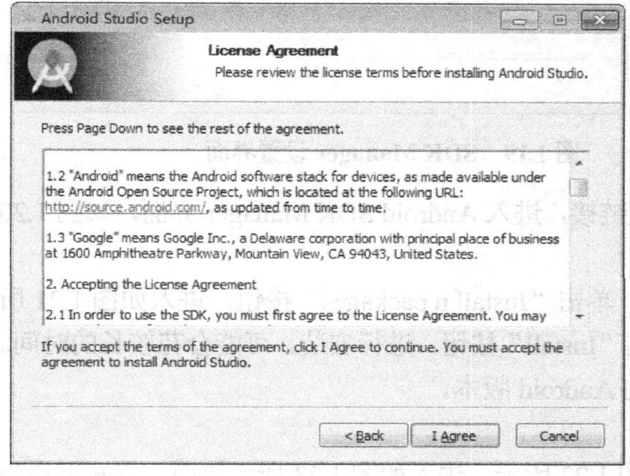

图 1.14 同意许可证协议

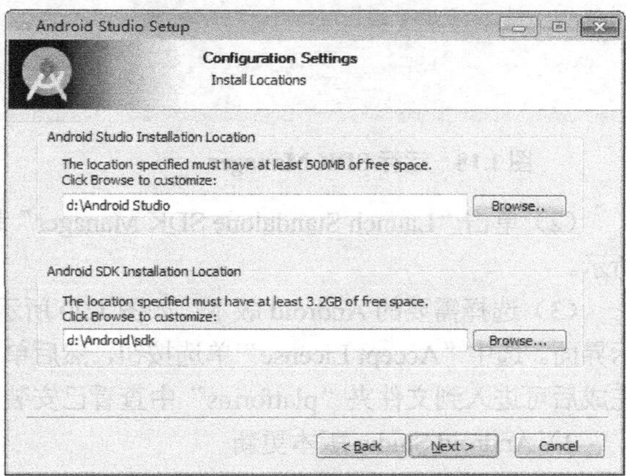

图 1.15 选择安装目录

（4）安装完成，进入 Android Studio 安装成功界面，如图 1.17 所示。

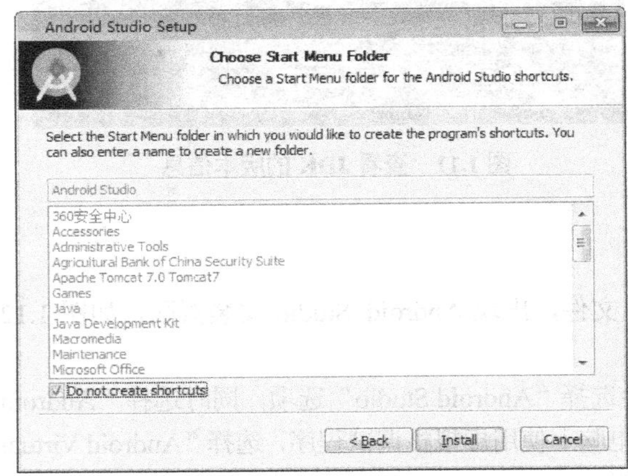

图 1.16　开始安装

图 1.17　安装成功

1.4.3　Android Studio 的基本配置

启动 Android Studio，有必要进行一些基本配置，才能够顺利运行项目。

1）安装 Android SDK

（1）打开 Android Studio 程序，选择“Tools→Android→SDK Manager”选项，如图 1.18 所示。或者单击工具栏中的 SDK Manager 按钮，进入 SDK Manager 界面，会列出已有的 Android 版本，如图 1.19 所示。

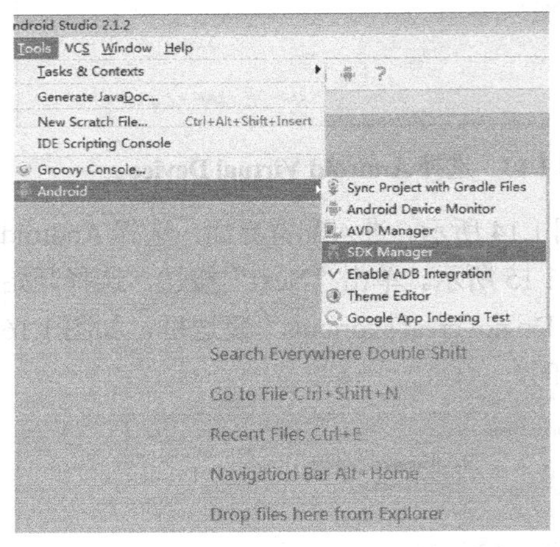

图 1.18　运行 SDK Manager

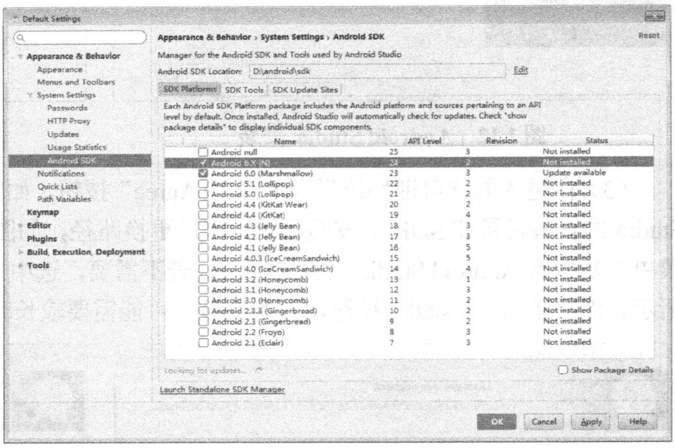

图 1.19　SDK Manager 设置界面

（2）单击“Launch Standalone SDK Manager”超链接，进入 Android SDK Manager 界面，如图 1.20 所示。

（3）选择需要的 Android 版本，如图 1.20 所示。单击“Install n packages”按钮，进入如图 1.21 所示界面。选中“Accept License”单选按钮，然后单击“Install”按钮，进行安装，可能会花较长的时间。完成后可进入到文件夹“platforms”中查看已安装的 Android 版本。

2）Android Studio 版本更新

（1）选择“Help→Check for Update”选项，如图 1.22 所示，进入如图 1.23 所示界面，告知用户当前版本以及最新版本。

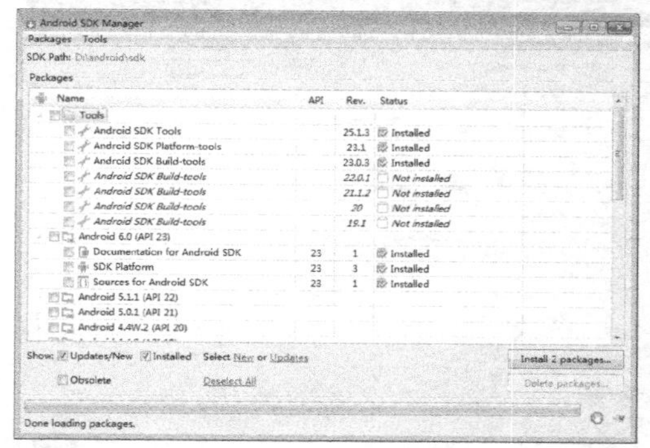

图 1.20 Android SDK Manager

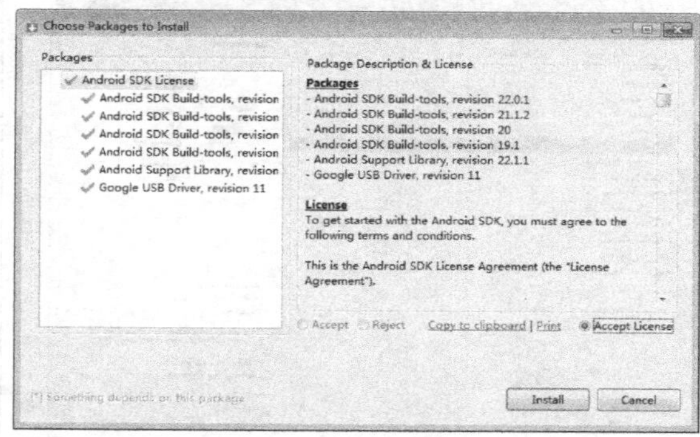

图 1.21 接受许可协议

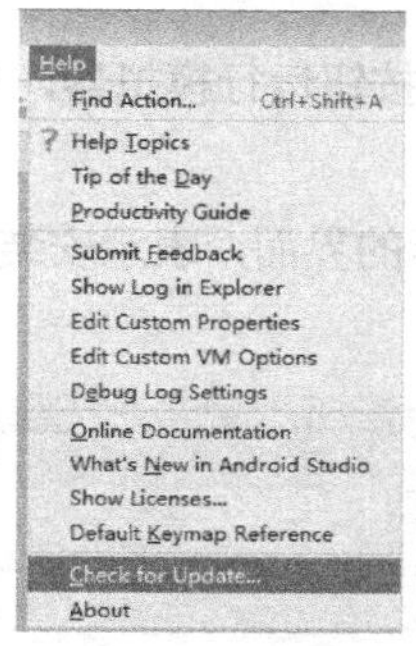

图 1.22 检查更新

图 1.23 版本信息是否更新

（2）单击“Download”按钮，进行下载更新。

3）创建 Android 虚拟设备 AVD

（1）选择“Tools→Android→AVD Manager”选项，或者单击工具栏中的 AVD Manager 按钮，进入 AVD Manager 界面，如图 1.24 所示。

（2）单击对话框中的“Create Virtual Device”按钮，进入如图 1.25 所示界面。选择 AVD 屏幕尺寸，然后单击“Next”按钮，进入如图 1.26 所示的界面。选择一个系统镜像，再单击“Next”按钮，进入 AVD 设备信息综合界面，可更改设备名称、设备版本、设备镜像、横竖屏、是否使用框架等信息，确认无误后单击“Finish”按钮，进行安装，如图 1.27 所示。

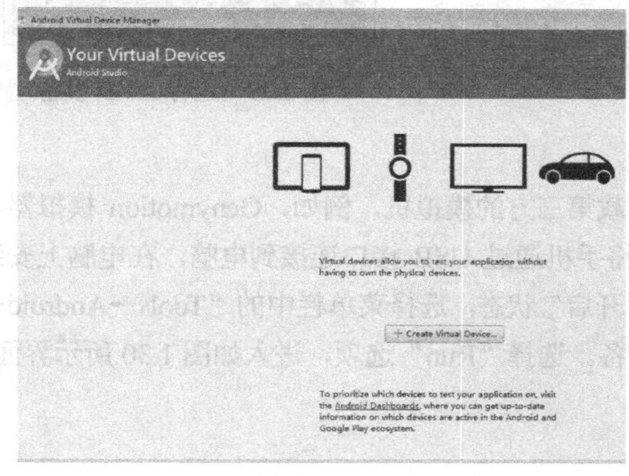

图 1.24 进入 AVD Manager 界面

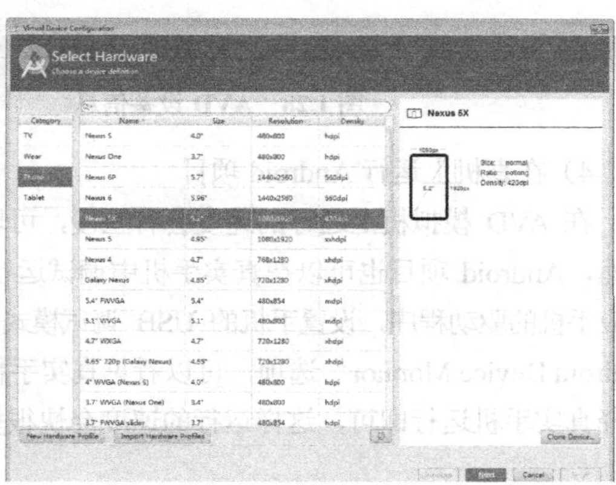

图 1.25 选择设备尺寸

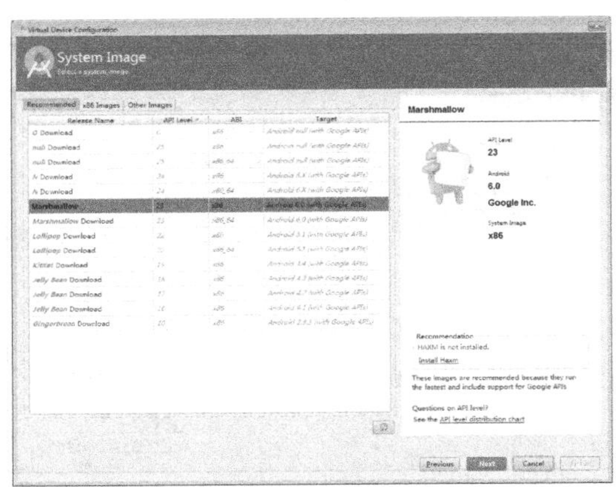

图 1.26　系统镜像选择

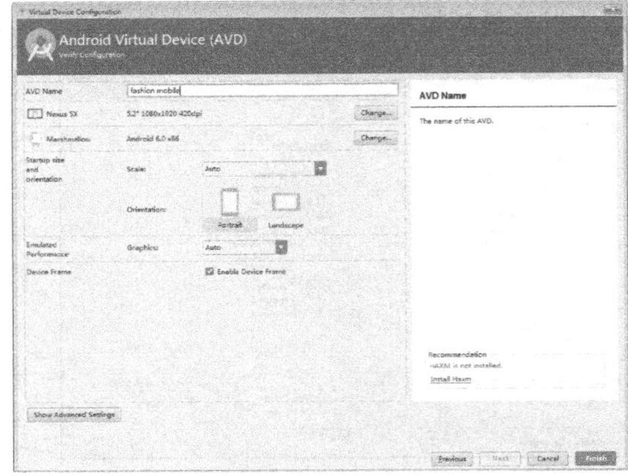

图 1.27　设备信息确定

（3）可以看到刚才创建的设备信息，可单击设备边的绿色三角图标按钮运行模拟器，单击笔图标按钮编辑 AVD 设备信息，如图 1.28 所示。

（4）单击绿色三角图标按钮，等待模拟器运行，进入如图 1.29 所示的界面，表示安装 AVD 设备成功。

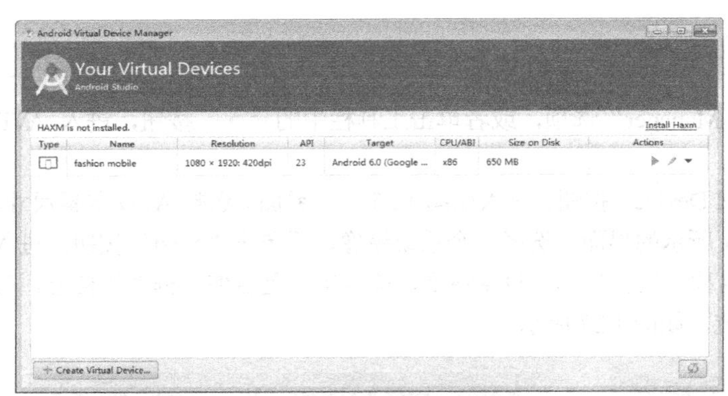

图 1.28　AVD 设备信息

图 1.29　运行 AVD 设备

4）在真机上运行 Android 项目

在 AVD 模拟机上运行的速度会有些慢，可以下载第三方的模拟机，例如，Genymotion 模拟器。当然，Android 项目也可以在真实手机中调试运行。将手机通过 USB 接口连接到电脑，在电脑上安装相应手机的驱动程序，设置手机的 USB 调试模式为“开启”状态，选择菜单栏中的“Tools→Android→Android Device Monitor”选项，可以看见真实手机设备。选择“Run”选项，进入如图 1.30 所示界面，选择真实手机运行即可，这样运行的速度会快很多。

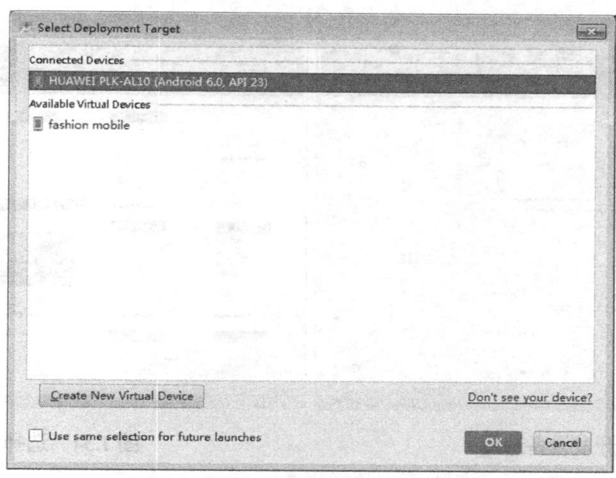

图 1.30 在真实手机上运行程序

1.5 开发第一个 Android 项目

经过基本设置，就可以通过 Android Studio 来建立一个 Android 项目了，大致流程如下。

（1）选择“File→New→New Project”选项来创建一个 Android 项目，如图 1.31 所示。

（2）在弹出的对话框中，输入项目的名称，如“FirstDemo”，注意，开头字母要大写，否则会有错误提示。公司域名可以自己输入（如“org.hnist.cn”），包名称会根据项目名称和公司域名自动生成。下面是指定应用存放的目录（如：“d:\androidspace\FirstDemo”），如图 1.32 所示，然后单击“Next”按钮。

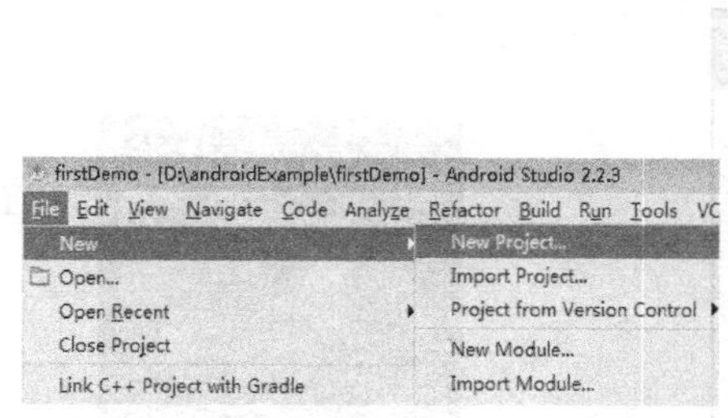

图 1.31 创建项目

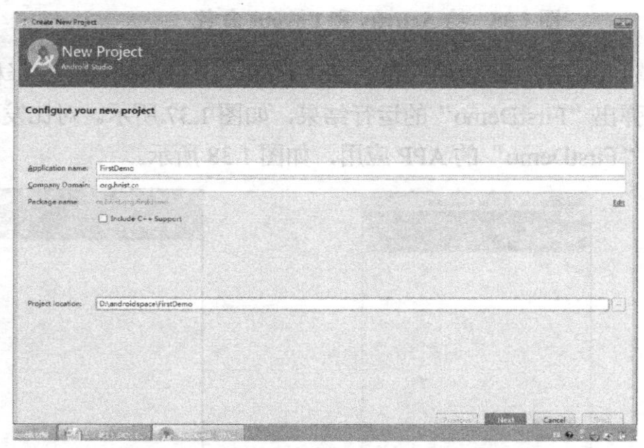

图 1.32 给项目命名、指定存放位置

（3）在弹出的对话框中，如图 1.33 所示。选择需要开发的设备，可选“手机和平板”、“穿戴（Wear）”、“TV”、“车载(Android Auto)”上的应用，以及确定最小 SDK 的版本号。单击“Help me choose”按钮，可查看每个版本的分布图表和描述。

（4）单击“Next”按钮，选择要制作的模板样式，如图 1.34 所示。这里选择“Empty Activity”模板，单击“Next”按钮。进入如图 1.35 所示界面，该界面需要给 Activity 和 Layout 命名，这里采用默认名。单击“Finish”按钮，Android Studio 开始创建应用，创建成功的界面如图 1.36 所示。

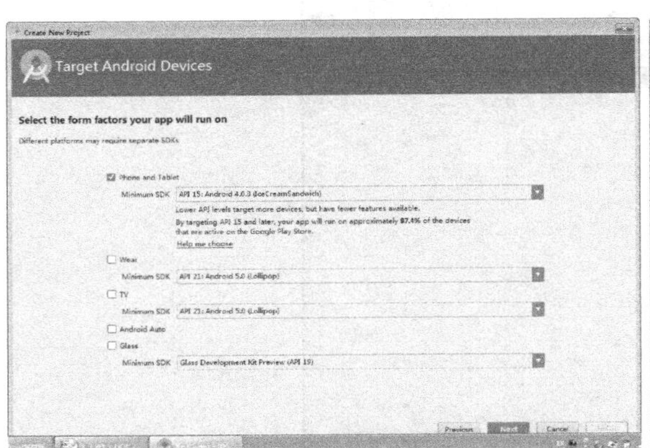

图 1.33　开发设备选择

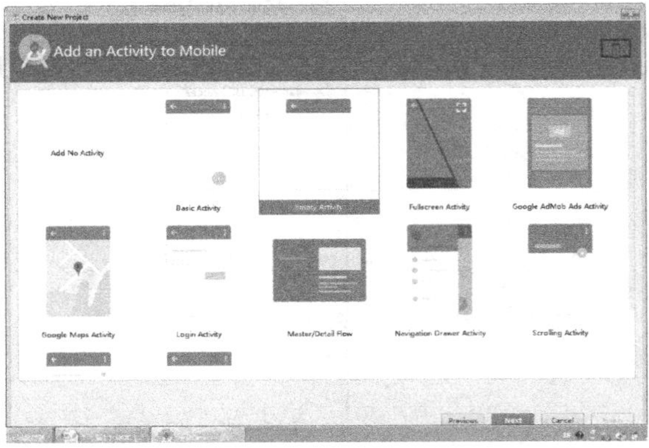

图 1.34　选择设计模板

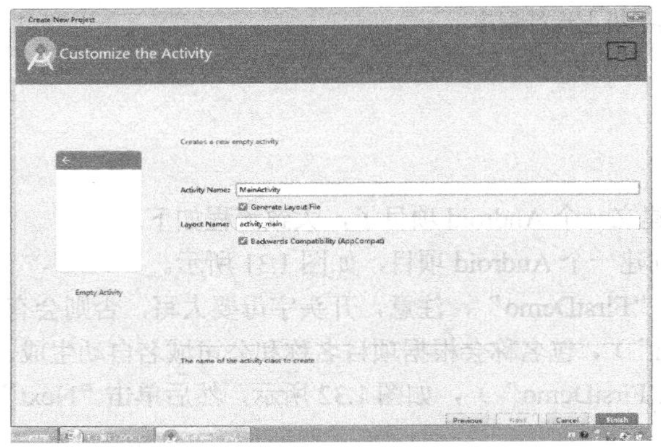

图 1.35　给 Activity 和 Layout 命名

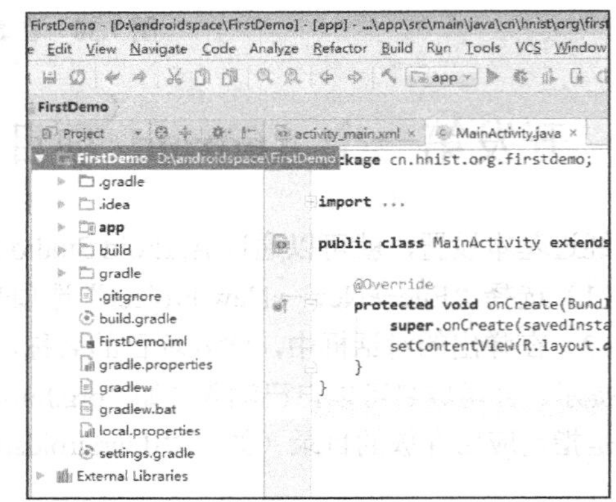

图 1.36　项目创建成功

单击菜单栏中的绿色三角形按钮，运行该项目，选择模拟器或者真实手机，等待一段时间后，将会弹出“FirstDemo”的运行结果，如图 1.37 所示。对比发现真机上运行要快得多，并且真机上已经有了“FirstDemo”的 APP 应用，如图 1.38 所示。

（a）模拟器上运行

（b）真机上运行

图 1.37　FirstDemo 运行结果

图 1.38　FirstDemo 应用

1.6 打包签名第一个 Android 项目

Android 程序开发完成后，为了方便用户使用，还需要将程序封装后上传至 Google Play 应用程序商店，经过商店审核通过后才能向用户提供下载和安装服务。下面介绍如何进行 Android 程序的封装。

（1）在 Android Studio 中，选择“Build→ Generate Signed APK...”选项，如图 1.39 所示，进入如图 1.40 所示界面。

（2）在没有 Key 时，可以先单击“Create new...”按钮来创建一个 Key，填写路径密码等信息，如图 1.41 所示。有 Key 时，可直接单击“Choose existing…”按钮进行选择。

（3）信息填写完毕后单击“OK”按钮，图 1.40 界面中的信息已自动填写完，如图 1.42 所示。单击“Next”按钮，进入如图 1.43 所示界面，选择要存放的 APK 位置和文件类型。单击“Finish”按钮，待主页面右上角弹出如图 1.44 所示的小提示时，项目打包签名完毕。

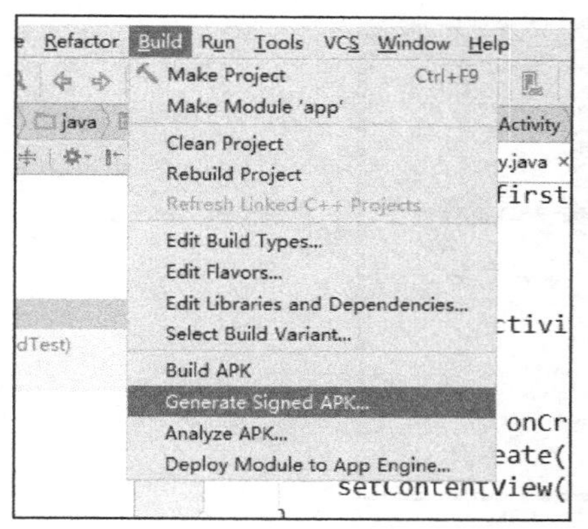

图 1.39　打包签名选项

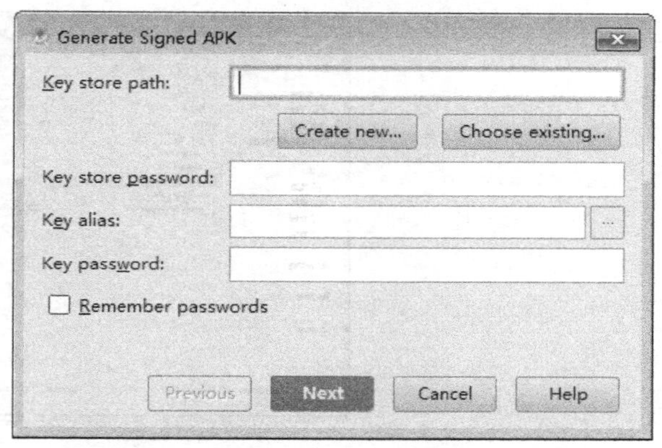

图 1.40　创建或者选择 Key

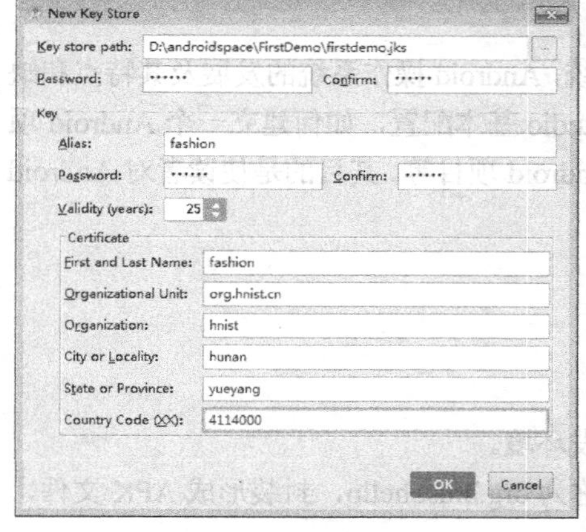

图 1.41　创建 Key 页面

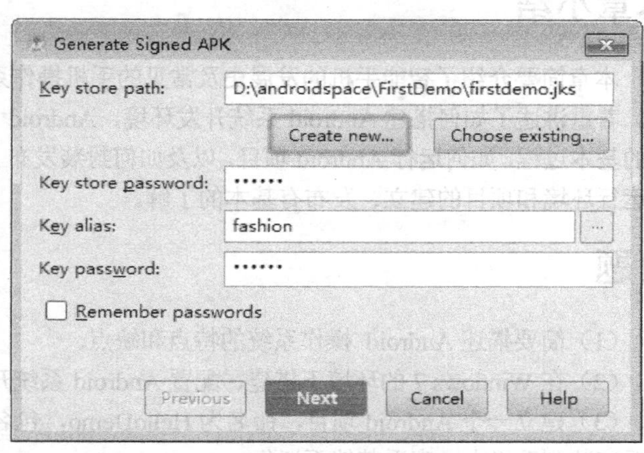

图 1.42　信息输入

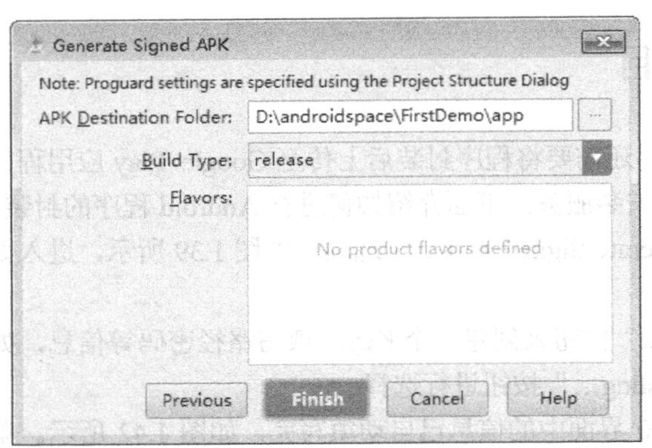

图 1.43　APK 保存位置

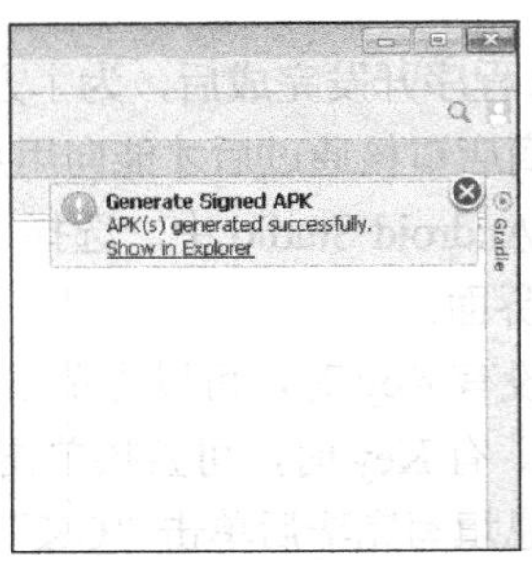

图 1.44　APK 打包成功

（4）单击图 1.44 中的“Show in Explorer”按钮，可查看签名好的 APK 文件，如图 1.45 所示。这个 APK 文件即可安装在手机上运行，也可以在 Android Market 注册后发布。

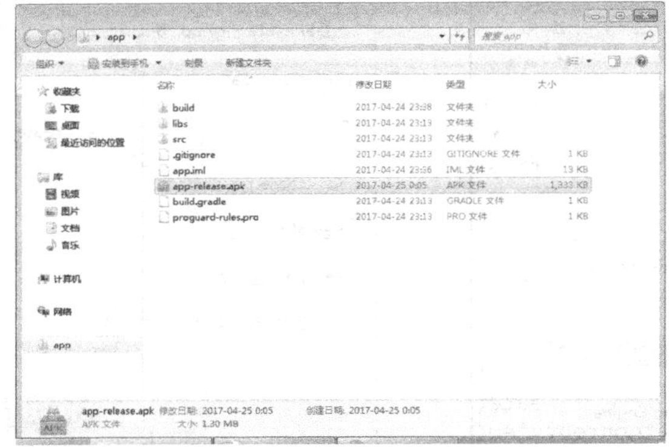
图 1.45　文件夹中的 APK 文件

本章小结

本章简要介绍了智能手机的发展史及常见的手机操作系统，Android 操作系统的发展及其特点和缺点，着重讲述了如何搭建 Android 系统开发环境，Android Studio 基本配置，如何建立一个 Android 项目的基本过程，如何运行 Android 项目，以及如何封装发布 Android 项目等。其目的是使读者对 Android 的运行环境和项目的建立、发布有基本的了解。

习题

（1）简要描述 Android 操作系统的特点和缺点。

（2）在 Windows 7 的环境下搭建、配置 Android 系统开发环境。

（3）建立一个 Android 项目，命名为 HelloDemo，包名称为 org.hnist.hello，封装形成 APK 文件，然后安装到手机上，查看其能否运行。

第 2 章　Android 中的项目

学习目标：

- 了解 Android 项目结构。
- 掌握 Android 项目各个常用部分的相互关系。
- 了解 Activity 基础知识。
- 了解 Android 项目中的开发包。
- 了解 Android 项目中的大致开发流程。
- 了解 Android 项目中常见的文件。

第 1 章中在没有编写一条代码的情况下就建立了一个 Android 项目，并能够在屏幕上显示“Hello World！”，项目建立后在 Android Studio 界面的左边导航栏出现了很多文件和文件夹，它们都是用来干什么的？显示的“Hello World！”能不能换成别的字符？本章对此会有详细的介绍。

2.1　Android 项目的组成

在 Android Studio 中创建的项目提供了多种项目结构类型，如 Android、Project、Packages 等结构类型，默认值是 Android，如图 2.1 所示。比较常用的是 Project 和 Android 结构类型，如图 2.2、图 2.3 所示。

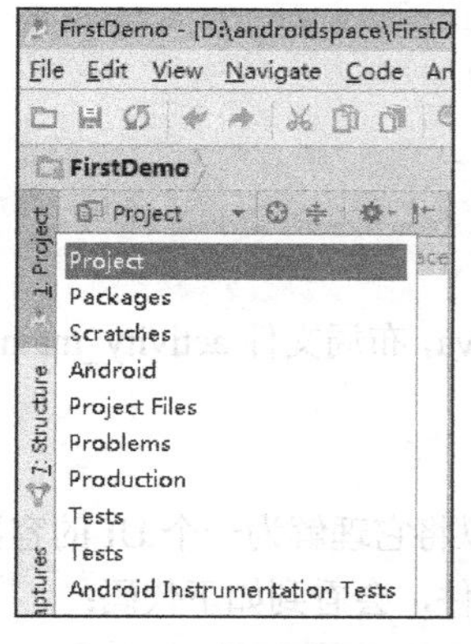

图 2.1　项目结构类型

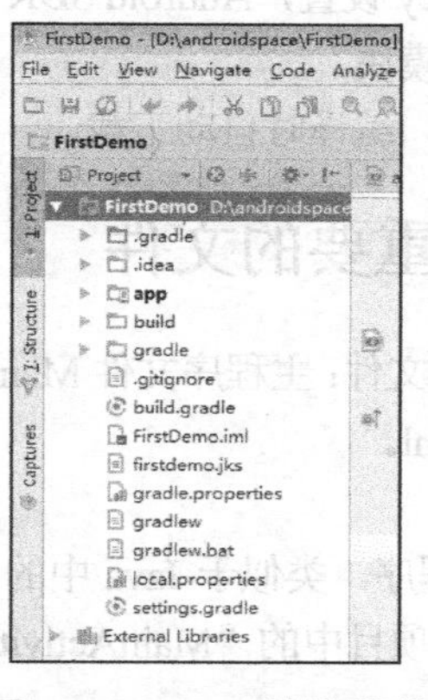

图 2.2　Project 结构类型

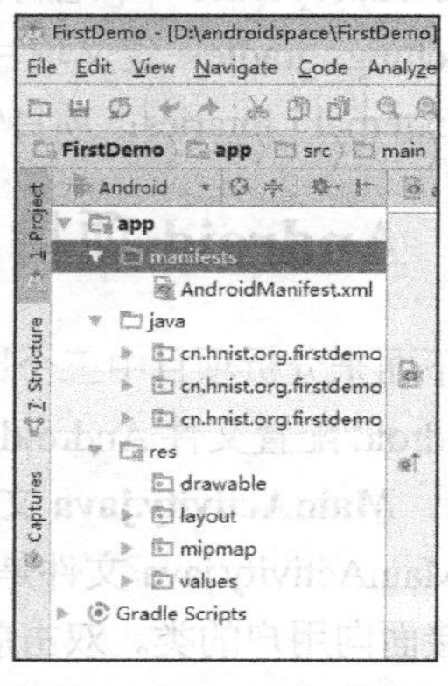

图 2.3　Android 结构类型

Android 结构类型文件及文件夹介绍如下。

（1）app/manifests/AndroidManifest.xml 配置文件，用来存储一些关于 Android 项目的配置数据。

（2）app/java 源码文件夹，所有的 Android 的 Java 源代码和测试代码都保存在这个文件夹中，其中的 Activity 程序 MainActivity.java 比较关键，类似于 Java 中的主类，是直接面向用户的类。

（3）app/res 资源文件夹，所有的资源文件都可以在这里找到，里面有如下文件夹。

① drawabel：包含应用程序要用到的图标文件（*.png、*.jpg 等）。

② layout：存放界面布局文件，默认为 activity_main.xml 文件，界面布局文件主要用于放置不同的显示组件。在 MainActivity.java 中通过 setContentView(R.layout.activity_main)语句来调用布局文件 activity_main.xml。

③ mipmap：存储原生图片资源文件，如图标文件。

④ values：该文件夹中可以有多个 XML 文件，以便存放不同类型的数据。例如，字符串（string.xml）、颜色文件（colors.xml）、尺寸文件（dimens.xml）和类型文件（styles.xml）等。

注意：在项目中尽量使用 dp 作为空间大小单位，sp 作为和文字相关大小单位。

（4）Gradle Scripts 存放 gradle 编译相关的脚本，其中 build.gradle 为项目的 gradle 配置文件。

Project 结构类型体现的是项目在电脑上真实的结构，其文件及文件夹形式如下。

（1）.gradle 文件夹：存放 gradle 编译系统，版本由 wrapper 指定。

（2）.idea 文件夹：存放 Android Studio IDE 所需要的文件。

（3）app 文件夹：存放应用相关的文件及目录。

（4）build 文件夹：存放编译后产生的相关文件，例如，最后生成的 APK 文件就在 outputs\apk 文件夹下，这里是 app-debug.apk。

（5）libs 文件夹：存放相关依赖库。

（6）src 文件夹：存放项目源代码和测试代码。

（7）main 文件夹：Java 文件夹存放项目源代码，Res 文件夹存放项目资源文件。其中有如下文件夹。

.gitignore：git 版本管理忽略文件，标记出哪些文件不用进入 git 库。

build.gradle：模块的 gradle 相关配置。

MyApplication.iml：项目的配置文件。

MyApplication.jks：项目生成 APK 文件时的 JKS 签名文件。

gradle.propertie：gradle 相关的全局属性设置。

gradlew：编译脚本，可以在命令行下执行打包封装 Gradle。

local.properties：本地属性设置（Key 设置，Android SDK 位置等属性）。

settings.gradle：定义项目包含哪些模块。

External Libraries：项目依赖的 Lib，编译时自动下载。

2.2 Android 项目中三个重要的文件

下面来分析项目中三个比较重要的文件：主程序文件 MainActivity.java，布局文件 activity_main.xml 和 Android 配置文件 AndroidManifest.xml。

1．MainActivity.java 文件

MainActivity.java 文件是 Activity 程序，类似于 Java 中的主类。可以将它理解为一个 UI 的容器，是直接面向用户的类。双击前面建立的项目中的“MainActivity.java”文件，会看到如下代码：

```
package cn.hnist.org.firstdemo;          //程序所在包为cn.hnist.org.firstdemo
import android.support.v7.app.AppCompatActivity;//导入Activity支持类
import android.os.Bundle;                //导入支持包
public class MainActivity extends AppCompatActivity {//定义MainActivity类
   @Override
   protected void onCreate(Bundle savedInstanceState)  {//覆写onCreate方法
      super.onCreate(savedInstanceState); //调用父类的onCreate()方法
      setContentView(R.layout.activity_main);}}//调用布局文件activity_main.xml
```

AppCompatActivity 类：定义的 Activity 程序 MainActivity 类是自动继承 AppCompatActivity 类（最后还是继承 Activity 类），Activity 类包含与用户交互的属性和方法，可以用方法 setContentView（View）将自己的用户界面（UI）放在其中。

以下两个方法是几乎所有的 Activity 子类都要实现的。

onCreate(Bundle)：初始化 Activity 程序。

setContentView(int)：指定由哪个文件指定布局（如 activity_main.xml），可以将这个界面显示出来，然后进行相关操作。

2．activity_main.xml 文件

Layout 文件夹下有界面布局文件“activity_main.xml”，界面布局文件主要用于放置不同的显示组件。在 MainActivity.java 中通过 setContentView(R.layout.activity_main)语句来调用它，显示在手机屏幕上。

一个具体的 activity_main.xml 的代码如下，注意，后面的汉字是对本行代码的解释，实际的程序中不能有这些解释的汉字。

```
    <?xml version="1.0" encoding="utf-8"?>
    <RelativeLayout                                        //采用的相对布局模式
        xmlns:android="http://schemas.android.com/apk/res/android"
        xmlns:tools="http://schemas.android.com/tools"
        android:id="@+id/activity_main"                    //设置此布局文件的 ID 值
        android:layout_width="match_parent"                //布局管理器的宽度为屏幕宽度
        android:layout_height="match_parent                //布局管理器的高度为屏幕高度
        android:paddingBottom="@dimen/activity_vertical_margin"  /*指控件中内容
        距离控件底边距离为 dimen.xml 中变量 activity_vertical_margin 设定的值*/
        android:paddingLeft="@dimen/activity_horizontal_margin"  /*指控件中内容
        距离控件左边距离为 dimen.xml 中变量 activity_horizontal_margin 设定的值*/
        android:paddingRight="@dimen/activity_horizontal_margin"  /*指控件中
        内容距离控件右边距离为 dimen.xml 中变量 activity_horizontal_margin 设定的值*/
        android:paddingTop="@dimen/activity_vertical_margin"      /*指控件中
        内容距离控件上边距离为 dimen.xml 中变量 activity_vertical_margin 设定的值*/
        tools:context="cn.hnist.org.firstdemo.MainActivity"> /*说明当前的布局文件
所在的对象是 MainActivity 对应的那个 Activity。*/
        <TextView                                          //设置一个文本显示组件
            android:id="@+id/textView1"                    //设置此文本显示组件的 ID
            android:layout_width="wrap_content"            //组件的宽度为文字的宽度
            android:layout_height="wrap_content"           //组件的高度为文字的高度
            android:text="Hello World" />                  //文本显示的内容为 Hello World
    </RelativeLayout>                                      //相对布局模式结束
```

打开一个 XML 布局文件，选择左侧的“Design”选项卡，如图 2.4 所示，即可进入预览界面，还可以选择手机的尺寸进行预览，在右边还可以选择某个组件进行设置，此项功能大大提高了开发效率。

values 文件夹里有尺寸文件 dimens.xml，一个 dimen.xml 文件的代码如下：

```
<resources>
   <!-- Default screen margins, per the Android Design guidelines. -->
   <dimen name="activity_horizontal_margin">16dp</dimen>
                                //定义 activity_horizontal_margin 的值为 16dp
   <dimen name="activity_vertical_margin">16dp</dimen>
                                //定义 activity_vertical_margin 的值为 16dp
</resources>
```

Android 系统中定义的单位有以下几种类型。

px（Pixels，像素）：屏幕上的实际像素点。例如，320*480 的屏幕在横向有 320 个像素，在纵向有 480 个像素。

in（Inches，英寸）：屏幕物理长度单位，1 英寸等于 2.54 厘米。例如，形容手机屏幕大小，可以用 3.2（英）寸、3.5（英）寸、4（英）寸来表示。这些尺寸是指屏幕的对角线长度。

mm（Millimeters，毫米）：屏幕物理长度单位。

pt（Points，磅）：屏幕物理长度单位，大小为 1 英寸的 1/72。

dp（与密度无关的像素）：逻辑长度单位，在 160 dpi 屏幕上，1dp=1px=1/160 英寸。随着密度变化，对应的像素数量也会变化，但并没有直接的变化比例。

dip：与 dp 相同，多用于 Google 示例中。

sp（与密度和字体缩放度无关的像素）：与 dp 类似，但是可以根据用户的字体大小首选项进行缩放。

注意：在项目中尽量使用 dp 作为空间大小单位，sp 作为文字相关大小单位。

values 文件夹里还有 strings.xml 文件的代码，如下所示：

```
<resources>
    <string name="app_name">FirstDemo</string>//定义 app_name 的值为 FirstDemo
</resources>
```

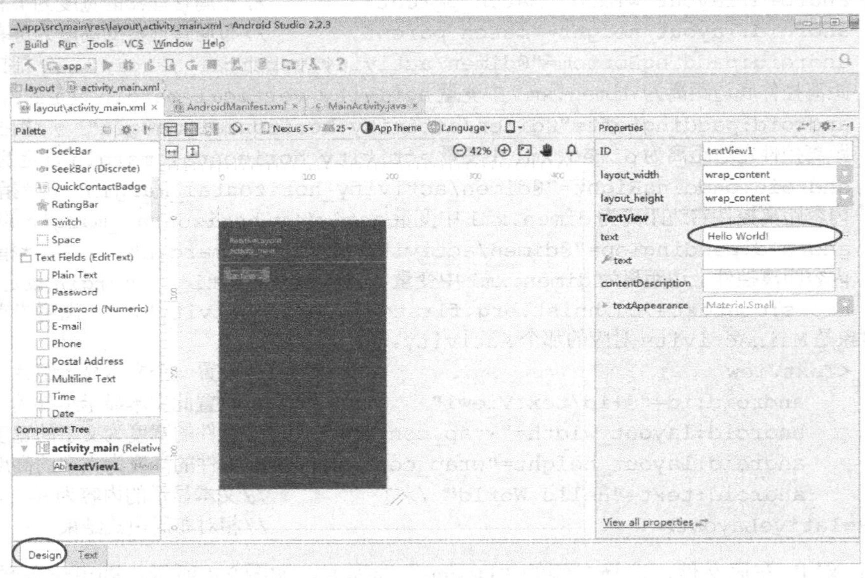

图 2.4　显示预览界面

3．AndroidManifest.xml 文件

每个应用程序都有 AndroidManifest.xml 文件，此文件提供了关于这个应用程序的基本信息，例如，要开发 Activity、Broadcast、Service 等应用都要在 AndroidManifest.xml 中进行定义；要使用系统自带的服务，如拨号服务、GPRS 服务等都必须在 AndroidManifest.xml 中声明权限；新添加一个 Activity 的时候，也需要在这个文件中进行相应配置，只有配置好后，才能调用此 Activity。

AndroidManifest.xml 文件用来存储一些关于 Android 项目的配置数据，主要包含以下功能。

① 命名应用程序的 Java 应用包，这个包名用来唯一标识应用程序。

② 描述应用程序的组件——活动、服务、广播接收者、内容提供者；对实现每个组件和公布其功能的类进行命名。这些声明使得 Android 系统了解这些组件以及它们在什么条件下可以被启动。

③ 决定应用程序组件运行在哪个进程中。

④ 声明应用程序所必须具备的权限，用以访问受保护的部分 API 以及和其他应用程序的交互。

⑤ 声明应用程序其他的必备权限，用以组件之间的交互。

⑥ 列举测试设备 Instrumentation 类，用来提供应用程序运行时所需的环境配置及其他信息，这些声明只在程序开发和测试阶段存在，发布前将被删除。

一般 AndroidManifest.xml 包含如下设置：application、Activities、intent filters 等。

例如，一个 AndroidManifest.xml 文件的代码如下：

```
<?xml version="1.0" encoding="utf-8"?>
<manifest xmlns:android=http://schemas.android.com/apk/res/android /*定义 Android
项目的命名空间*/
    package="cn.hnist.org.firstdemo">      //指定本应用内 Java 主程序的包名
    <application  //声明了每一个应用程序的组件及其属性
        android:allowBackup="true"          //允许备份文件
        android:icon="@mipmap/ic_launcher" //应用程序图标文件的位置
        android:label="@string/app_name"    //应用程序的名称，也就是安装到手机上的名称
        android:supportsRtl="true"          //声明 application 是否支持从右到左的布局
        android:theme="@style/AppTheme">   //默认 Theme 样式
          <activity android:name=".MainActivity">     //默认启动的 Activity
            <intent-filter>
                <action android:name="android.intent.action.MAIN" />//主程序
                <category android:name="android.intent.category.LAUNCHER" />
                       //应用程序是否显示在程序列表中
            </intent-filter>
        </activity>
    </application>
</manifest>
```

2.3 扩充 FirstDemo 项目

下面对项目 FirstDemo 用几种不同方法进行扩充，在现有的基础上增加了一个文本显示组件和按钮组件，通过这些不同的实例，来说明这些文件之间的一些关系。

（1）修改布局文件 activity_main.xml，增加一个文本显示组件和按钮组件。

在原有代码基础上，增加如下代码：

```
    <TextView                                           //定义一个文本显示组件
        android:id="@+id/textView2"                     //组件的名称叫 textView2
        android:layout_width="wrap_content"             //定义组件的宽度
        android:layout_height="wrap_content"            //定义组件的高度
        android:layout_below="@+id/textView1"           //位于 textView1 组件的下方
        android:layout_centerHorizontal="true"          //垂直居中
        android:layout_marginTop="20dp"                 //与顶部边界的距离为 20dp
        android:text="欢迎您使用本系统！" />              //显示的文字为"欢迎您使用本系统"
    <Button                                             //定义一个按钮组件
        android:id="@+id/mybutton"                      //定义组件名称为 mybutton
        android:layout_width="fill_parent"              //组件的宽度为屏幕的宽度
        android:layout_height="wrap_content"            //定义组件的高度
        android:layout_alignLeft="@+id/textView1"    //位于 textView1 组件的左边
        android:layout_alignParentBottom="true"        //贴紧父元素的下边缘
        android:layout_marginBottom="260dp"             //与底部边界的距离为 260dp
        android:text="我是按钮！" />                     //按钮上的文字为"我是按钮！"
</RelativeLayout>
```

其他不做任何修改，保存并运行该项目，结果如图 2.5 所示。

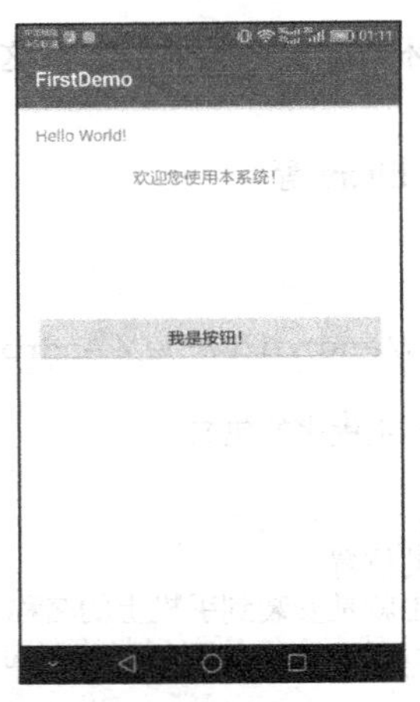

图 2.5　运行结果

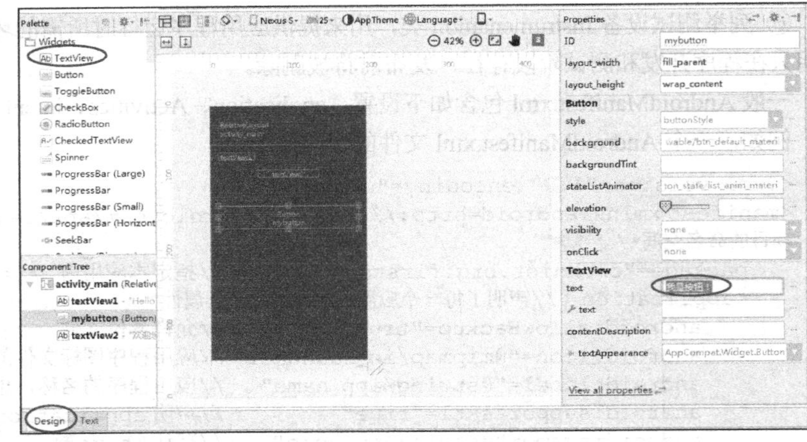

图 2.6　组件拖动方式

布局管理文件除了像上面那样通过输入代码的形式增加组件外，还可以使用拖动的方式增加组件。

① 双击打开“Activity_main.xml”文件，选择“Design”选项卡，进入显示预览界面，如图 2.6 所示。

② 选择“Widgets”选项，选择要加入的组件，拖动到右侧的手机屏幕上，将它放在合适的地方，并设置属性值。

③ 保存后，再双击“activity_main.xml”文件，进入该文件界面，发现组件相应的代码已经自动添加。

（2）修改布局文件和 string.xml 文件，增加一个文本显示组件和按钮组件。

在 Activity_main.xml 文件中修改如下代码：

```
    <TextView                                    //定义一个文本显示组件
        …
        android:text="@string/hello_world"  />/*显示的文字从文件 string.xml 中的
hello_world 变量获得*/
    <TextView                                    //增加一个文本显示组件
        …
        android:text="@string/txt"/> //显示的文字从文件 string.xml 中的 txt 变量获得
    <Button                                      //增加一个按钮组件
        …
       android:text=""@string/but"" /> //显示的文字从文件 string.xml 中的 but 变量获得
  </RelativeLayout>
```

对 strings.xml 代码做修改，如下所示：

```
  <?xml version="1.0" encoding="utf-8"?>
  <resources>
     <string name="app_name">FirstDemo</string> //定义 app_name 的值为 FirstDemo
     <string name="hello_world">Hello world!</string>   //定义 hello_world 的值
     <string name="txt">欢迎您使用本系统! </string>//定义 txt 的值为"欢迎您使用本系统! "
     <string name="but">我是按钮! </string>      //定义 but 的值为"我是按钮! "
  </resources>
```

保存并运行该项目，结果如图 2.5 所示。再对 strings.xml 文件代码做修改，如下所示：

```
  <?xml version="1.0" encoding="utf-8"?>
  <resources>
```

```
    <string name="app_name">我的第一个项目演示</string>    //定义 app_name 的值
    <string name="hello_world">你好!</string>          //定义 hello_world 的值
    <string name="txt">欢迎您使用本系统! </string>        //定义 txt 的值
    <string name="but">我是按钮! </string>              //定义 but 的值
</resources>
```

其他的文件不做修改，保存并运行该项目，看看与图 2.5 结果有什么不同？

（3）修改布局文件和 MainActivity.java 文件，增加一个文本显示组件和按钮组件。

在上面的修改基础上对 Activity_main.xml 文件进行修改，代码如下：

```
…
    <TextView                                          //定义一个文本显示组件
        android:id="@+id/textView2"                    //组件的名称叫 textView2
        android:layout_width="wrap_content"            //定义组件的宽度
        android:layout_height="wrap_content"           //定义组件的高度
        android:layout_below="@+id/textView1"          //位于 textView1 组件的下方
        android:layout_centerHorizontal="true"         //垂直居中
        android:layout_marginTop="20dp"                //与顶部边界的距离为 20dp
    <Button                                            //增加一个按钮组件
        android:id="@+id/mybutton"                     //定义组件名称为 mybutton
        android:layout_width="fill_parent"             //组件的宽度为屏幕的宽度
        android:layout_height="wrap_content"           //定义组件的高度
        android:layout_alignLeft="@+id/textView1"      //位于 textView1 组件的左边
        android:layout_alignParentBottom="true"        //贴紧父元素的下边缘
        android:layout_marginBottom="260dp"            //与底部边界的距离为 260dp
</RelativeLayout>
```

注意：这里没有对文本显示组件和按钮组件设置 Text 的值，需要在 MainActivity.java 程序中设置，在原有 MainActivity.java 基础上对文件进行如下修改。

```
…
    public void onCreate(Bundle savedInstanceState) { //覆写 onCreate 方法
        super.onCreate(savedInstanceState);     //调用父类的 onCreate()方法
        setContentView(R.layout.activity_main);  //调用布局文件 activity_main.xml
        this.txt=(TextView)super.findViewById(R.id.textView2); //取得 txt 组件
        txt.setText("欢迎您使用本系统! ");          //设置 txt 组件显示的内容
        this.but=(Button)super.findViewById(R.id.mybutton); //取得 but 按钮组件
        but.setText(super.getString(R.string.but)); } } //设置 but 按钮上显示的
                 内容来源于 string.xml 中 but 定义的值
```

其他的文件不做修改，保存并运行该项目，结果如图 2.5 所示。

这种方法是初学者常使用的方法，比较简单、直观。首先在布局管理器中定义要使用的组件，然后在 Activity 程序中通过 setContentView()方法调用这个布局管理文件，布局管理文件里的组件则通过 findViewById()方法来获得，然后对组件进行相应的操作，如通过 setText()方法设置在组件上要显示的内容。

事实上，也可以不使用布局管理文件，直接在 MainActivity.java 上添加组件，设置参数。

（4）修改 MainActivity.java 文件，增加一个文本显示组件。

对 MainActivity.java 文件做如下修改：

```
…
    public void onCreate(Bundle savedInstanceState){//覆写 onCreate 方法
```

```
        super.onCreate(savedInstanceState);                    //调用父类的 onCreate()方法
        //setContentView(R.layout.activity_main);   //注意，这里没有调用布局文件
        TextView txt=new TextView(this);                      //定义一个文本显示组件
        txt.setText(super.getString(R.string.txt));  //设置文本显示组件显示的文字
        super.setContentView(txt); }  }                      //设置显示该组件
```

这里直接通过程序的方式生成组件，因此不需要加载布局管理文件，要注意的是组件生成之后要利用 setContentView()方法将组件显示出来。

上面代码中的语句“txt.setText(super.getString(R.string.txt))”通过调用 string.xml 文件中的变量 txt 的值来设置文本显示组件上显示的文字，也可以直接在这里使用 txt.setText（“欢迎您使用本系统!”），运行效果一样。这种方法比较灵活，但是利用这种方式所生成的组件，每次只能显示一个组件，虽然布局管理文件中存在按钮组件，但是没有在 Activity 程序中调用布局文件，因此没有显示按钮。在后续的课程中可以定义一个布局管理器对象，然后在其中添加多个组件。

2.4　Activity

Android 项目中包括 4 个应用程序组件（Component），一个 Android 应用程序是一个包（Package），包中可能包含下面的一个或者多个 Android 组件。

（1）活动（Activity）：Activity 是最基本的 Android 应用程序组件，在应用程序中，一个 Activity 通常就是一个单独的用户界面。每一个 Activity 都可以视为一个独立的类，并且从 Activity 基类继承而来，Activity 类将会显示由视图（View）控件组成的用户接口，并对事件（Event）做出响应。大多数的应用程序会有多个用户界面，因此便会有多个相应的 Activity 程序。

一个活动一般对应界面中的一个屏幕显示，可以理解成一个界面，每一个 Activity 在界面上可以包含按钮、文本框等多种可视的 UI 元素。

（2）广播接收器（BroadcastReceiver）：广播接收器用于让应用程序对一个外部事件做出响应。例如，电话呼入事件、数据网络可用通知或者指定时间进行通知等。

（3）服务（Service）：服务是具有一段较长生命周期但没有用户界面的程序。例如，一个正在从播放列表中播放歌曲的媒体播放器会在后台运行。

（4）内容提供者（Content Provider）：应用程序能够将它们的数据保存到文件或 SQLite 数据库中，甚至是任何有效的设备中。当需要将数据与其他的应用共享时，内容提供者组件将会很有用。一个内容提供者类实现了一组标准的方法，从而能够让其他应用程序保存或读取此内容提供者处理的各种数据类型。

简单来说，Activity 就是一个用户所能看到的屏幕，是应用程序的界面容器，可以放置各种各样的控件、设置处理事件（如按键事件、触摸屏事件等）、为用户显示指定的 View、启动其他 Activity 等。Activity 中所有操作都与用户密切相关，是负责与用户交互的组件，所有应用的 Activity 都继承于 android.app.Activity 类。

创建一个 Activity，要注意以下 4 个方面。

（1）一个 Activity 就是一个类，并且这个类要继承 Activity 或者 Activity 支持类。

```
import android.support.v7.app.AppCompatActivity;//导入 Activity 支持类
import android.app.Activity;                    //导入 Activity 类
```

（2）需要覆写 onCreate 方法（应用程序启动后第一个运行的函数，由 Android 框架决定）。

```
public void onCreate(Bundle savedInstanceState) {//覆写 onCreate 方法
    super.onCreate(savedInstanceState);          //调用父类的 onCreate()方法
```

```
        setContentView(R.layout.activity_main); }  //调用布局文件 activity_main.xml
```

（3）每一个 Activity 都需要在 AndroidManifest.xml 文件中进行配置。

```
<activity
    android:name="cn.hnist.org.firstdemo.MainActivity"//设置默认启动 Activity
    android:label="@string/app_name" >                  //定义 Activity 的名称
     <intent-filter>
        <action android:name="android.intent.action.MAIN" />    //主程序
      //表示放到手机应用程序的列表中
        <category android:name="android.intent.category.LAUNCHER" />/
    </intent-filter>
</activity>
```

（4）根据需求为 Activity 添加必要的控件。

通过 findViewById(控件 ID)方法可以得到所要显示的控件。例如：

```
this.txt=(TextView)super.findViewById(R.id.textView2); //取得 txt 文本显示组件
```

Activity 的启动可以通过 Launcher 组件，也可以通过 Activity 内部调用 startActivity 接口来启动。Activity 之间的数据传递可通过 Intent 来进行参数的传递，传递数据的过程可以是双向的，这些将在本书的第 6 章中进行详细介绍。

2.5 Android 中的常用包

在 Android 应用程序开发中，使用的是 Java 语言，除了需要熟悉 Java 语言的基础知识之外，还需要了解 Android 提供的扩展的 Java 功能，Android 提供了一些扩展的 Java 类库，类库又分为若干个包，每个包中包含若干个类。Android Java API 包含 40 多个包和 700 多个类，这些为编写 Android 应用程序提供了一个功能丰富的平台。

下面对一些常用的重要包进行了简要的描述，如表 2-1 所示。

表 2-1 重要包描述

包名称	功能描述
android.app	实现 Android 的应用程序模型，提供基本的运行环境
android.bluetooth	提供一些类来处理蓝牙功能
android.content	包含各种的对设备上的数据进行访问和发布的类
android.database	通过内容提供者浏览和操作数据库
android.database.sqlite	实现 android.database 包，将 SQLite 用做物理数据库
android.graphics	底层的图形库，包含画布、颜色过滤、点、矩形，可以将它们直接绘制到屏幕上
android.graphics.drawable	实现绘制协议和背景图像，支持可绘制对象动画
android.graphics.drawable.shapes	实现各种形状
android.hardware	实现与物理照相机相关的类
android.location	定位和相关服务的类
android.media	提供一些类管理多种音频、视频的媒体接口
android.net	提供帮助网络访问的类，超过通常的 java.net.*接口
android.net.wifi	管理 WiFi 连接
android.opengl	提供 OpenGL 的工具
android.os	表示可通过 Java 编程语言访问的操作系统服务

续表

包名称	功能描述
android.provider	提供类访问 Android 的内容提供者
android.speech	包含用于语音识别的常量。这个包只在 1.5 版和更新版本中提供
android.speech.tts	提供从文本到语音转换的支持
android.telephony	提供与拨打电话相关的 API 交互
android.telephony.gsm	可用于根据基站来收集手机位置
android.text	包含文本处理类
android.view	提供基础的用户界面接口框架
android.util	涉及工具性的方法，如时间日期的操作
android.webkit	默认浏览器操作接口
android.widget	包含通常派生自 View 类的所有 UI 控件。主要的部件包括 Button、Checkbox、Chronometer、AnalogClock、DatePicker、DigitalClock、EditText、ListView、FrameLayout、GridView、ImageButton、MediaController、ProgressBar、RadioButton、RadioGroup、RatingButton、Scroller、ScrollView、Spinner、TabWidget、TextView、TimePicker、VideoView 和 ZoomButton

例如，希望在 Android 程序中用 Color.BLACK 来表示黑色，就要调用包含颜色的包 android.graphics.Color，通常使用 import android.graphics.Color；语句进行引用，引用后就可以直接使用 Color.BLACK 来表示黑色，Color.RED 来表示红色，等等。如果事先不进行引用，直接使用 Color.BLACK 来表示黑色，则程序将会报错。

2.6 Android 项目的大致开发流程

在开发一个 Android 项目前，先要对这个项目进行一些最基本的分析，规划好项目的基本开发步骤，确保项目顺利开发。一般来说，Android 项目开发步骤包括以下几个方面。

（1）对项目进行分析：了解项目的主要功能、有哪些必需的界面及界面之间的跳转关系、需要的数据及数据的来源、是否需要服务器的支持、是否需要后台服务等。

（2）架构设计：将整个项目进行分解，确定在 Activity 里设计的项目有哪些、哪个进行网络连接、哪个进行数据库处理等。

（3）界面设计：确定程序的主界面，各模块界面，列表，查看、编辑界面，菜单，按钮、对话框、提示信息的设计，界面总体颜色设计，使项目更加美观统一。

（4）数据操作和存储：确定项目需要的数据的来源、如何存储和读取等。

（5）代码的编写：对分解的各个子模块进行代码编写，包括控件、事件、菜单、页面跳转等。

（6）程序调试。

当然，这里只是一个简单的介绍，如果真正开发一个较大的 Android 项目，可能还会考虑得更多。例如，可行性分析、用户需求分析、项目进度设计、项目的总体设计、详细设计等，读者可以参考软件工程的相关书籍进行了解。

本章小结

本章着重介绍了 Android 项目的目录结构、主要文件夹和文件的用途，通过四种不同的方法对第一个项目 FirstDemo 进行扩充，加深对项目各组成部分之间的关系理解，简单介绍了 Activity 及建立 Activity 要注意的几个方面，简单介绍了 Android 项目中的开发包，项目开发的大致流程等。其目的是让用户深入了解 Android 项目各组成部分之间的关系，对 Android 项目开发有一个大致了解。

习题

（1）简要描述 Android 项目开发的大致开发流程。

（2）在网络上下载 Android API 文档，了解其用途。

（3）新建一个项目，命名为 HelloDemo，包名称为 org.hnist.hello，运行项目，在屏幕上显示一行文字“信息提交！”，显示两个按钮，一个为“确定”按钮，另一个为“取消”按钮。要求用以下两种方式实现：调用布局管理文件和不调用布局管理文件。

（4）如果要修改 APP 项目的标题为“成绩管理系统”，应该怎么修改？

（5）如果要修改 APP 项目显示在手机上的图标，应该怎么修改？

（6）要在 res\layout 文件夹下建立一个 XML 文件，应该如何操作？

（7）要在 src 文件夹下建立一个 Java 文件，应该如何操作？

第3章　Android常用基本组件

学习目标：

- 了解Android中的View类。
- 掌握Android中的TextView等常用基本组件及操作方法。
- 掌握Android中的布局管理器的使用。

在软件开发过程中，界面设计和功能开发同样重要，界面美观可以大大增强用户的体验感，还能吸引更多的新用户。而Android提供了大量的UI开发组件，只要合理地使用它们，就能编写出各种各样美观实用的界面。

接下来就介绍一些常见的UI控件和四大布局管理器的调用及其基本操作。

3.1　Android平台中的View类

前面章节介绍的按钮和文本显示组件都是View类的子类，Android平台中的android.view.View类包含了大多数的图形显示组件，其层次关系如下：

```
java.lang.Object
  android.view.View
```

要在Android程序中使用View类，必须在程序中使用下面的语句，否则会报错。

```
import android. android.view.View;                //导入android.view.View类
```

除了按钮和文本显示组件外，android.view.View类中还定义了许多图形组件，如表3-1所示，这些组件都在android.widge包中定义。

表3-1　部分常见图形组件

组件名称	类名称	描述
TextView	android.widget. TextView	文本的显示组件
Button	android.widget. Button	普通按钮组件
EditText	android.widget. EditText	编辑文本框组件
CheckBox	android.widget. CheckBox	表示复选框组件
RadioGroup	android.widget. RadioGroup	表示单选按钮组件
Spinner	android.widget. Spinner	下拉列表框
ImageView	android.widget. ImageView	图片显示组件
ImageButton	android.widget. ImageButton	图片按钮组件
Toast	android.widget. TextView	信息提示框组件

View组件有相应的属性和方法，参数的值可以在布局管理中设置，也可以在Activity程序中通过代码来设置。表3-2列出了View组件常用属性及对应方法。

表3-2　View组件常用属性及对应方法

属性名称	方法名称	描述
android:background	public void setBackgroundResource (int resid)	设置组件背景
android:clickable	public void setClickable (boolean clickable)	是否可以产生单击事件

续表

属性名称	方法名称	描述
android:contentDescription	public void setContentDescription (CharSequence contentDescription)	定义视图的内容描述
android:drawingCacheQuality	public void setDrawingCacheQuality (int quality)	设置绘图时所需要的缓冲区大小
android:focusable	public void setFocusable (boolean focusable)	设置是否可以获得焦点
android:focusableInTouchMode	public void setFocusableInTouchMode (boolean focusableInTouchMode)	在触摸模式下配置是否可以获得焦点
android:id	public void setId (int id)	设置组件 ID
android:longClickable	public void setLongClickable (boolean longClickable)	设置长按事件是否可用
android:minHeight		定义视图的最小高度
android:minWidth		定义视图的最小宽度
android:padding	public void setPadding (int left, int top, int right, int bottom)	填充所有的边缘
android:paddingBottom	public void setPadding (int left, int top, int right, int bottom)	填充下边缘
android:paddingLeft	public void setPadding (int left, int top, int right, int bottom)	填充左边缘
android:paddingRight	public void setPadding (int left, int top, int right, int bottom)	填充右边缘
android:paddingTop	public void setPadding (int left, int top, int right, int bottom)	填充上边缘
android:scaleX	public void setScaleX (float scaleX)	设置 X 轴缩放
android:scaleY	public void setScaleY (float scaleY)	设置 Y 轴缩放
android:scrollbarSize		设置滚动条大小
android:scrollbarStyle	public void setScrollBarStyle (int style)	设置滚动条样式
android:visibility	public void setVisibility (int visibility)	设置是否显示组件
android:layout_width		定义组件显示的宽度
android:layout_height		定义组件显示的长度
layout_toRightOf		位于组件的右边
layout_above		位于组件的上方
layout_below		位于组件的下方
layout_toLeftOf		位于组件的左边
android:layout_gravity		组件文字的对齐位置
android:layout_margin		设置文字的边距
android:layout_marginTop		上边距
android:layout_marginBottom		下边距
android:layout_marginLeft		左边距
android:layout_marginRight		右边距

在 Android 平台中 View 类是一个包含子类最多的一个类，常见的组件有以下几种。

文本类组件：TextView、EditText、…

按钮类组件：Button、ImageButton、…

选择类组件：RadioButton、CheckBox、…

列表类组件：Spinner、ListView、…

图像类组件：ImageView、Gallery、…

时间类组件：DatePicker、TimePicker、…

布局类组件：RelativeLayout、LinearLayout、…

提示类组件：Toast、…

菜单类组件：Menu、…

…

下面介绍 View 类中几个常见的组件，以加深对 Activity 的认识。

3.2 文本显示组件 TextView

在 Android 中，文本框使用 TextView 表示，主要用于在界面中显示一段文本信息，如 Hello World！Android 中的文本框组件既可以显示单行文本，也可以显示多行文本，还可以显示带图像的文本。其层次关系如下：

```
java.lang.Object
    android.view.View
        android.widget.TextView
```

直接子类：Button、CheckedTextView、Chronometer、DigitalClock、EditText。

间接子类：AutoCompleteTextView、ExtractEditText、MultiAutoCompleteTextView、RadioButton、ToggleButton、CheckBox、CompoundButton。

要在 Android 程序中使用 TextView 组件必须在程序中使用下面的语句。

```
import android.widget.TextView;                    //导入 widget.TextView 类
```

1. TextView 组件常见的属性和方法

TextView 组件继承了 View 类，所以前面介绍的 View 类的属性它都具备，除此之外，该组件还有其他属性，如表 3-3 所示。表 3-4 列出了该组件的常用方法。

表 3-3　TextView 组件常用属性

属性名称	值	描述
android:text	自定义	设置显示文本
android:textColor	自定义	设置文本颜色
android:textColorLink	自定义	文字链接的颜色
android:textSize	自定义	设置文字大小，推荐度量单位“sp”，如“15sp”
android:height	自定义	设置文本区域的高度，支持度量单位：px(像素)/dp/dip/sp/in/mm，常用“wrap_content”、“fill_parent”等设置
android:width	自定义	设置文本区域的宽度，支持度量单位：px(像素)/dp/dip/sp/in/mm，常用“wrap_content”、“fill_parent”等设置
android:layout_marginTop	自定义	设置文本上边距，支持度量单位：px(像素)/dp/dip/sp/in/mm
android:layout_marginRight	自定义	设置文本右边距，支持度量单位：px(像素)/dp/dip/sp/in/mm
android:layout_marginBottom	自定义	设置文本下边距，支持度量单位：px(像素)/dp/dip/sp/in/mm
android:layout_marginLeft	自定义	设置文本左边距，支持度量单位：px(像素)/dp/dip/sp/in/mm
android:layout_centerHorizontal	true/false	默认为 false，设置为 true 时，组件水平居中显示
android:layout_centerVertical	true/false	默认为 false，设置为 true 时，组件垂直居中显示
android:shadowColor	自定义	指定文本阴影的颜色，需要与 shadowRadius 一起使用
android:shadowRadius	自定义	设置阴影的半径。设置为 0.1 就会变成字体的颜色，一般而言，设置为 3.0 的效果比较好
android:autoLink	“none”、“web”、“Email”、“phone”、“map”、“all”	设置是否当文本为 URL 链接/E-mail/电话号码/map 时，文本显示为可单击的链接
android:maxLines	自定义	设置文本最多显示行数

表 3-4 TextView 组件常用方法

方法	描述
public void setText(CharSquence str)	设置组件显示文字
public String setText getText()	获得 TextView 对象的文本
public int length()	获得 TextView 中的文本长度
public void getEditableText()	取得文本的可编辑对象，通过这个对象可对 TextView 的文本进行操作，如在光标之后插入字符
public void getAutoLinkMask()	返回自动连接的掩码
public void setTextColor()	设置文本显示的颜色
public void setHintTextColor()	设置提示文字的颜色
public void setLinkTextColor()	设置链接文字的颜色
public void setGravity()	设置当 TextView 超出了文本本身时横向及垂直对齐

2. TextView 组件的使用实例

前面已经介绍了 TextView 组件要显示在屏幕上可以调用布局管理文件中设置的组件，例如：

```
<TextView                                       //定义一个文本显示组件
   android:id="@+id/mytxt"                      //组件的名称为 mytxt
   android:layout_width="wrap_content"          //定义组件的宽度
   android:layout_height="wrap_content"         //定义组件的高度
   android:text="欢迎您使用本系统！"              //显示的文字为"欢迎您使用本系统"
   ... />
```

代码中“android:”后面的都是 TextView 组件的属性，这些属性可以参照表 3-3 进行设置，也可以在 Activity 程序中通过调用 TextView 相应的方法来实现，例如：

```
TextView txt=new TextView(this);               //定义一个文本显示组件
txt.setText("欢迎您使用本系统！");              //设置文本显示组件显示的文字
super.setContentView(txt);                     //设置显示该组件
```

实例 3-1：TextView 的使用

按照如下步骤新建一个 Android 项目。

（1）打开程序 Android Studio，选择“File→New→New Project”选项。

（2）在弹出的窗口中，输入项目的名称 Exam3_1，公司域名 cn.hnist.org，单击“Next”按钮。

（3）在后续弹出的窗口中，所有选项采用默认值（也可根据自身情况做出相应的修改），直到单击“Finish”按钮完成创建。

（4）在 res 文件夹下选择 layout，双击“activity_main.xml”文件，打开布局管理文件（当然，为了加快代码编写的速度也可以结合拖动的方式进行），进行如下修改：

```
<?xml version="1.0" encoding="utf-8">
<RelativeLayout                                      //相对布局开始
   xmlns:android="http:  //schemas.android.com/apk/res/android"
   xmlns:tools="http:    //schemas.android.com/tools"
   android:layout_width="fill_parent"                //布局管理器的宽度为屏幕宽度
   android:layout_height="fill_parent"               //布局管理器的高度为屏幕高度
   android:background="#fff">                        //设置布局管理器背景颜色为白色
<TextView                                            //定义一个文本显示组件
      android:id="@+id/mytxt1"                       //组件的名称为 mytxt1
      android:layout_width="wrap_content"            //组件的宽度为文字的宽度
```

```
        android:layout_height="wrap_content"    //组件的高度为文字的高度
        android:layout_centerHorizontal="true" //设置文字水平居中显示
        android:background="#FFFF00"           //设置文字背景颜色
        android:textSize="20sp"                //设置文字的大小为 20sp
        android:text="@string/mytxt" />  //设置文字调用 strings.xml 中定义的文本
    <TextView                                  //定义一个文本显示组件
        android:id="@+id/mytxt2"               //组件的名称为 mytxt2
        android:layout_width="wrap_content"    //组件的宽度为文字的宽度
        android:layout_height="wrap_content"   //组件的高度为文字的高度
        android:layout_below="@+id/mytxt1"     //该组件位于组件 mytxt1 的下方
        android:layout_centerHorizontal="true" //设置文字水平居中显示
        android:layout_marginTop="25dp"        //距离上面组件 25dp
        android:textColor="#00FF00"            //设置文字的颜色
        android:textSize="24sp"                //设置文字的大小为 24sp
        android:textStyle="bold"               //设置文字加粗显示
        android:text="@string/mytxt" />  //设置文字调用 strings.xml 中定义的文本
    <TextView                                  //定义一个文本显示组件
        android:id="@+id/mytxt3"               //组件的名称为 mytxt3
        android:layout_width="wrap_content"    //组件的宽度为文字的宽度
        android:layout_height="wrap_content"   //组件的高度为文字的高度
        android:layout_below="@+id/mytxt2"     //该组件位于组件 mytxt2 的下方
        android:layout_marginTop="35dp"        //距离上面组件 35dp
        android:autoLink="all"                 //设置超级链接
        android:textSize="20sp"                //设置文字的大小为 20sp
        android:text="@string/mytxt"/>  //设置文字调用 strings.xml 中定义的文本
    <TextView                                  //定义一个文本显示组件
        android:id="@+id/mytxt4"               //组件的名称为 mytxt4
        android:layout_width="wrap_content"    //组件的宽度为文字的宽度
        android:layout_height="wrap_content"   //组件的高度为文字的高度
        android:layout_alignParentRight="true" //设置文字相对父控件右对齐显示
        android:layout_below="@+id/mytxt3"     //该组件位于组件 mytxt3 的下方
        android:layout_marginTop="20dp"        //距离上面组件 20dp
        android:textSize="20sp"                //设置文字的大小为 20sp
        android:text="@string/mytxt"/> //设置文字调用 strings.xml 中定义的文本
    <TextView                              //定义一个文本显示组件
        android:id="@+id/mytxt5"               //组件的名称为 mytxt5
        android:layout_width="wrap_content"    //组件的宽度为文字的宽度
        android:layout_height="wrap_content"   //组件的高度为文字的高度
        android:layout_above="@+id/mytxt6"     //该组件位于组件 mytxt6 的上方
        android:layout_marginBottom="16dp"     //距离下面组件 16dp
        android:layout_centerHorizontal="true" //组件水平居中
        android:textColor="#000000"            //设置文字的颜色
        android:shadowColor="#FF0000"          //设置文本阴影的颜色
        android:shadowRadius="3.0"             //设置阴影的半径为 3.0
        android:textSize="22sp"                //设置文字的大小为 22sp
        android:text="@string/mytxt"/>   //设置文字调用 strings.xml 中定义的文本
    <TextView                                  //定义一个文本显示组件
        android:id="@+id/mytxt6"               //组件的名称为 mytxt6
        android:layout_width="80dp"            //组件的宽度为 80dp
        android:layout_height="wrap_content"   //组件的高度为文字的高度
        android:layout_centerHorizontal="true" //组件水平居中
        android:layout_centerVertical="true"   //组件垂直居中
```

```
        android:textSize="28sp"                   //设置文字的大小为 28sp
        android:ellipsize="marquee"               //设置为滚动的文字
        android:focusable="true"                  //设置可以获得焦点
        android:focusableInTouchMode="true"       //设置触摸模式下可以获得焦点
        android:gravity="center"                  //设置文本居中显示
        android:marqueeRepeatLimit="marquee_forever"//设置滚动的次数为无限次
        android:maxLines="1"               //设置文字内容最多显示为 1 行
        android:text="@string/mytxt"/>     //设置文字调用 strings.xml 中定义的文本
</RelativeLayout>                          //相对布局结束
```

（5）在 res\values 文件夹下选择“strings.xml”文件，双击打开，定义相应的字符串和数值，例如：

```
<resources>
    <string name="app_name">Exam3_1</string>
    <string name="mytxt">www.hnist.cn</string>  //设置相应的文本 ID 要显示的文本
</resourse>
```

保存文件，程序运行结果如图 3.1 所示。

单击带有链接的“www.hnist.cn”能够链接到 www.hnist.cn 网页。

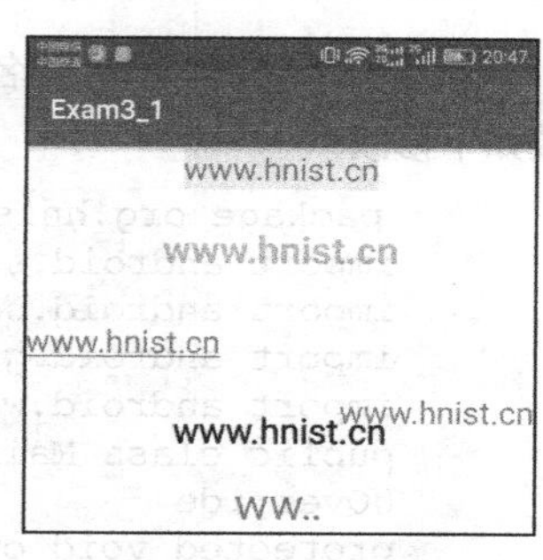

图 3.1　Exam3_1 运行结果

实例 3-2：使用样式文件简化 XML 代码

程序中的文字如果使用某种指定格式的效果，可以建立样式文件，以便于维护和使用。样式文件格式与 strings.xml 文件类似，一般是在“res\values”文件夹中的“styles.xml”文件中建立。

按照前面的步骤建立一个项目，项目命名为 Exam3_2，公司域名为 cn.hnist.org。

在“res\values”文件夹中双击“styles.xml”文件，打开的文件中输入如下代码：

```
    <resources>
       <style name="AppTheme" parent="Theme.AppCompat.Light.DarkActionBar"    >
</style>
       <style name="my_styles" >                                //定义 my_styles 样式
        <item name="android:layout_width">wrap_content</item>//组件的宽度为文字的宽度
        <item name="android:layout_height">wrap_content</item>//组件的高度为文字的高度
        <item name="android:layout_centerHorizontal">true</item>//组件水平居中
        <item name="android:textSize">20sp</item>               //设置文字的大小为 20sp
        <item name="android:textColor">#890004 </item>          //设置文字颜色
        <item name="android:textStyle">bold</item>              //设置文字的加粗显示
        <item name="android:shadowColor">#FFFFFF</item>         //设置文字阴影颜色
        <item name="android:shadowRadius">3.0</item>            //设置文字阴影半径
      </style>                                                  //结束样式定义
    </resources>
```

一旦样式文件建立，被调用后，显示出来的文字就是前面定义好的样式。例如：

```
    …
        <TextView                                      //定义一个文本显示组件
           android:id="@+id/mytxt1"                    //组件的名称为 mytxt1
           style="@style/my_styles"                    //文字的显示样式调用 my_styles 样式
           android:background="@drawable/a1"           //设置背景图像为 a1.jpg（将 a1.jpg
文件先复制到 drawable 文件夹中，否则程序会报错）
```

```
        android:gravity="center"                    //文字居中显示
        android:autoLink="phone"                    //设置电话号码链接
        android:text="拨打手机：13207304569" />      //设置显示文字
    <TextView                                       //定义一个文本显示组件
        android:id="@+id/mytxt2"                    //组件的名称为 mytxt2
        style="@style/my_styles"                    //文字的显示调用 my_styles 样式
        android:layout_below="@+id/mytxt1"          //组件位于 mytxt1 组件的下方
        android:background="#000000"                //设置文字背景颜色
        android:text="湖南理工学院" />               //设置显示文字的为湖南理工学院
</RelativeLayout>
```

保存文件，程序运行结果如图 3.2 所示。虽然在布局文件里没有对文字的颜色字号等属性进行设置，但是因为调用了样式文件，所以显示的文字在颜色、阴影等方面都有变化。因为设置了电话号码链接，单击号码，可以进入电话拨打界面。

实例 3-3：在 Activity 程序中动态创建 TextView 组件

新建一个项目，项目命名为 Exam3_3，公司域名为 cn.hnist.org，在项目中对 MainActivity.java 文件做如下修改：

```
package org.hnist.cn.Exam3_3;      //包名
import android.support.v7.app.AppCompatActivity;
import android.os.Bundle;          //导入 os.Bundle 类
import android.graphics.Color;     //导入 graphics.Color 类，就能识别 RED 等
import android.widget.TextView;    //导入 widget.TextView 类
public class MainActivity extends AppCompatActivity {
@Override
protected void onCreate(Bundle savedInstanceState) { //覆写 onCreate()方法
    super.onCreate(savedInstanceState);              //调用父类的 onCreate()方法
  // setContentView(R.layout.activity_main);         //注意，这里没有调用布局文件
    TextView txt=new TextView(this);                 //定义一个文本显示组件
    txt.setText("www.hnist.cn");                     //设置组件文字为 www.hnist.cn
    txt.setBackgroundColor(Color.GRAY);              //设置背景颜色为灰色
    txt.setTextColor(Color.GREEN);                   //设置文字颜色为绿色
    txt.setTextSize(30);                             //设置文字大小为 30dp
    super.setContentView(txt);                       //设置 txt 组件显示
}}
```

保存文件，程序运行结果如图 3.3 所示。

图 3.2　Exam3_2 运行结果

图 3.3　Exam3_3 运行结果

3.3 按钮组件 Button

按钮组件是在人机交互时使用较多的组件，当用户进行某些选择的时候，就可以通过按钮的操作来接收用户的选择。Button 组件的层次关系如下：

```
java.lang.Object
    android.view.View
        android.widget.TextView
              android.widget.Button
```

直接子类：CompoundButton。

间接子类：CheckBox、RadioButton、ToggleButton。

要在 Android 程序中使用 Button 组件，必须在程序中使用下面的语句：

```
import android.widget.Button;                    //导入 widget. Button 类
```

1. Button 组件常见的属性和方法

由于 Button 组件继承了 TextView 类，TextView 类的属性，Botton 组件都可以使用，常见属性如表 3-3 所示。

Button 组件一般用来做事件处理，用于接收用户的选择并执行相应的程序或操作，Button 组件常用的方法如表 3-5 所示。

表 3-5　Button 组件常用的方法

方法	描述
public Boolean onKeyDown()	当用户按键时，该方法被调用
public Boolean onKeyUp()	当用户按键弹起后，该方法被调用
public Boolean onKeyLongPress()	当用户保持按键时，该方法被调用
public Boolean onKeyMultiple()	当用户多次调用时，该方法被调用
public void invalidateDrawable()	刷新 Drawable 对象
public void scheduleDrawable()	定义动画方案的下一帧
public void unscheduleDrawable()	取消 scheduleDrawable 定义的动画方案
public Boolean onPreDraw()	设置视图显示
public void sendAccessibilityEvent()	发送事件类型指定的 AccessibilityEvent。发送请求之前，需要检查 Accessibility 是否打开
public void sendAccessibilityEventUnchecked()	发送事件类型指定的 AccessibilityEvent。发送请求之前，不需要检查 Accessibility 是否打开
public void setOnKeyListener()	设置按键监听

2. Button 组件使用实例

因为 Button 组件是 TextView 的子类，如果将上面例子中的“TextView”换成“Button”也可以顺利执行。参照前面的例子将“TextView”换成“Button”，查看运行结果，注意，这里的按钮暂时单击没有效果，因为里面没有添加事件处理。限于篇幅，这里没有给出代码，附带的源代码 Exam3_4 中有示例，可以导入查看学习。需要说明的是，按钮一般不用来显示漂亮的文字效果，而是主要用来做事件处理，接收用户的选择后执行其他的程序或操作，通俗地说，就是当单击或双击该按钮后要进行的操作。

由 Button 派生而来 ToggleButton(开关按钮)是 Android 系统中比较常用的一个组件，是一个具有选中和未选择状态的按钮，而且在按中时和未按中的时候分别可以显示不同的文本，ToggleButton 常用于切换程序中的某种状态。

实例 3-4：ToggleButton 的使用

新建一个项目，项目命名为 Exam3_4.1，公司域名为 cn.hnist.org，在项目中对 activity_main.xml 文件做如下修改：

```
…
<TextView
    android:id="@+id/net_3g"
    android:text="启用 3G"
    android:layout_margin="10dp"
    android:layout_width="wrap_content"
    android:layout_height="wrap_content"/>
<ToggleButton
    android:id="@+id/toggleButton_3g"
    android:layout_width="wrap_content"
    android:layout_height="wrap_content"
    android:layout_marginRight="10dp"
    android:layout_alignParentRight="true"
    android:layout_alignBaseline="@id/net_3g"
    android:textOff="关闭"
    android:textOn="开启" />
</RelativeLayout>
```

保存文件，程序运行结果如图 3.4 所示，单击按钮，按钮上的文字会在设置的文字之间交替变化。

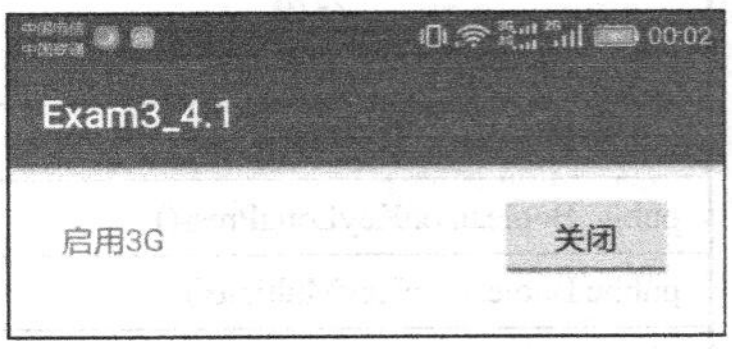

图 3.4　Exam3_4.1 运行结果

3.4　编辑框组件 EditText

在 Android 中，编辑框用 EditText 表示，用于在屏幕上显示文本编辑框，它既可以编辑单行文本，也可以编辑多行文本，如图 3.5 所示。除此之外，还可以输入指定格式的文本（如密码、电话号码、邮箱地址等），有了它就等于有了一扇和 Android 应用传输的窗口，通过它用户可以将数据传给 Android 应用，然后得到想要的数据。EditText 组件也是 TextView 的一个子类，其层次关系如下：

```
java.lang.Object
   android.view.View
      android.widget.TextView
         android.widget.EditText
```

直接子类：AutoCompleteTextView，ExtractEditText。

间接子类：MultiAutoCompleteTextView。

要在 Android 程序中使用 EditText 组件，必须在程序中使用下面的语句。

```
import android.widget.EditText;          //导入 widget.EditText 类
```

1. EditText 组件常见的属性和方法

EditText 组件继承自 TextView 类，除了具备 TextView 类的属性外，其他常用属性如表 3-6 所示。

表 3-6 EditText 组件常用的属性

属性名称	值	描述
android:textColor	自定义	设置字体颜色
android:textColorHint	自定义	设置提示信息文字的颜色，默认为灰色
android:capitalize	“sentences”、“words”、“characters”	设置英文字母大写类型。例如，“sentences”为仅第一个字母大写；“words”为每一个单词首字母大写，用空格区分单词；“characters”为每一个英文字母都大写
android:cursorVisible	true/false	设定光标为显示/隐藏，默认显示。如果设置为 false，则即使选中了也不显示光标栏
android:digits	自定义	设置允许输入哪些字符，如“1234567890.+-*/%\n()”
android:editable	true/false	设置是否可编辑。若值为“false”，则仍然可以获取光标，但是无法输入
android:hint	true/false	是否在 Text 为空时显示文字提示信息，可通过 textColorHint 设置提示信息的颜色
android:imeOptions	“normal”、“actionNext”、“actionDone”、“actionSearch”	设置软键盘的 Enter 键，部分输入法对此的支持可能不够好
android:includeFontPadding	true/false	设置文本是否包含顶部和底部额外空白，默认为 true
android:inputType	“phone”、“date”、“time”、“textPassword”等	设置文本的类型，用于帮助输入法显示合适的键盘类型，可以用|选择多个，取值可参考 android.text.InputType 类。取值包括“phone”、“date”、“time”等，分别对应“拨号键盘”、“日期键盘”、“时间键盘”等
android:ems	自定义	设置 TextView 的宽度为 N 个字符的宽度
android:maxLength	自定义	限制输入字符数。如设置为 5，那么仅可以输入 5 个汉字/数字/英文字母
android:lines	自定义	设置文本的行数，设置两行就显示两行，即使第二行没有数据
android:numeric	“integer”、“signed”、“decimal”	如果被设置，该 TextView 有一个数字输入法。例如，integer 正整数、signed 带符号整数、decimal 带小数点浮点数
android:selectAllOnFocus	true/false	是否自动选取该组件内的所有文本内容（前提是 EditText 有内容）
android:typeface	“normal”、“sans”、“serif”、“monospace”	设置字体

EditText 组件的属性可以在 XML 布局文件中使用上面的参数进行设置，也可以在 Activity 文件中通过调用方法来实现，表 3-7 给出了 EditText 组件常用方法。

表 3-7 EditText 组件常用的方法

方法	描述
public void setImeOptions()	设置软键盘的 Enter 键
public Charsequence getImeActionLable()	设置 IME 动作标签
public Boolean getDefaultEditable()	获取是否默认可编辑
public void setEllipse()	设置文件过长时控件的显示方式
public void setFreeezesText()	设置保存文本内容及光标位置
public Boolean getFreeezesText()	获取保存文本内容及光标位置
public void setGravity()	设置文本框在布局中的位置
setTransformationMethod()	设置隐藏密码或者显示密码
public int getGravity()	获取文本框在布局中的位置
public void setHint()	设置文本框为空时，文本框默认显示的字符
public Charsequence getHint()	获取文本框为空时，文本框默认显示的字符
public void setIncludeFontPadding()	设置文本框是否包含底部和顶端的额外空白
public void setMarqueeRepeatLimit()	在 ellipsize 指定 marquee 的情况下，设置重复滚动的次数，当设置为 marquee_forever 时表示无限次

2. EditText 组件使用实例

实例 3-5：EditText 的使用

新建一个项目，项目命名为 Exam3_5，公司域名为 cn.hnist.org，在 res 文件夹下选择 layout 文件夹，双击 activity_main.xml 文件，将代码做如下修改，当然，为了加快代码编写的速度，也可以结合拖动的方式进行修改，如图 3.5 所示。

```
...
    <EditText
        android:id="@+id/edit1"
        android:layout_width="fill_parent"
        android:layout_height="wrap_content"
        android:text="请输入您所在的单位："
        android:textSize="20dp" />
    <!-- 用于不可编辑的文本框 -->
    <EditText
        android:id="@+id/edit2"
        android:layout_width="fill_parent"
        android:layout_height="wrap_content"
        android:layout_below="@+id/edit1"          //设置组件位于 edit1 下方
        android:editable="false                    //设置文本框不可编辑
        android:text="www.hnist.cn"
        android:textSize="20dp" />
    <!-- 用于输入数字的文本框 -->
    <EditText
        android:id="@+id/edit3"
        android:layout_width="fill_parent"
        android:layout_height="wrap_content"
        android:layout_below="@+id/edit2"
        android:text="请输入电话号码"
        android:inputType="phone"                  //设置文本框只能输入电话号码
        android:textColorHint="#238745"            //设置文字颜色
        android:selectAllOnFocus="true"/>          //设置选中全部内容
    <!-- 用于输入密码的文本框 -->
    <EditText
        android:id="@+id/edit4"
        android:layout_width="fill_parent"
        android:layout_height="wrap_content"
        android:layout_below="@+id/edit3"
        android:hint="请输入密码"
        android:inputType="textPassword"           //设置文本框输入为密文显示
        android:textColorHint="#238745" />
</RelativeLayout>
```

保存文件，程序运行结果如图 3.6 所示。

可以在编辑框中输入数据，默认状态是输入英文，长按“请输入电话号码”编辑框，弹出数字键盘，在“请输入密码”输入框内输入字符，显示为密文。注意，第一个框输入时提示信息还在，输入电话号码和密码时提示信息就没有了，仔细体会代码的作用。

Text Fields (EditText)
Plain Text
Password
Password (Numeric)
E-mail
Phone
Postal Address
Multiline Text
Time
Date
Number
Number (Signed)
Number (Decimal)
AutoCompleteTextView
MultiAutoCompleteTextView

图 3.5 Android Studio 中的 EditText

图 3.6 Exam3_5 运行结果

实例 3-6：在 Activity 文件中动态创建 EditText 组件

（1）在 activity_main.xml 文件中添加如下代码：

```
<EditText
    android:id="@+id/edit5"
    android:layout_width="fill_parent"
    android:layout_height="wrap_content"
    android:layout_below="@+id/edit4"/>
```

保存文件，程序运行结果如图 3.7 所示，增加了一个没有内容的编辑框。

（2）对 MainActivity.java 文件做如下修改：

```
package org.hnist.cn.Exam3_5;
import android.support.v7.app.AppCompatActivity;
import android.os.Bundle;
import android.widget.EditText;             //导入 widget.EditText 类
public class MainActivity extends AppCompatActivity {
private EditText myedit=null;               //定义一个编辑框组件
@Override
protected void onCreate(Bundle savedInstanceState) {
    super.onCreate(savedInstanceState);
    setContentView(R.layout.activity_main);                  //调用布局文件
    this.myedit=(EditText)super.findViewById(R.id.edit5);  //获取编辑框组件
    this.myedit.setText("请输入您的姓名"); } } //设置编辑框中的内容
```

保存文件，程序运行结果如图 3.8 所示，增加的编辑框中有了提示内容。

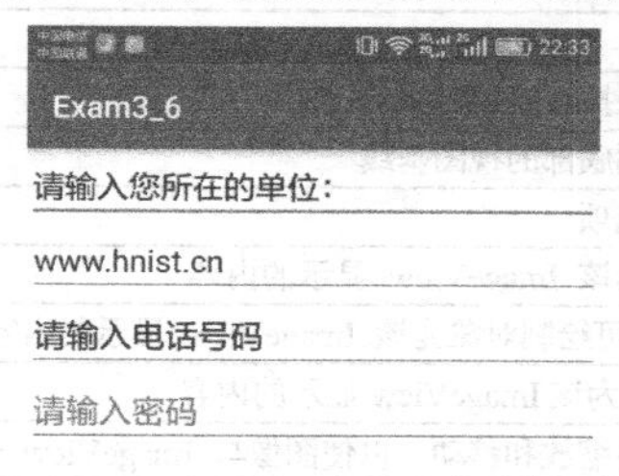

图 3.7 添加了新编辑框的 Exam3_6 运行结果

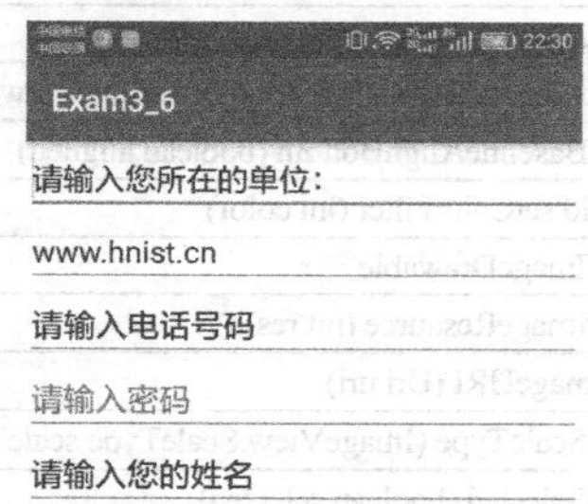

图 3.8 添加了新编辑框内容的 Exam3_6 运行结果

自己试着用修改代码的方式或者拖动的方式在最下面添加一个“提交”按钮。

3.5 图片视图组件 ImageView

ImageView 组件的主要功能是为图片展示提供一个容器，它可以加载各种来源（如资源或图片库）的图片文件，并提供缩放和着色等各种显示选项。其层次关系如下：

```
java.lang.Object
   android.view.View
      android.widget.ImageView
```

直接子类：ImageButton。

间接子类：ZoomButton。

要在 Android 程序中使用 ImageView 组件，必须在程序中使用下面的语句：

```
import android.widget.ImageView;                    //导入 widget.ImageView 类
```

1. ImageView 组件常用的属性和方法

ImageView 组件是 View 的子类，所以 View 的属性和方法它都具备，还具有以下常见的属性和方法，如表 3-8 和表 3-9 所示。

表 3-8 ImageView 常见属性

属性名称	描述
android:adjustViewBounds	设置该属性为真，可以在 ImageView 调整边界时保持图片的纵横比例（需要与 maxWidth、maxHeight 一起使用）
android:baseline	视图内基线的偏移量
android:baselineAlignBottom	如果为 true，图像视图将基线与父控件底部边缘对齐
android:cropToPadding	如果为真，会剪切图片以适应内边距的大小（单独设置无效果，需要与 scrollY 一起使用）
android:maxHeight	为视图提供最大高度的可选参数（单独使用无效，需要与 setAdjustViewBounds 一起使用）
android:maxWidth	为视图提供最大宽度的可选参数
android:scaleType	控制为了使图片适合 ImageView 的大小，相应地变更图片大小或移动图片
android:src	设置可绘制对象作为 ImageView 显示的内容
android:tint	为图片设置着色颜色

表 3-9 ImageView 常见方法

方法	描述
public void setMaxHeight (int maxHeight)	定义图片的最大高度
public void setMaxWidth (int maxWidth)	定义图片的最大宽度
public void setImageResource (int resId)	定义显示图片的 ID
public void setImageBitmap (Bitmap bm)	定义显示图片
public void setAlpha (int alpha)	设置透明度
public void setAdjustViewBounds (boolean adjustViewBounds)	调整边框时是否保持可绘制对象的比例
public void setBaselineAlignBottom (boolean aligned)	设置是否设置视图底部的视图基线
public final void setColorFilter (int color)	为图片设置着色选项
public void setImageDrawable	设置可绘制对象为该 ImageView 显示的内容
public void setImageResource (int resId)	通过资源 ID 设置可绘制对象为该 ImageView 显示的内容
ublic void setImageURI (Uri uri)	设置指定的 URI 为该 ImageView 显示的内容
public void setScaleType (ImageView.ScaleType scaleType)	控制图像应该如何缩放和移动，以使图像与 ImageView 一致
public void setSelected (boolean selected)	改变视图的选中状态

2. ImageView 组件使用实例

实例 3-7：ImageView 的使用

新建一个项目，项目命名为 Exam3_7，公司域名为 cn.hnist.org，将已有图片“test1.jpg”和“test2.jpg”文件复制到“\res\drawable”文件夹下，如图 3.9 所示。修改布局管理器文件使之加载“test1.jpg”图片文件，然后修改 Main_Activity.java 文件加载“test2.jpg”图片文件。

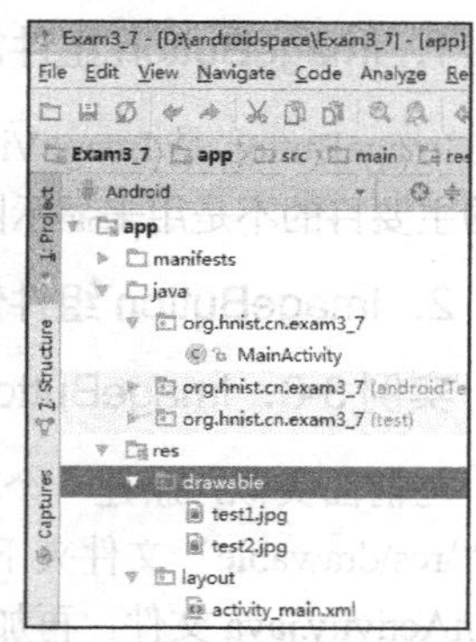

图 3.9　复制图片到 drawable 中

（1）修改 activity_main.xml 文件：

```
…
 <ImageView
    android:layout_width="wrap_content"
    android:layout_height="wrap_content"
    android:src="@drawable/test1" />
<ImageView
    android:id="@+id/myimg"
    android:layout_width="wrap_content"
    android:layout_height="wrap_content"
    android:layout_centerHorizontal="true"
    android:layout_marginTop="160dp"/>
</RelativeLayout>
```

注意：代码中第一个 ImageView 组件中加载了图片 test1，第二个 ImageView 组件中没有加载图片。

（2）修改 MainActivity.java 文件：

```
…
import android.widget.ImageView;          //导入widget. ImageView类
public class MainActivity extends AppCompatActivity {
private ImageView img= null;              //定义一个ImageView对象img
@Override
protected void onCreate(Bundle savedInstanceState)
 {
    super.onCreate(savedInstanceState);
    setContentView(R.layout.activity_main);
    this.img=(ImageView)
super.findViewById(R.id.myimg); //获得ImageView组件
    this.img.setImageResource(R.drawable.test2);
//设置显示的图片为test2.jpg
  }  }
```

保存文件，程序运行结果如图 3.10 所示，程序界面中显示了 2 张图片，其中“test1.jpg”是在布局管理器中加载的，“test2.jpg”是在 Main_Activity 中加载的。

图 3.10　Exam3_7 运行结果

3.6 图片按钮组件 ImageButton

ImageButton 是带有图标的按钮，与 Button 组件类似，可以直接使用 ImageButton 定义，其层次关系如下：

```
java.lang.Object
android.view.View
     android.widget.ImageView
         android.widget.ImageButton
```

要在 Android 程序中使用 ImageButton 组件，必须在程序中使用下面的语句：

```
import android.widget.ImageButton;        //导入widget.ImageButton类
```

1．ImageButton 组件常用的属性和方法

ImageButton 是 ImageView 的子类，因此 ImageView 的属性和方法它都具备。和 Button 组件类似，它的主要目的不是用来显示图片，而是用做事件处理，接收用户的选择后执行其他的程序或操作。

2．ImageButton 组件使用实例

实例 3-8：ImageButton 的使用

与前面类似，新建一个项目，项目命名为 Exam3_8，将图片“test1.jpg”和“test2.jpg”文件复制到“\res\drawable”文件夹下，修改布局管理器文件，加载“test1.jpg”图片文件到按钮上，然后修改 MainActivity.java 文件，再加载“test2.jpg”图片文件到按钮上。

（1）修改 activity_main.xml 文件，代码如下：

```
…
<ImageButton                                            //定义一个 ImageButton 组件
        android:layout_width="300dp"                    //定义组件宽度
        android:layout_height="200dp"                   //定义组件高度
        android:layout_centerHorizontal="true"          //水平居中
        android:src="@drawable/test1" />                //图片为 test1.jpg
<ImageButton                                            //定义一个 ImageButton 组件
        android:id="@+id/myimg"                         //命名为 myimg
        android:layout_width="300dp"                    //定义组件宽度
        android:layout_height="200dp"                   //定义组件高度
        android:layout_centerHorizontal="true"          //水平居中
        android:layout_marginTop="160dp"/>              //图片与顶部的距离
</RelativeLayout>
```

（2）修改 MainActivity.java 文件，代码如下：

```
…
import android.widget. ImageButton;          //导入 widget.ImageButton 类
public class MainActivity extends AppCompatActivity {
 private ImageButton img= null;              //定义一个 ImageButton 对象 img
    @Override
    protected void onCreate(Bundle savedInstanceState) {
      super.onCreate(savedInstanceState);
      setContentView(R.layout.activity_main);
      this.img=(ImageButton)super.findViewById(R.id.myimg);
                                                       //获得 ImageButton 组件
      this.img.setImageResource(R.drawable.test2);}}//设置按钮上的图片为 test2
```

保存文件，运行结果显示了 2 个图片按钮，ImageButton 与 ImageView 显示在屏幕上时区别不大。

3.7 单选按钮组件 RadioGroup

单选按钮组件在开发中提供了一种多选一的操作模式，例如，在性别选择时，用户需要从多个选择中选定一个选项，实现这个需求可在 Android 程序中使用 RadioGroup 组件来定义单选按钮并提供多个选项存放的容器，RadioGroup 组件的层次关系如下：

```
java.lang.Object
    android.view.View
        android.view.ViewGroup
            android.widget.LinearLayout
```

```
android.widget.RadioGroup
```

RadioGroup 组件提供的只是一个单选按钮的容器，在 Android 平台中提供了 RadioButton 类，可通过该类配置多个选项，设置单选按钮的内容，RadioButton 组件的层次关系如下：

```
RadioGroupjava.lang.Object
 android.view.View
      android.widget.TextView
         android.widget.Button
             android.widget.CompoundButton
                 android.widget.RadioButton
```

要在 Android 程序中使用 RadioGroup 组件，必须在程序中使用下面的语句：

```
import android.widget.RadioGroup;                 //导入 widget.RadioGroup 类
```

1. RadioGroup 组件常见的属性

RadioGroup 组件是 TextView 的子类，所以 TextView 组件的属性和方法它都继承了，其属性如表 3-3 所示，方法如表 3-4 所示，RadioGroup 组件其他常见方法如表 3-10 所示。

表 3-10 RadioGroup 组件常见方法

方法	描述
public void check (int id)	设置要选中的单选按钮编号
public void clearCheck ()	清空选中状态
public int getCheckedRadioButtonId()	取得选中按钮的 RadioButton 的 ID
public void setOnCheckedChangeListener(RadioGroup.OnCheckedChangeListener listener)	设置单选按钮选中的操作事件
public void setOnHierarchyChangeListener (ViewGroup.OnHierarchyChangeListener listener)	注册一个当该单选按钮组中的选中状态发生改变时所要调用的回调函数

2. RadioGroup 组件使用实例

实例 3-9：RadioGroup 组件的使用

新建一个项目，项目名称为 Exam3_9，该程序的目标是显示一个单项选择题的界面。

（1）修改 activity_main.xml 文件，定义一个 RadioGroup 组件，里面包含四个 RadioButton 组件，其中 2 个选项有文字提示信息，另外 2 个没有，具体代码如下：

```
...
  <TextView
       android:layout_width="wrap_content"
       android:layout_height="wrap_content"
       android:textSize="20sp"
       android:text="下面的计算哪个是正确的：" />
  <RadioGroup                                  //定义一个 RadioGroup 组件
       android:id="@+id/Radio1"                //给组件命名为 Radio1
       android:layout_width="wrap_content"
       android:layout_height="wrap_content"
       android:orientation="vertical"> //RadioButton 对齐方式为垂直对齐
 <RadioButton                                  //定义一个 RadioButton 组件
       android:id="@+id/RBut1"                 //命名为 RBut1
       android:layout_width="wrap_content"
       android:layout_height="wrap_content"
       android:layout_marginTop="30dp"
       android:textSize="20sp"
       android:text="  1+2=1"/>                //显示内容为 1+2=1
 <RadioButton                                  //定义一个 RadioButton 组件
```

```
            android:id="@+id/RBut2"               //命名为RBut2
            android:layout_width="wrap_content"
            android:layout_height="wrap_content"
            android:textSize="20sp"
            android:text="    1+2=2"/>            //显示内容为1+2=1
      <RadioButton                                //定义一个RadioButton组件
            android:id="@+id/RBut3"               //命名为RBut3
            android:layout_width="wrap_content"
            android:layout_height="wrap_content"
            android:textSize="20sp"/>
      <RadioButton                                //定义一个RadioButton组件
            android:id="@+id/RBut4"               //命名为RBut4
            android:layout_width="wrap_content"
            android:layout_height="wrap_content"
            android:textSize="20sp"/>
    </RadioGroup>
</RelativeLayout>
```

注意：这里的选项3和选项4没有显示值，运行结果如图3.11(a)所示。

（2）修改MainActivity.java文件，在RBut3和RBut4选项处也显示提示信息，代码如下：

```
...
import android.widget.RadioButton;        //导入widget. RadioButton类
public class MainActivity extends AppCompatActivity {
   private RadioButton rbut1=null;        //定义一个RadioButton
   private RadioButton rbut2=null;        //定义一个RadioButton
   @Override
   protected void onCreate(Bundle savedInstanceState) {
      super.onCreate(savedInstanceState);
      setContentView(R.layout.activity_main);
      this.rbut1=(RadioButton) super.findViewById(R.id.RBut3);  //获得RadioButton
      this.rbut2=(RadioButton) super.findViewById(R.id.RBut4); //获得RadioButton
      this.rbut1.setText("    1+2=3");     //设置RadioButton的内容为“1+2=3”
      this.rbut2.setText("    1+2=4");}}   //设置RadioButton的内容为“1+2=4”
```

保存文件，程序运行结果如图3.11(b)所示。当然，各选项的内容也可以在string.xml文件中定义好，然后在Main_Activity.java文件和activity_main.xml文件中进行调用，请读者自行完成。

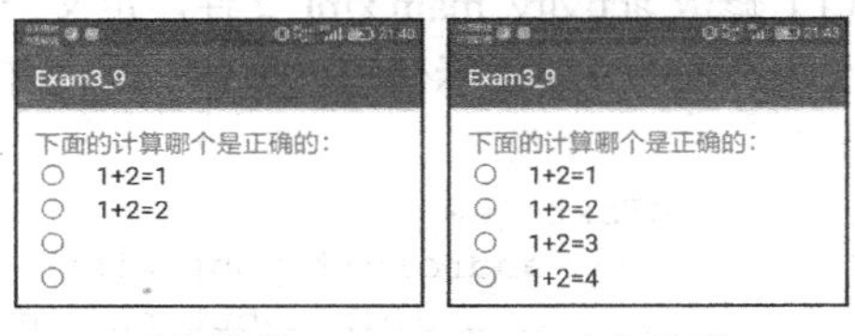

（a）选项3　　（b）选项4

图3.11　Exam3_8运行结果

3.8　复选框组件CheckBox

CheckBox组件的主要功能是完成多项选择的操作，例如，在选择兴趣爱好时可能会存在多种选择的情况，Android平台提供的CheckBox组件可以实现这样的功能，该组件层次关系如下：

```
java.lang.Object
    android.view.View
      android.widget.TextView
        android.widget.Button
          android.widget.CompoundButton
             android.widget.CheckBox
```

要在Android程序中使用CheckBox组件，必须在程序中使用下面的语句：

```
import android.widget.CheckBox;                //导入 widget.CheckBox 类
```

1. CheckBox 组件常见的属性和方法

CheckBox 组件是 TextView 的子类，其属性如表 3-3 所示。另外，CheckBox 组件还有以下常用的方法，如表 3-11 所示。

表 3-11 CheckBox 常见方法

方法	描述
public CheckBox(Context context)	实例化 CheckBox 组件
public void setChecked (boolean checked)	设置默认选中
public int getCheckedRadioButtonId()	取得选中按钮的 RadioButton 的 ID
public Boolean isChecked()	判断组件状态是否选中
public void onRestoreInstanceState()	设置视图恢复以前的状态
public Boolean iperformClick()	执行 click 动作触发事件监听器
public void setButtonDrawable()	根据 Drawable 对象设置组件的背景
public void setChecked()	设置组件的状态
public void setOnCheckedChangeListener()	设置事件监听器
public void toggle()	改变按钮当前的状态
public CharSequence onCreateDrawableState()	获取文本框为空时文本框中的内容
public int onCreateDrawableState()	为当前视图生成新的 Drawable 状态

2. CheckBox 组件使用实例

实例 3-10：CheckBox 的使用

新建一个项目，项目名称为 Exam3_10，该项目用于用户进行个人爱好的选择。修改布局管理器文件 activity_main.xml，进行一个选项的定义，其他两个选项的数据来源于 string.xml 文件。

（1）修改 activity_main.xml 文件，代码如下：

```
...
    <TextView
        android:id="@+id/txt1"
        android:layout_width="wrap_content"
        android:layout_height="wrap_content"
        android:text="您的喜爱的体育运动有："
        android:textSize="18dp" />                //设置文字的大小为 18dp
    <CheckBox                                     //定义一个 CheckBox
        android:id="@+id/mybox1"                  //命名为 mybox1
        android:layout_width="wrap_content"
        android:layout_height="wrap_content"
        android:layout_below="@+id/txt1"
        android:layout_marginTop="10dp"
        android:text="@string/hobby1" />  //显示的内容来源于 string.xml 中定义的 hobby1
    <CheckBox                                 //定义一个 CheckBox
        android:id="@+id/mybox2"              //命名为 mybox2
        android:layout_width="wrap_content"
        android:layout_height="wrap_content"
        android:layout_alignBottom="@+id/mybox1"
        android:layout_toRightOf="@+id/mybox1"
        android:text="@string/hobby2"  />//显示的内容来源于 string.xml 中定义的 hobby2
    <CheckBox                                 //定义一个 CheckBox
```

```
        android:id="@+id/mybox3"        //命名为mybox3
        android:layout_width="wrap_content"
        android:layout_height="wrap_content"
        android:layout_alignBottom="@+id/mybox2"
        android:layout_toRightOf="@+id/mybox2"
        android:checked="true"                  //处于被选中状态
        android:text="羽毛球" />               //显示的内容为"羽毛球"
</RelativeLayout>
```

（2）修改 string.xml 文件，代码如下：

```
<resources>
    <string name="app_name">Exam3_9</string>
    <string name="action_settings">Settings</string>
    <string name="hello_world">Hello world!</string>
    <string name="hobby1">篮球</string>      //定义 hobby1 的值为篮球
    <string name="hobby2">排球</string>      //定义 hobby2 的值为排球
</resources>
```

保存文件，程序运行结果如图 3.12 所示。

当然，也可以在 Activity 的 MainActivity.java 文件中利用表 3-11 提供的方法来定义 CheckBox，设置各个选项的内容，请读者自行完成。

图 3.12　Exam3_10 运行结果

3.9　下拉列表框组件 Spinner

Spinner 组件的功能类似于 RadioGroup 组件，它可以为用户提供列表的选择方式，一个 Spinner 对象包含多个子项，子项来源于与之相关联的适配器，每个子项只有两种状态——选中或未被选中。在 Android 平台中可以使用 android.widget.Spinner 类实现该组件。它的层次关系如下：

```
java.lang.Object
    android.view.View
        android.view.ViewGroup
            android.widget.AdapterView<T extends android.widget.Adapter>
                android.widget.AbsSpinner
                    android.widget.Spinner
```

要在 Android 程序中使用 Spinner 组件，必须在程序中使用下面的语句：

```
import android.widget.Spinner;                    //导入widget.Spinner类
```

1．Spinner 组件常见的属性和方法

Spinner 组件是 View 的子类，所以 Spinner 组件的属性如表 3-3 所示。此外，Spinner 组件还有一个常用的属性——android:prompt，该属性在下拉列表对话框显示时显示，也就是显示对话框的标题。Spinner 组件常用方法如表 3-12 所示。

表 3-12　Spinner 常见方法

方法	描述
public CharSequence getPrompt ()	取得提示文字
public void setPrompt (CharSequence prompt)	设置组件的提示文字
public void setAdapter (SpinnerAdapter adapter)	设置下拉列表项
public CharSequence getPrompt()	得到提示信息

续表

方法	描述
public void setOnItemClickListener(AdapterView.OnItemClickListener l)	设置选项单击事件
public void onClick(DialogInterface dialog, int which)	当单击弹出框中的项时，此方法将被调用
public Boolean performClick()	如果它被定义，则调用此视图的 OnClickListener
public void setPromptId(CharSequence prompt)	设置弹出视图的标题字
public ArrayAdapter (Context context, int textViewResourceId, List<T> objects)	定义 ArrayAdapter 对象，传入一个 Activity 实例、列表项的显示风格、List 集合数据
public ArrayAdapter (Context context, int textViewResourceId, T[] objects)	定义 ArrayAdapter 对象，传入一个 Activity 实例、列表项的显示风格、数组数据
public static ArrayAdapter<CharSequence> createFromResource (Context context, int textArrayResId, int textViewResId)	通过静态方法取得 ArrayAdapter 对象，传入 Activity 实例、资源文件的 ID、列表项的显示风格
public void setDropDownViewResource (int resource)	设置下拉列表项的显示风格

2．Spinner 组件使用实例

在 Android 平台中，可以在 activity_main.xml 文件中定义 Spinner 组件，但是不能在该文件中直接设置其显示的列表项，下拉列表框中的列表项可用以下几种方式进行配置。

方式一：直接通过资源文件配置。

方式二：通过 android.widget.ArrayAdapter 类读取资源文件的数据。

方式三：通过 android.widget.ArrayAdapter 类设置具体的数据。

实例 3-11：Spinner 组件的使用

新建一个项目，项目的名称为 Exam3_11，该项目显示的界面是让用户选择学历、专业和班级，分别用上面的三种方式来设置具体的列表项数据。

（1）修改 activity_main.xml 文件，具体代码如下：

```
…
    <TextView
        android:id="@+id/txt1"
        android:layout_width="wrap_content"
        android:layout_height="wrap_content"
        android:textSize="20dp"
        android:text="您的学历是" />
    <Spinner                                          //定义一个 Spinner 组件
        android:id="@+id/spin1"                       //命名为 spin1
        android:layout_width="wrap_content"
        android:layout_height="wrap_content"
        android:layout_below="@+id/txt1"
        android:prompt="@string/eduprompt" //设置提示信息，在 string.xml 中定义
        android:entries="@array/edu"       //选项的列表项来源于 edu.xml 文件
        android:spinnerMode="dialog" />    //设置为对话框模式时，prompt 才有效

    <TextView
        android:id="@+id/txt2"
        android:layout_width="wrap_content"
        android:layout_height="wrap_content"
        android:layout_below="@+id/spin1"
        android:textSize="20dp"
```

```
        android:text="您的专业是"/>
    <Spinner                                        //定义一个 Spinner 组件
        android:id="@+id/spin2"                     //命名为 spin2
        android:layout_width="wrap_content"
        android:layout_height="wrap_content"
        android:layout_below="@+id/txt2" />
    <TextView
        android:id="@+id/txt3"
        android:layout_width="wrap_content"
        android:layout_height="wrap_content"
        android:layout_below="@+id/spin2"
        android:textSize="20dp"
        android:text="您的班级是"/>
    <Spinner                                        //定义一个 Spinner 组件
        android:id="@+id/spin3"                     //命名为 spin3
        android:layout_width="wrap_content"
        android:layout_height="wrap_content"
        android:layout_below="@+id/txt3"/>
</RelativeLayout>
```

（2）打开 string.xml 文件，在</resources>标签前面添加一行语句：“<string name="eduprompt">请选择您的学历：</string>”。具体代码如下：

```
<resources>
    <string name="app_name">Exam3_10</string>
    <string name="eduprompt">请选择您的学历：</string>
</resources>
```

（3）代码中的“android:entries="@array/edu"”语句表示这个选项的列表数据项来源于 edu.xml 文件，edu.xml 文件要事先建立完成，以免程序报错。

① 找到 res/values 文件夹，右击该文件夹，选择“New→ XML→ Values XML File”选项，如图 3.13 所示。

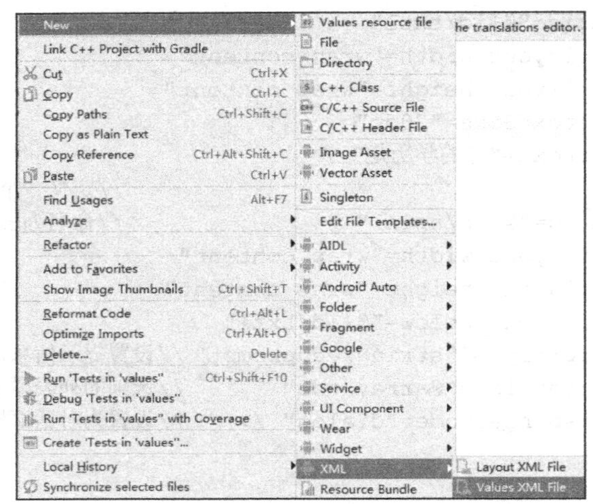

图 3.13　新建一个 XML 文件

② 在弹出的对话框中的 Values File Name 文本框中输入“edu”，然后单击“Finish”按钮，如图 3.14 所示，进入 edu.xml 文件的编辑状态，输入如下的内容：

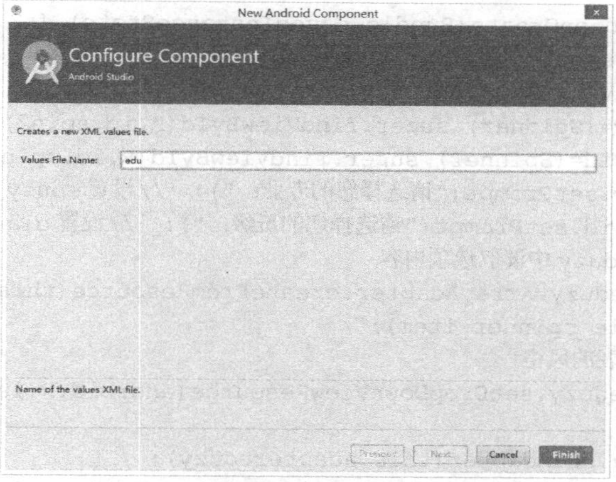

图 3.14 为新建的 XML 文件命名

```
<?xml version="1.0" encoding="utf-8"?>
<resources>
    <string-array name="edu">
        <item>高中</item>
        <item>大专</item>
        <item>本科</item>
        <item>硕士</item>
        <item>博士</item>
    </string-array>
</resources>
```

类似的，新建 eduzy.xml 文件，文件内容如下：

```
<?xml version="1.0" encoding="utf-8"?>
<resources>
    <string-array name="eduzy">
        <item>信息工程</item>
        <item>通    信</item>
        <item>电子信息</item>
        <item>机械电子</item>
    </string-array>
</resources>
```

（4）通过 android.widget.ArrayAdapter 类读取资源文件 eduzy.xml 中的数据，并通过该类设置具体的数据，对 MainActivity.java 文件进行修改，代码如下所示：

```
…
import android.widget.Spinner;                 //导入 widget.Spinner 类
import android.widget.ArrayAdapter;            //导入 widget.ArrayAdapter 类
import java.util.ArrayList;                    //导入 java.util.ArrayLis 类
import java.util.List;                         //导入 java.util.List 类
public class MainActivity extends AppCompatActivity {
private Spinner eduzy=null;                    //定义表示专业的 Spinner
private Spinner classid=null;                  //定义表示班级的 Spinner
private ArrayAdapter<CharSequence> adaptereduzy=null; //定义下拉列表适配器
private ArrayAdapter<CharSequence> adapterclassid=null;//定义下拉列表适配器
private List<CharSequence> dataclassid=null;   //定义 List 集合来保存下拉列表选项
@Override
```

```
    protected void onCreate(Bundle savedInstanceState) {
        super.onCreate(savedInstanceState);
        setContentView(R.layout.activity_main);
        this.eduzy=(Spinner) super.findViewById(R.id.spin2);  //获得 Spinner 组件
        this.classid=(Spinner) super.findViewById(R.id.spin3);//获得 Spinner 组件
        this.eduzy.setPrompt("请选择您的专业: ");  //设置 eduzy 下拉列表的提示信息
        this.classid.setPrompt("请选择您的班级: ");  //设置 classid 下拉列表的提示信息
        从资源文件 eduzy 中读取选项内容
      this.adaptereduzy=ArrayAdapter.createFromResource(this,R.array.eduzy,andro
id.R.layout.simple_spinner_item);
       //设置下拉列表的风格
      this.adaptereduzy.setDropDownViewResource(android.R.layout.simple_spinner_
dropdown_item);
        this.eduzy.setAdapter(this.adaptereduzy);              //设置下拉列表的选项内容
        this.dataclassid=new ArrayList<CharSequence>();      //实例化 List 集合
        this.dataclassid.add("信工 12-2BF");              //向 List 集合中添加选项内容
        this.dataclassid.add("通信 12-2BF");              //向 List 集合中添加选项内容
        this.dataclassid.add("电信 12-2BF");              //向 List 集合中添加选项内容
        this.dataclassid.add("机电 12-2BF");              //向 List 集合中添加选项内容
        //获得一个下拉列表适配器
        this.adapterclassid=new ArrayAdapter<CharSequence>(this,android.R.layout.
              simple_spinner_item,this.dataclassid);
        //设置下拉列表的风格
        this.adapterclassid.setDropDownViewResource(android.R.layout.simple_spi
nner_dropdown_item);
        this.classid.setAdapter(this.adapterclassid);}} //设置下拉列表选项内容
```

上述代码中设置了 3 个下拉列表，其中，表示学历的下拉列表数据来源于 edu.xml；表示专业的下拉列表在 Activity 程序中引用，数据来源于 eduzy.xml；表示班级的下拉列表在 Activity 程序中引用，数据也在 Activity 程序中添加。保存所有文件，程序运行后，可以进行学历、专业和班级的选择，如图 3.15 所示。

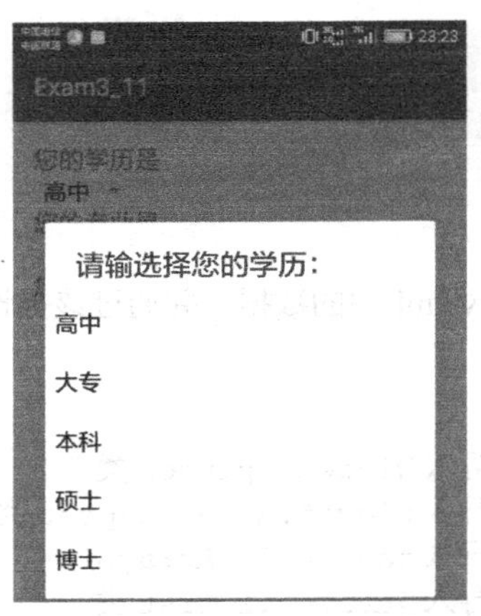

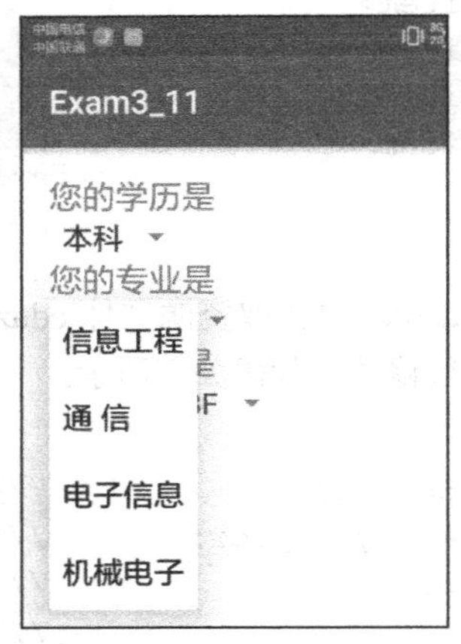

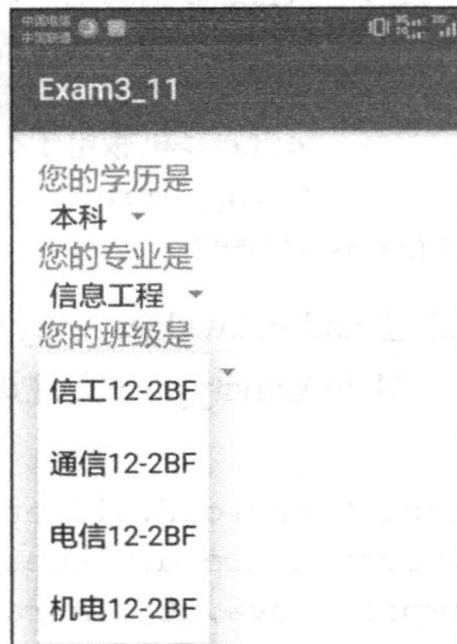

图 3.15　Exam3_11 运行结果

3.10　信息提示框组件 Toast

Android 平台中提供了提示界面效果，这种提示不会打断用户的正常操作，这就是信息提示框组件

——Toast，它以浮于应用程序之上的形式呈现给用户。因为它并不获得焦点，即使用户正在进行内容输入也不会受到影响。它的目标是尽可能以不显眼的方式，使用户看到提供的信息。该组件直接继承java.lang.Object，其类层次结构如下：

```
java.lang.Object
    android.widget.Toast
```

要在 Android 程序中使用 Toast 组件，必须在程序中使用下面的语句：

```
import android.widget.Toast;                //导入 widget.Toast 类
```

1．Toast 组件常见的属性和方法

信息提示框 Toast 组件的属性不多，Toast 组件常用的方法如表 3-13 所示。

表 3-13　Toast 组件常见的方法

方法	描述
public static final int LENGTH_LONG	常量，持续显示视图或文本提示较长时间
public static final int LENGTH_SHORT	常量，显示时间较短，为默认值
public Toast(Context context)	创建 Toast 对象
public static Toast makeText(Context context, int resId, int duration)	创建一个 Toast 对象，并指定显示文本资源的 ID 和信息的显示时间
public static Toast makeText(Context context, CharSequence text, int duration)	创建一个 Toast 对象，并指定显示文本资源和信息的显示时间
public void show()	显示信息
public void setDuration(int duration)	设置显示的时间
public void setView(View view)	设置显示的 View 组件
public void setText(int resId)	设置显示的文字资源 ID
public void setText(CharSequence s)	直接设置要显示的文字
public void setGravity(int gravity, int xOffset, int yOffset)	设置组件的对齐方式
public View getView()	取得内部包含的 View 组件
public int getXOffset()	返回组件的 X 坐标位置
public int getYOffset()	返回组件的 Y 坐标位置
public void cancel()	取消显示

2．Toast 组件使用实例

实例 3-12：Toast 的使用

新建一个项目，项目命名为 Exam3_12，该程序的目标是显示一个长时间提示信息。

（1）修改 activity_main.xml 文件，代码如下：

```
...
    <TextView
        android:id="@+id/txt1"
        android:layout_width="wrap_content"
        android:layout_height="wrap_content"
        android:layout_centerHorizontal="true"
        android:textSize="20dp"
        android:text="Toast 演示实例" />
</RelativeLayout>
```

（2）修改 MainActivity.java 文件，代码如下：

```
…
import android.widget.Toast;                                //导入 widget.Toast 类
public class MainActivity extends AppCompatActivity {
@Override
protected void onCreate(Bundle savedInstanceState) {
    super.onCreate(savedInstanceState);
    setContentView(R.layout.activity_main);
    //创建 Toast，设置提示信息和显示时间长短
    Toast myToast = Toast.makeText(this, "长时间显示的 Toast 信息提示框",Toast.
LENGTH_LONG) ;
    myToast.setGravity(BIND_AUTO_CREATE, 5, 10);  //设置组件的对齐方式
    myToast.show();}}//显示提示信息
```

保存文件，默认的提示信息显示在屏幕的底部，可以通过 setGravity()方法修改提示信息的位置，程序运行结果如图 3.16 所示，程序界面显示几秒钟后，下面的提示信息会自动消失。

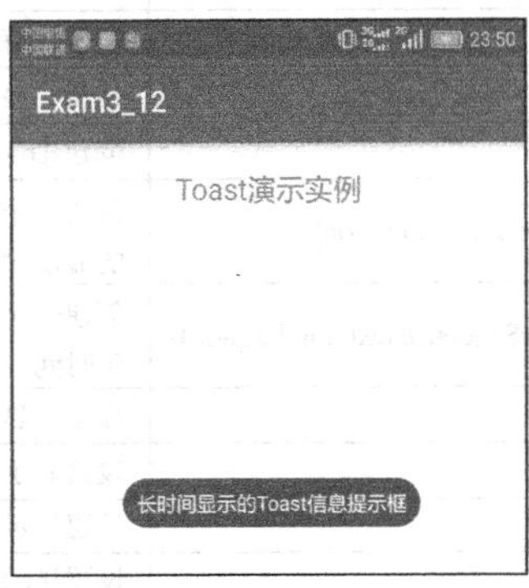

图 3.16　Exam3_12 运行结果

3.11　布局编辑器

Android Studio 中的布局编辑器组件如图 3.17 所示，在 Android Studio 布局编辑器中，可以快速地通过将控件拖入编辑器组件来代替手写 XML 来创建布局。

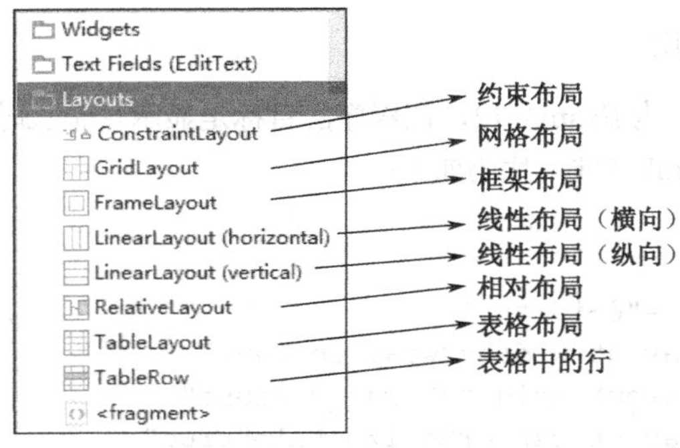

图 3.17　Android Studio 中的布局编辑器组件

打开一个布局 XML 文件时布局编辑器也会被打开。编辑器中的各部分如图 3.18 所示。

可以通过工具栏改变布局的样式，工具栏上的按钮显示彩色的布局预览界面，蓝图界面按钮显示了每个组件的轮廓图。按钮并排查看蓝图界面与设计界面。按钮改变横向或纵向手机界面。

Nexus4 按钮设置设备类型和分辨率，也可以通过拖拽布局的右下角来调整设备的大小。25 按钮选择在哪个版本的 Android 版本中预览界面。AppTheme 按钮选择预览界面的主题。Language 按钮选择界面所要显示的语言。按钮从可选的布局中选择一个，或者创建一个新的布局变量。

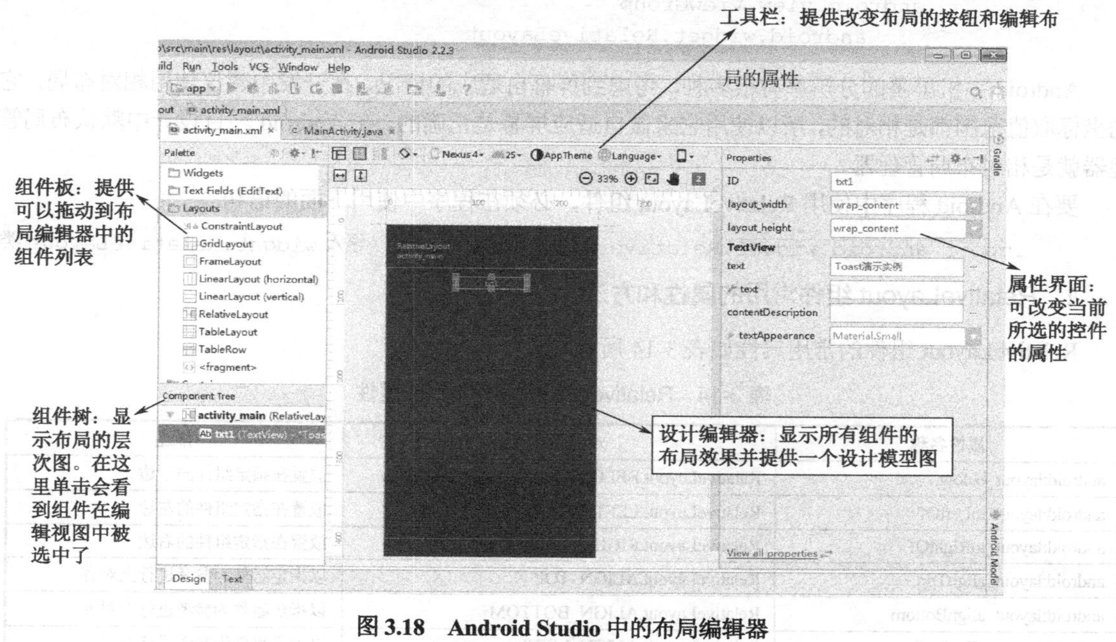

图 3.18　Android Studio 中的布局编辑器

当创建一个新的 APP 项目时，可能需要多个布局文件，默认在 res/layout 目录下只有一个布局文件 activity_main.xml，如果要新建一个布局文件，则可以通过以下步骤完成：右击要添加布局的项目，选择“New→XML→ Layout XML File”选项，在弹出的窗口中，输入要创建文件的名称，选择根布局的标签和布局所属的来源，然后单击“Finish”按钮。

如果要为已存在的布局创建横竖屏切换时可选的布局，可以通过布局变量来实现，步骤如下：打开布局文件，选择“Design”标签，单击工具栏中的按钮，建议直接选择“Create Landscape Variant”选项，在的下拉列表中选择来进行横竖屏切换即可。

布局编辑器通过在设计编辑器中拖动控件和属性窗口的简化属性完成了大量的工作，使界面设计更加快捷，应该说明的是，设计编辑器中显示的只是预览，可能与实际显示有差别。

Android Studio 平台中的布局管理器组件常见的有以下几类。

LinearLayout 组件：线性布局管理器，分为水平和垂直两种，只能进行单行布局。

FrameLayout 组件：所有的组件放在左上角，一个覆盖另一个。

TableLayout 组件：任意行和列的表格布局管理器，其中 TableRow 代表一行，可以向行中增加组件。

RelativeLayout 组件：相对布局管理器，根据最近的一个视图组件，或顶层父组件来确定下一个组件的位置。它是 Android 平台建议采用的布局管理器。

ConstraintLayout 组件：约束布局管理器，Android Studio 2.2 版本提供了约束布局，不需要多个布局的嵌套就可以实现复杂的用户界面。

3.12　相对布局管理器组件 RelativeLayout

相对布局管理器是指参考其他控件进行组件的放置，可以将组件摆放在一个指定参考组件的上、下、左、右等位置，也可以直接通过各个组件提供的属性完成。组件的层次关系如下：

```
java.lang.Object
    android.view.View
        android.view.ViewGroup
            android.widget.RelativeLayout
```

Android 手机屏幕的分辨率有很多种，考虑到屏幕自适应的情况，在开发中建议使用相对布局，它的坐标取值范围都是相对的，所以使用它来做自适应屏幕是正确的。在 Android 项目设计中默认布局管理器就是相对布局管理器。

要在 Android 程序中使用 RelativeLayout 组件，必须在程序中使用下面的语句：

```
import android.widget.RelativeLayout;          //导入 widget.RelativeLayout 类
```

1. RelativeLayout 组件常用的属性和方法

RelativeLayout 组件的常用属性如表 3-14 所示。

表 3-14　RelativeLayout 组件常用的属性

属性名称	对应的常量	描述
android:layout_below	RelativeLayout.BELOW	放置在指定组件的下边
android:layout_toLeftOf	RelativeLayout.LEFT_OF	放置在指定组件的左边
android:layout_toRightOf	RelativeLayout.RIGHT_OF	放置在指定组件的右边
android:layout_alignTop	RelativeLayout.ALIGN_TOP	以指定组件为参考进行上对齐
android:layout_alignBottom	RelativeLayout.ALIGN_BOTTOM	以指定组件为参考进行下对齐
android:layout_alignLeft	RelativeLayout.ALIGN_LEFT	以指定组件为参考进行左对齐
android:layout_alignRight	RelativeLayout.ALIGN_RIGHT	以指定组件为参考进行右对齐

如果要在程序中控制 RelativeLayout 组件的操作，则需要对一些布局参数进行配置，布局参数保存在 RelativeLayout.LayoutParams 类中。组件的层次关系如下：

```
java.lang.Object
    android.view.ViewGroup.LayoutParams
        android.view.ViewGroup.MarginLayoutParams
            android.widget. RelativeLayout.LayoutParams
```

RelativeLayout.LayoutParams 类提供了以下操作方法，如表 3-15 所示。

表 3-15　RelativeLayout.LayoutParams 类常用的方法

方法	描述
public RelativeLayout.LayoutParams (int w, int h)	指定 RelativeLayout 组件布局的宽度和高度
public void addRule (int verb, int anchor)	增加指定的参数规则
public int[] getRules ()	取得一个组件的全部参数规则

2. RelativeLayout 组件使用实例

实例 3-13：RelativeLayout 的使用

前面实例中的布局文件中的布局管理器采用 RelativeLayout 新建了一个项目，项目命名为 Exam3_13，将已有图片“a1.jpg”和“a2.jpg”复制到\res\drawable 文件夹中，利用相对布局管理器对组件进行放置。

（1）修改 activity_main.xml 文件，代码如下：

```
…
    <ImageView                                           //定义一个 ImageView 组件
        android:id="@+id/img1"                           //命名为 img1
        android:src="@drawable/a2"                       //图片来源于 a2.jpg
        android:layout_width="match_parent"              //设置图片宽度为屏幕宽度
        android:layout_height="100dp" />                 //组件高度为 100dp
    <TextView
        android:id="@+id/mytxt"
        android:layout_width="match_parent"
        android:layout_height="wrap_content"             //组件高度为 100dp
        android:textSize="30dp"                          //设置文字大小为 30dp
        android:gravity="center"                         //设置文字居中显示
        android:layout_below="@+id/img1"                 //设置文字位于 img1 下方
        android:text="湖南理工学院"/>
    <ImageView
        android:id="@+id/img2"
        android:src="@drawable/a1"
        android:layout_width="match_parent"
        android:layout_height="wrap_content"
        android:layout_centerVertical="true" />          //设置图片垂直居中显示
    <Button
        android:text="学生入口"
        android:id="@+id/butstu"
        android:layout_width="wrap_content"
        android:layout_height="wrap_content"
        android:layout_marginRight="27dp"                //设置按钮与右边间距为 27dp
        android:layout_alignBaseline="@+id/buttea"       //设置与 buttea 底部对齐
        android:layout_alignParentEnd="true" />
    <Button
        android:text="教师入口"
        android:id="@+id/buttea"
        android:layout_width="wrap_content"
        android:layout_height="wrap_content"
        android:layout_marginLeft="27dp"                 //设置按钮与左边间距为 27dp
        android:layout_marginTop="25dp"                  //设置按钮与上面组件间距为
25dp
        android:layout_below="@+id/img2" />
</RelativeLayout>
```

保存文件，程序运行结果如图 3.19 所示。

也可以在 Activity 中动态生成 RelativeLayout 组件，这里不再给出代码，感兴趣的读者可以下载 Exam3_14 进行学习。

图 3.19　Exam3_13 运行结果

3.13　线性布局管理器组件 LinearLayout

线性布局的形式可以分为两种：第一种是横向线性布局，第二种是纵向线性布局。它们以线性的形式一个个排列出来的，纯线性布局的缺点是不方便修改控件的显示位置，在开发中一般会以线性布局与相对布局嵌套的形式设置布局。

LinearLayout 层次关系如下：

```
java.lang.Object
    android.view.View
         android.view.ViewGroup
              android.widget.LinearLayout
```

要在 Android 程序中使用 LinearLayout 组件，必须在程序中使用下面的语句。

```
import android.widget.LinearLayout;           //导入 widget.LinearLayout 类
```

1. LinearLayout 组件常用的方法

LinearLayout 组件常用的方法如表 3-16 所示。

表 3-16　LinearLayout 组件常用的方法

方法及常量	描述
public static final int HORIZONTAL	常量，设置水平对齐
public static final int VERTICAL	常量，设置垂直对齐
public LinearLayout(Context context)	创建 LinearLayout 类的对象
public void addView(View child, ViewGroup.LayoutParams params)	增加组件并且指定布局参数
public void addView(View child)	增加组件
protected void onDraw(Canvas canvas)	用于图形绘制的方法
public void setOrientation(int orientation)	设置对齐方式

如果要在程序中控制 LinearLayout 的操作，则需要对一些布局参数进行配置，布局参数保存在 LinearLayout.LayoutParams 类中。其层次关系如下：

```
java.lang.Object
     android.view.ViewGroup.LayoutParams
          android.view.ViewGroup.MarginLayoutParams
               android.widget.LinearLayout.LayoutParams
```

LinearLayout.LayoutParams 类提供了以下构造方法。

public LinearLayout.LayoutParams (int width, int height)，创建布局参数时需要传递布局参数的宽度和高度，这两个参数可以通过下面的两个常量来提供。

```
public static final int FILL_PARENT，填充全部父控件
public static final int WRAP_CONTENT，包裹自身控件
```

2. LinearLayout 组件使用实例

实例 3-14：LinearLayout 的使用

新建一个项目，项目命名为 Exam3_15，利用线性布局管理器对组件进行放置。

（1）建立一个布局文件 activity_main.xml 文件，代码如下：

```
<?xml version="1.0" encoding="utf-8"?>
<LinearLayout                                       //采用线性布局管理器
    xmlns:android="http://schemas.android.com/apk/res/android"
    android:orientation="vertical"                  //组件垂直放置
```

```
        android:layout_width="fill_parent"          //此布局管理器填充整个屏幕宽度
        android:layout_height="fill_parent">        //此布局管理器填充整个屏幕高度
        <TextView                                   //定义一个文本显示组件
            android:id="@+id/mytxt"                 //命名为 mytxt
            android:layout_width="fill_parent"      //文本显示组件的宽度为屏幕宽度
            android:layout_height="wrap_content"    //文本显示的高度为文字的高度
            android:text="更多精彩单击按钮去看看..." //文本显示的内容
            android:textSize="25sp" />              //文字显示的大小为 25sp
        <Button                                     //定义一个按钮组件
            android:id="@+id/mybut"                 //命名为 mybut
            android:layout_width="fill_parent"      //按钮组件的宽度为屏幕宽度
            android:layout_height="wrap_content"    //按钮组件的高度为文字的高度
            android:text="去看看"                    //设置按钮上的文字
            android:textSize="20sp"/>               //按钮上文字大小为 20sp
    </LinearLayout>                                 //结束线性布局管理器
```

保存文件，代码中就没有了组件之间的位置关系的语句，会将组件一行一行地显示出来，程序运行结果如图 3.20 所示。

图 3.20　Exam3_15 运行结果

也可以在 Activity 中动态地生成 LinearLayout 组件，这里就不给出代码了，感兴趣的读者可以下载 Exam3_16 进行学习。

3.14　表格布局管理器组件 TableLayout

TableLayout 采用表格的形式对组件的布局进行管理，在表格布局中使用 TableRow 进行表格行的控制，通过设置 TableRow 组件，可以设置表格中每一行显示的内容及位置等，所有的组件的增加操作都要在 TableRow 组件中进行。TableLayout 组件层次关系如下：

```
java.lang.Object
    android.view.View
        android.view.ViewGroup
            android.widget.LinearLayout
                android.widget.TableLayout
```

要在 Android 程序中使用 TableLayout 组件，必须在程序中使用下面的语句：

```
import android.widget.TableLayout;          //导入 widget.TableLayout 类
```

1. TableLayout 组件常用的属性和方法

TableLayout 组件是 LinearLayout 的子类，因此 LinearLayout 组件的属性和方法它都具备。另外，它还有如表 3-17 所示的属性和如表 3-18 所示的方法。

表 3-17　TableLayout 组件的常用属性

属性名称	对应方法	描述
android:collapseColumns	setColumnCollapsed(int,boolean)	隐藏从 0 开始的索引列。列之间必须用逗号隔开，如 1, 2, 5
android:shrinkColumns	setColumnCollapsed(int,boolean)	收缩从 0 开始的索引列。列之间必须用逗号隔开，如 1, 2, 5
android:stretchColumns	setColumnCollapsed(int,boolean)	拉伸从 0 开始的索引列。列之间必须用逗号隔开，如 1, 2, 5

表 3-18 TableLayout 组件常用的方法

方法	描述
public TableLayout (Context context)	创建表格布局
public void addView (View child)	添加子视图
public void addView (View child, int index)	添加子视图
public void addView (View child, int index, ViewGroup.LayoutParams params)	根据指定参数，添加子视图
public TableLayout.LayoutParams generateLayoutParams (AttributeSet attrs)	返回一组基于提供的属性集合的布局参数集合
public boolean isColumnCollapsed (int columnIndex)	返回指定列的折叠状态
public boolean isColumnShrinkable (int columnIndex)	返回指定的列是否可收缩
public boolean isColumnStretchable (int columnIndex)	返回指定的列是否可拉伸
public boolean isShrinkAllColumns ()	指示是否所有的列都是可收缩的
public boolean isStretchAllColumns ()	指示是否所有的列都是可拉伸的

例如，表格中，要在一行显示一个文本，其后接着显示一个文本编辑框，代码如下：

```
<TableRow>                                              //增加一行
    <TextView                                           //在此行添加一个文本组件
        android:id="@+id/myinput1"                      //给文本组件命名
        android:layout_width="wrap_content"             //组件宽度为文字宽度
        android:layout_height="wrap_content"            //组件高度为文字高度
        android:text="请输入您的姓名："/>                 //设置显示内容
    <EditText                                           //在此行添加一个文本编辑框组件
        android:id="@+id/myinput2"                      //给文本编辑框组件命名
        android:layout_width="wrap_content"             //组件宽度为文字宽度
        android:layout_height="wrap_content"            //组件高度为文字高度
</TableRow>                                             //一行结束
```

有时文本内容太长，会有一些内容看不到，可以将长的列定义为可收缩列，它会根据文字信息调整显示格式，可用如下代码定义可收缩列：

```
<TableLayout                                        //采用表格布局管理器
    xmlns:android="http://schemas.android.com/apk/res/android"
    android:id="@+id/Layout01"                      //定义名称为 Layout01，程序中使用
    android:layout_width="fill_parent"              //布局管理器宽度为屏幕宽度
    android:layout_height="fill_parent"             //布局管理器高度为屏幕高度
    android:shrinkColumns="2">                      //设置第二列为可收缩列
```

在表格布局管理器内还可以设置某些列不显示，需要注意的是，列数的计数是从第 0 列开始的，可用如下代码设置第 1 列和第 3 列不显示：

```
<TableLayout                                        //采用表格布局管理器
    xmlns:android="http://schemas.android.com/apk/res/android"
    android:id="@+id/Layout01"                      //定义名称为 Layout01，程序中使用
    android:layout_width="fill_parent"              //布局管理器宽度为屏幕宽度
    android:layout_height="fill_parent"             //布局管理器高度为屏幕高度
    android:collapseColumns="0,2">                  //设置第 1 列和第 3 列不显示
```

2. TableLayout 的使用举例

实例 3-15：TableLayout 的使用

新建一个项目，项目命名为 Exam3_17，利用 TableLayout 对组件进行放置。

建立一个布局文件 activity_main.xml 文件，代码如下：

```
<?xml version="1.0" encoding="utf-8"?>
<TableLayout                                         //采用表格布局管理器
xmlns:android="http://schemas.android.com/apk/res/android"
android:id="@+id/Layout01"                           //定义名称为 Layout01
android:layout_width="fill_parent"                   //布局管理器宽度为屏幕宽度
android:layout_height="fill_parent">                 //布局管理器高度为屏幕高度
<TextView
    android:layout_width="fill_parent"
    android:layout_height="wrap_content"
    android:textSize="24sp"
    android:text="个人信息登记表"
    android:gravity="center"                         //文本居中显示
    android:paddingBottom="12dp"  />                 //距离下边缘为 12dp
<TableRow>                                           //增加一行
    <TextView                                        //在此行添加一个文本组件
      android:id="@+id/myinput1"                     //给文本组件命名
      android:layout_width="wrap_content"
      android:layout_height="wrap_content"
      android:textSize="20sp"
      android:text="请输入您的姓名："/>
    <EditText                                        //在此行添加一个文本编辑框组件
      android:id="@+id/myinput2"
      android:layout_width="wrap_content"
      android:layout_height="wrap_content"/>
</TableRow>                                          //一行结束
 <View
    android:layout_height="1dip"                     //画一根高度为 1dip 的线
    android:background="#909090" />                  //设置线的颜色
<TableRow>                                           //增加一行
    <TextView                                        //在此行添加一个文本组件
      android:id="@+id/info"
      android:layout_width="wrap_content"
      android:layout_height="wrap_content"
      android:textSize="20sp"
      android:text="请选择您的性别："/>
    <RadioGroup                                      //在此行添加一个单选按钮组
      android:id="@+id/sex"
      android:orientation="horizontal"                  //组件水平摆放
      android:checkedButton="@+id/man">                 //设置被选中的选项
      <RadioButton                                      //设置单选按钮选项
          android:id="@+id/man"
          android:layout_width="wrap_content"
          android:layout_height="wrap_content"
          android:text=" 男"/>
      <RadioButton                                      //设置单选按钮选项
          android:id="@+id/woman"
          android:layout_width="wrap_content"
          android:layout_height="wrap_content"
          android:text="女"/>
    </RadioGroup>                                       //单选按钮组结束
</TableRow>                                             //一行结束
<View
```

```
        android:layout_height="1dip"
        android:background="#909090" />
        <Button                                      //添加一个按钮
            android:id="@+id/ok"
            android:layout_width="wrap_content"
            android:layout_height="wrap_content"
            android:text="确  定"/>
    </TableLayout>                                   //表格布局结束
```

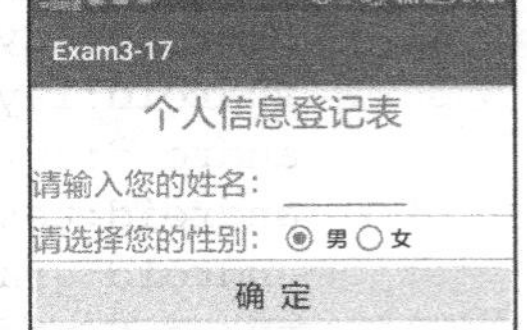

图 3.21 exam3_17 运行结果

保存文件，程序运行结果如图 3.21 所示。

和其他布局管理器一样，表格布局管理器也可以在 Activity 程序中通过调用表 3-18 提供的方法来实现，感兴趣的读者可以下载 Exam3_18 进行学习。

从上面的实例可以看出，采用布局管理器的方式来设置组件比在 Activity 程序中设置组件要轻松得多。虽然，布局管理器可以在 Activity 程序中生成，但是在设计时通常不建议这样做。一般的做法是，先在布局管理器中定义各个要显示的组件，如果需要较多数据，则要定义一个 XML 文件，从 XML 文件中获取数据，然后在 Activity 程序中调用这些组件并对它们进行操作。

在使用布局管理器进行组件布局时，也可以将各个布局管理器嵌套在一起使用，以达到更好的效果。把布局管理器当做一个类似于 TextView 的组件放到另一个布局管理中，就可以实现布局管理器的嵌套。感兴趣的读者可以下载 Exam3_19 学习利用布局管理器的嵌套对组件进行放置。

3.15 约束布局 Constraint Layout

Android Studio 2.2 版本实现了约束布局，不需要多个布局的嵌套就可以实现复杂的用户界面，极大地提高了布局的性能。建议将旧版本的布局转换成约束布局，选择要转换的布局文件并右击，然后选择 Convert layout to ConstraintLayout 选项即可。ConstraintLayout 与 RelativeLayout 很相似，所有在布局中的组件的布局方式取决于组件与组件之间的关系和父布局。但是它比 RelativeLayout 更灵活。

ConstraintLayout 的所有工作都可以使用布局编辑器的可视化工具完成，可以通过拖动的方式去构建一个使用了 ConstraintLayout 的布局，避免直接在 XML 中编辑，这样大大提高了开发效率。

1. 添加约束布局 ConstraintLayout

使用约束布局，必须确保有约束布局的库，如果没有，则可以通过如下步骤进行添加。

（1）选择“Tools→Android→SDK Manager”选项。如图 3.22 所示。

（2）在弹出的对话框中，选择“SDK Tools”选项卡。展开 Support Repository，然后选中“ConstraintLayout for Android”和“Solver for ConstraintLayout”。选中“Show Package Details”复选框，如图 3.23 所示，单击“OK”按钮。

新建的项目，自动生成的默认布局不会使用 ConstraintLayout，但是可以转换成 ConstraintLayout。打开布局文件，选择“Design”选项卡，在 Component Tree 窗体中，右击布局文件，然后选择“Convert Relative Layout to ConstraintLayout”选项，如图 3.24 所示。在弹出的窗口中单击“OK”按钮即可，转换成 ConstraintLayout 后如图 3.25 所示。

也可以新建一个约束布局。新建一个布局文件，输入布局文件的名称，将布局的根元素改为 android.support.constraint.ConstraintLayout，然后单击“完成”按钮即可。

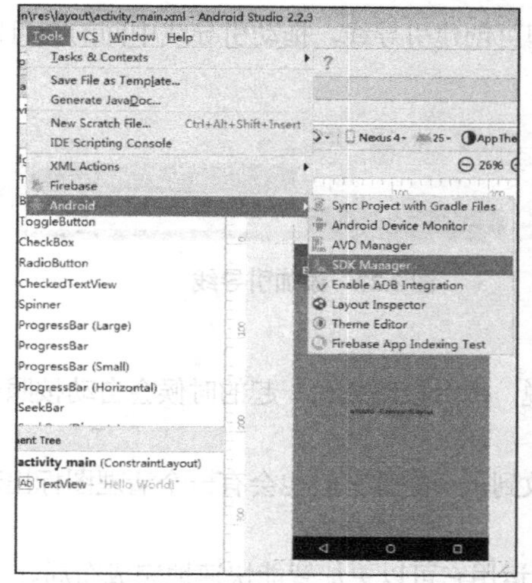

图 3.22 选择 SDK Manager

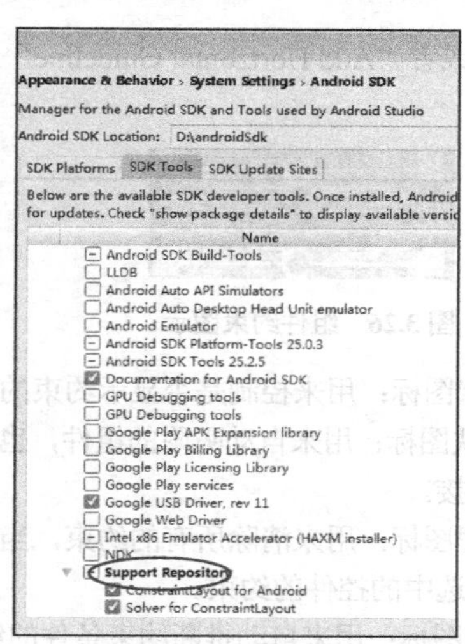

图 3.23 展开 Support Repository

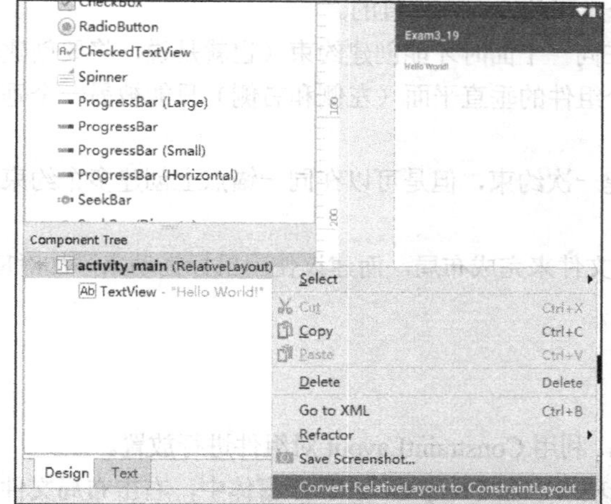

图 3.24 选择转换 ConstraintLayout

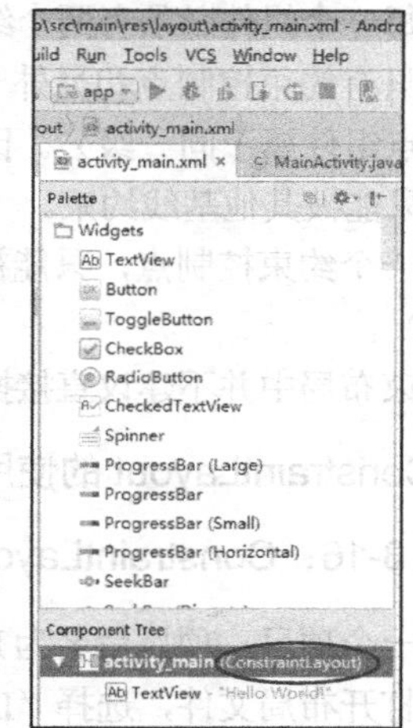

图 3.25 转换为 ConstraintLayout 后

2．约束的分类

拖动一个 View 组件到布局编辑器中，组件的四个角对应着的四个小矩形框用于控制大小，每一条边有四个圆形的约束控制点，如图 3.26 所示，对应着几种类型的约束：尺寸大小约束、边界约束、基准线约束、约束到一个引导线（辅助线）。

角上的实心方块是尺寸大小约束控制点，用来调整组件的大小，可以使用 Properties 窗口定义一个更具体的尺寸。

边上的空心圆圈是边界约束，用来建立组件与组件之间、组件与 Parent 边界之间的约束关系，实际上就是确定彼此的相对位置。选中一个组件，按住一个边界约束控制点并拖动这条线到另一个组件可用的锚点（其他组件的边缘或者引导线），松开鼠标左键，这个约束将会被创建，两个组件也将被默认的 margin 隔开。

中间的空心圆角矩形是基准线约束，用来让两个带有文本属性的组件进行对齐，需要注意的是，要把光标放在控件上，等基准线约束的图形亮了，才可以进行拖动。

约束到一个引导线（辅助线），可以添加一个水平和垂直方向上的引导线，当做附加约束。在布局内可以定位这个引导线，在工具栏中选择“Guidelines”选项，如图 3.27 所示，然后选择“Add Vertical

Guideline”或者“Add Horizontal Guideline”选项即可创建引导线，拖动引导线中间的圆可定位引导线的位置。

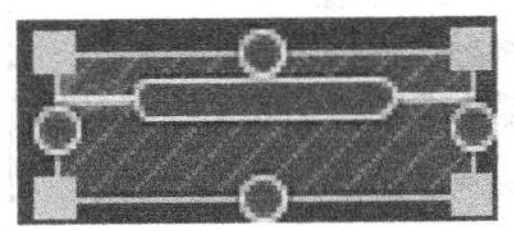

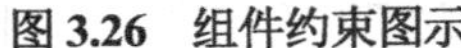

图 3.26　组件约束图示

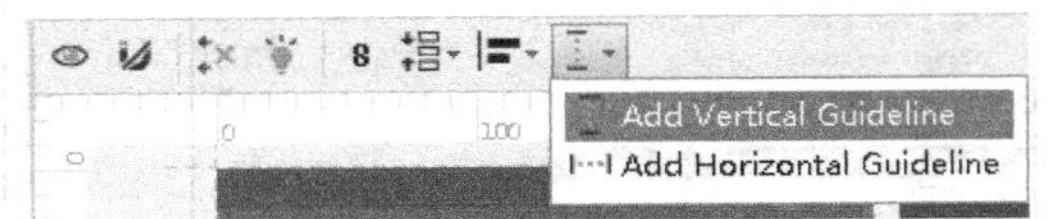

图 3.27　添加引导线

眼睛图标：用来控制是否显示约束的组件。

磁铁图标：用来自动吸附的组件，也就是说，两个按钮放在一起的时候会自动按照一定的约束条件进行连接。

清理图标：用来清除所有的约束，当光标放到一个控件上时也会有一个清理图标出现，单击可以清除当前选中的控件的约束。

灯泡图标：用来自动推断约束条件的组件，运用它可以更加智能快速地完成布局。

创建一个约束的时候，一般有下面几点规则。

（1）每一个组件必须有两个约束，一个水平的，一个垂直的。

（2）只有约束控制点和另外一个锚点在同一平面时才能创建约束（也就是说，将要创建的约束的组件和锚点组件属于同一级）。因此，一个组件的垂直平面（左侧和右侧）只能被另一个垂直平面约束，基线只能被其他基线约束。

（3）一个约束控制点，只能被用来创建一次约束，但是可以在同一锚点上创建多个约束（来自不同的组件）。

在约束布局中并不建议直接操作 XML 文件来完成布局，而建议使用鼠标拖动来添加对应的约束。

3. ConstraintLayout 的使用实例

实例 3-16：ConstraintLayout 的使用

新建一个项目，项目命名为 Exam3_20，利用 ConstraintLayout 对组件进行放置。

（1）打开布局文件，选择“Design”选项卡，在 Component Tree 窗体中，右击布局文件，然后选择“Convert Relative Layout to ConstraintLayout”选项，如图 3.24 所示。在弹出的对话框中单击“OK”按钮即可，转换成 ConstraintLayout 后如图 3.25 所示。

（2）打开布局文件“activity_main.xml”，找到“Images&Media”文件夹，找到其中的“ImageView”组件，拖动到手机界面中，如图 3.28 所示。在弹出的对话框中选择要添加的图片，如图 3.29 所示。

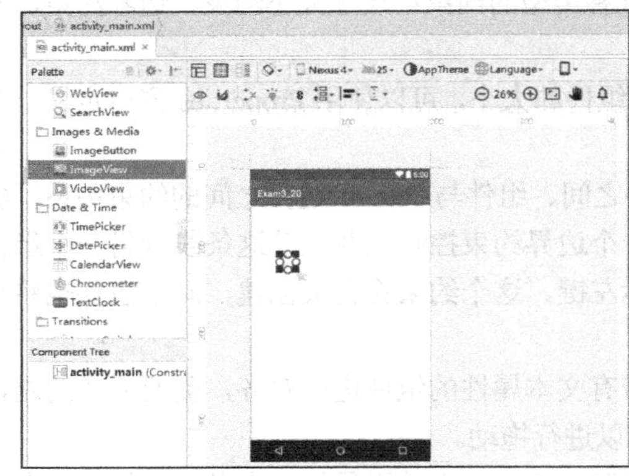

图 3.28　拖动 ImageView 组件

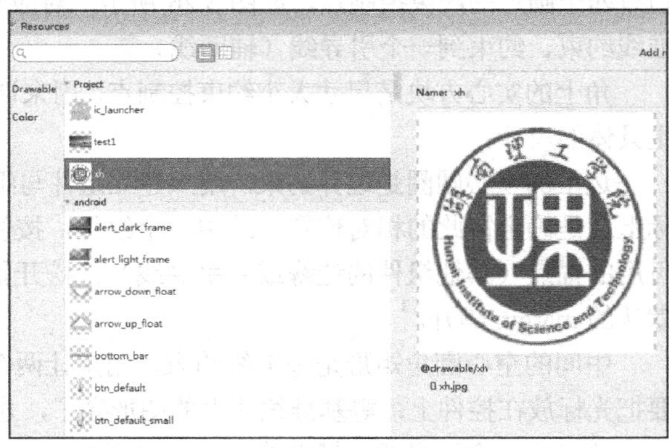

图 3.29　添加图片

（3）选择图片，找到左边竖线中的圆形图标，设置图片距左边的距离，然后找到右边竖线中的圆形图标，设置图片距右边的距离，将图片摆放在合适的位置，如图 3.30 所示。再找到其中的“TextView”组件，拖动到手机界面中，可以在右边的属性框中设置属性，如图 3.31 所示。

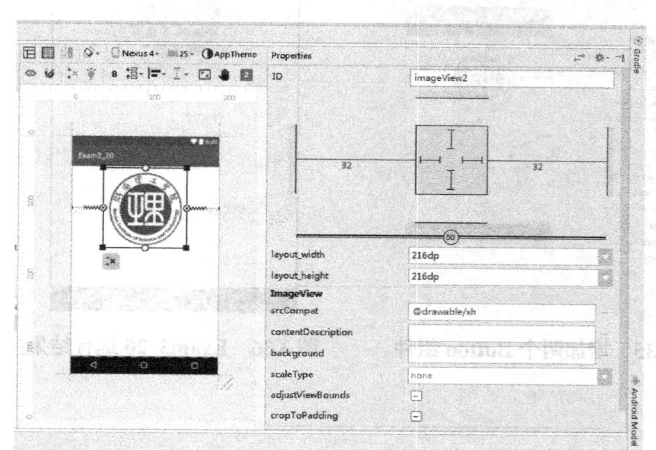

图 3.30 设置图片位置

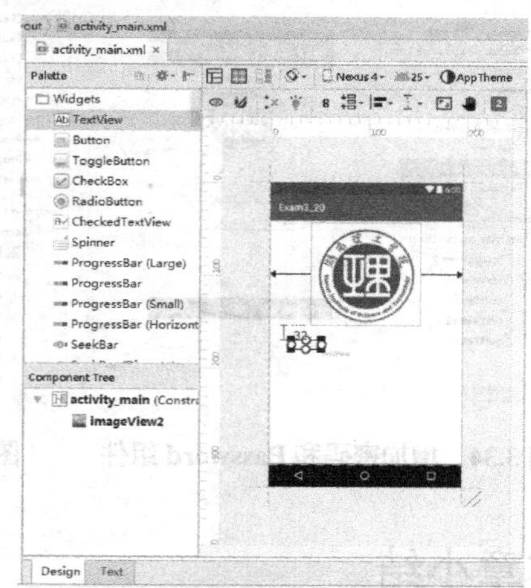

图 3.31 拖动 TextView 组件

（4）找到其中的“PlainText”组件，拖动到手机界面中，选择其中的“TextView”组件，选择组件左边线中间圆形图标，设置组件与左边界的距离，选择组件上边线中间圆形图标，设置组件与上面组件的距离，如图 3.32 和图 3.33 所示。

（5）类似的，再添加一个 TextView 组件和一个 Password 组件，设置它们的位置关系，注意为每个组件设置一个水平的、一个垂直的约束，如图 3.34 所示。

（6）类似的，添加两个按钮组件，并设置组件的位置，如图 3.35 所示。

保存文件，程序运行结果如图 3.36 所示。这里没有写一行代码，仅仅是拖动组件，设置组件的位置关系，就可以实现 Android 程序的界面设计。因此在实际设计中，界面设计采用这种形式可能会达到事半功倍的效果。

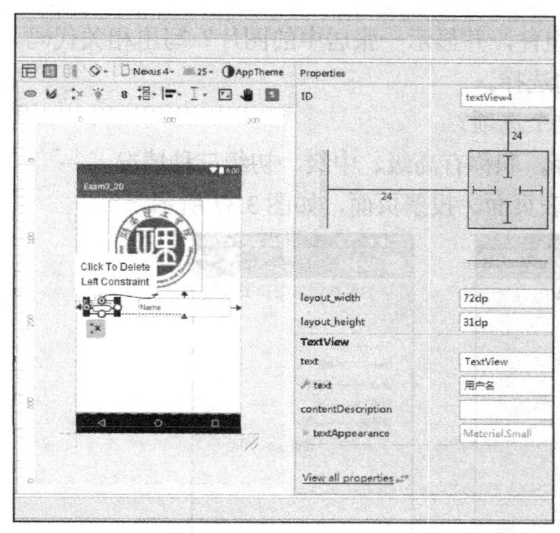

图 3.32 设置 TextView 组件位置

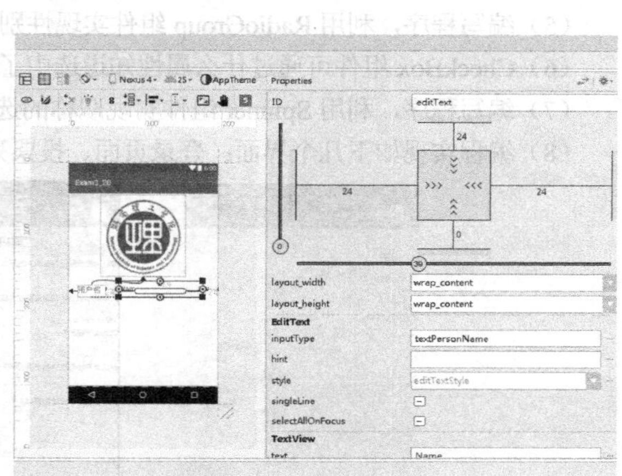

图 3.33 设置组件位置

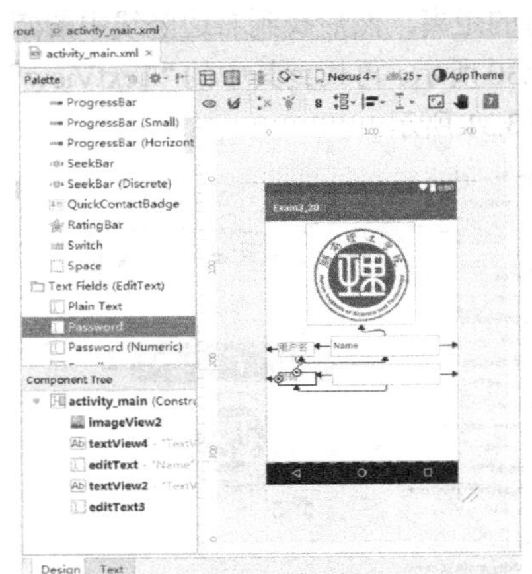
图 3.34　增加密码和 Password 组件

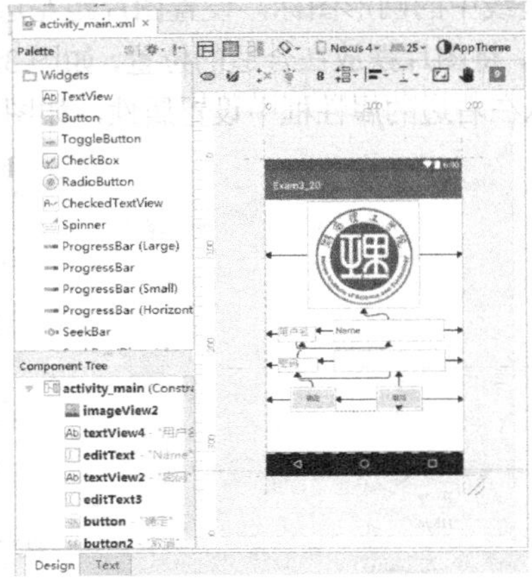
图 3.35　增加两个 Button 组件

图 3.36　Exam3_20 运行结果

本章小结

本章着重介绍了 Android 中常用的基本组件——View、TextView、EditText、Button、ImageButton、ImageView、RadioGroup、CheckBox、Spinner、Toast 及其基本操作，并详细介绍了 Android 中几种布局管理器及其基本使用，建议先在布局管理器中定义各个要显示的组件，然后在 Activity 程序中调用，这样设计可能会更加简单。

习题

（1）查找 Android　API，了解 View 组件的属性和方法，列举 10 种常用的属性和对应的方法。

（2）修改 EditText 组件的哪个属性值，能使 EditText 组件中的值不能编辑？

（3）是否可以在 Button 组件上同时显示一个图片和文字？如果可以，则该设置什么属性？写出相关代码。

（4）怎样在 Activity 程序中自动生成 ImageView 组件，并显示一张居中的图片？写出相关代码。

（5）编写程序，利用 RadioGroup 组件实现性别的选择。

（6）CheckBox 组件可通过什么属性知道选中了这个选项？

（7）编写程序，利用 Spinner 组件实现职称的选择，职称有高级、中级、初级三种情况。

（8）编程实现如下几个界面：登录页面、投票列表页面、投票页面，如图 3.37 所示。

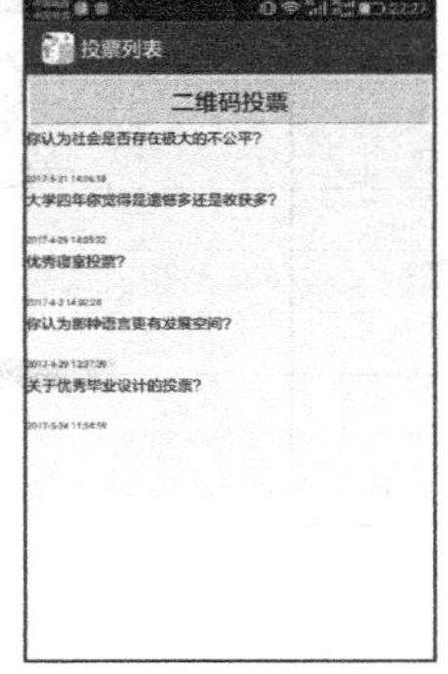

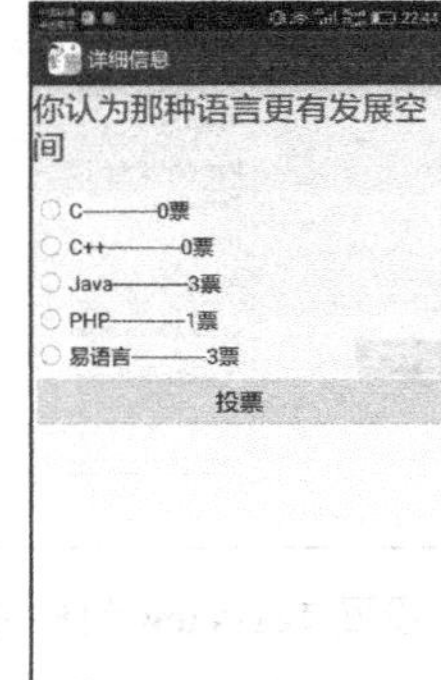

图 3.37　要实现的界面

第 4 章　Android 中的事件处理

学习目标：

- 了解 Android 中的事件处理的原理。
- 掌握 Android 中的主要事件——单击事件、长按事件、焦点改变事件、键盘事件、触摸事件、菜单事件的基本操作。

在前面的章节中介绍了组件的布局，事实上在图形界面的开发中，有两个非常重要的内容：一个是组件的布局，另一个是事件处理。本章主要对 Android 中几种不同的事件处理进行介绍。

4.1　Android 中的事件处理基础

Android 在事件处理过程中主要涉及以下 3 个概念。

（1）事件（Event）：在图形界面中操作的描述，通常是封装成各种类，如单击事件、触摸事件、键盘事件等。

（2）事件源（Event Source）：事件源是指事件发生的场所，通常是指各个组件，如 Button、EditText 等控件。

（3）事件监听器（Event Listener）：事件监听器是指接收事件对象并对其进行处理的对象，事件处理一般是一个实现某些特定接口类创建的对象。

例如，单击按钮后，屏幕显示“你好！”，在这个事件处理中，事件是“单击事件”，事件源是“按钮”，事件监听器是定义的“OnClickListener”对象，由它来实现具体的操作。

4.1.1　事件处理的过程

（1）为事件源对象添加监听，当某个事件被触发时，系统才会知道该通知谁来处理这个事件，如图 4.1（a）所示。

（2）当事件发生时，系统将事件封装成相应类型的事件对象，并发送给注册到事件源的事件监听器，如图 4.1（b）所示。

（3）当监听器对象接收到事件对象之后，系统会调用监听器中相应的事件处理方法来处理事件并给出响应，如图 4.1（c）所示。

事件处理有以下几个关键的操作：注册监听程序、根据指定的事件编写处理程序，在事件处理类中完成事件的处理操作。

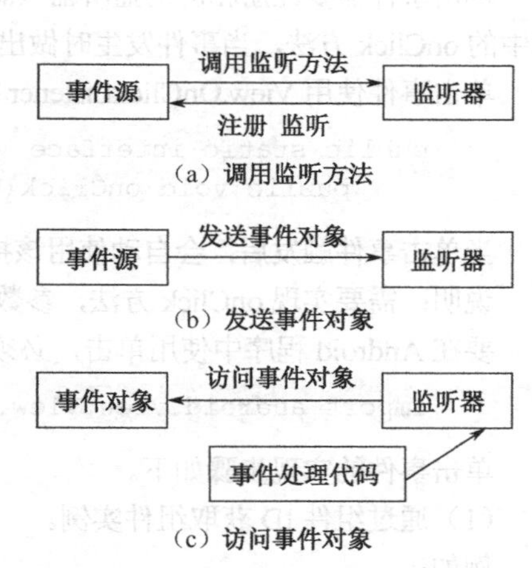

图 4.1　事件处理的过程示意图

4.1.2　事件处理模型

Android 中的事件处理模型常用的有：基于监听接口的事件处理、基于回调的事件处理。

1．基于监听接口的事件处理

基于监听的事件处理模型的编程步骤如下。

（1）获取普通界面组件。

（2）实现事件监听类，该监听类是一个特殊的 Java 类，必须实现一个 XXXListener 接口。

（3）调用事件源的 setXXXListener 方法注册事件监听器。

Android 中提供了以下几种基于监听接口的事件处理模型。

（1）OnClickListener 接口：单击事件。

（2）OnLongClickListener 接口：长按事件。

（3）OnFocusChangeListener 接口：焦点改变事件。

（4）OnKeyListener 接口：键盘事件。

（5）OnTouchListener 接口：触摸事件。

（6）OnCreateContextMenuListener 接口：上下文菜单事件。

2．基于回调机制的事件处理

在 Android 平台中，每个 View 类都有自己的处理事件的回调方法，可以通过重写 View 类中的回调方法来实现需要的响应事件，Android 类提供了以下回调方法供用户使用。

（1）onKeyDown：用来捕捉手机键盘被按下的事件。

（2）onKeyUp：用来捕捉手机键盘按键抬起的事件。

（3）onTouchEvent：用来处理手机屏幕的触摸事件。

（4）onTrackBallEvent：用来处理轨迹球事件。

（5）onFocusChanged：用来处理焦点改变的事件。

4.2　单击事件 OnClickListener

4.2.1　单击事件基础

单击事件需要注册相应的监听器（setOnClickListener）监听事件的来源，利用 OnClickListener 接口中的 onClick 方法，当事件发生时做出相应的处理。

单击事件使用 View.OnClickListener 接口进行事件的处理，此接口定义如下：

```
public static interface View.OnClickListener{
    public void onClick(View v) ;}
```

当单击事件触发后，会自动使用该接口中的 public void onClick(View v)方法进行事件处理。

说明：需要实现 onClick 方法，参数 v 为事件发生的事件源。

要在 Android 程序中使用单击，必须在程序中使用下面的语句导入相应类：

```
import android.view.View.OnClickListener;    //导入 View.OnClickListener 类
```

单击事件的实现步骤如下。

（1）通过组件 ID 获取组件实例。

例如：

```
this.mybut=(Button)super.findViewById(R.id.mybut); //获得按钮
```

（2）为该组件注册 OnClickListener 监听。

例如：

```
mybut.setOnClickListener(new ShowListener()) ; //注册监听
```

（3）实现 onClick 方法。

例如：

```
private class ShowListener implements OnClickListener {//定义监听处理程序
    public void onClick(View v) {                        //执行具体操作
            ...}}
```

4.2.2 单击事件实例

实例 4-1：单击事件举例

新建一个项目，项目命名为 Exam4_1，包名称为 cn.hnist.org，编程实现：手机界面显示 3 个组件 EditText、Button 和 TextView，单击“Button”按钮时会在手机上显示输入的信息。

（1）修改 activity_main.xml 文件，代码如下：

```
...
    <EditText
        android:id="@+id/myedit"
        android:layout_width="wrap_content"
        android:layout_height="wrap_content"
        android:textSize="20sp"
        android:phoneNumber="true"
        android:selectAllOnFocus="true"
        android:hint="请输入您的手机号码："/>
    <Button
         android:id="@+id/mybut"
         android:layout_width="wrap_content"
         android:layout_height="wrap_content"
         android:layout_below="@+id/myedit"
         android:textSize="20sp"
         android:layout_marginTop="20dp"
         android:text="显示您输入的号码" />
    <TextView
         android:id="@+id/mytxt"
         android:layout_width="wrap_content"
         android:layout_height="wrap_content"
         android:layout_below="@+id/mybut"
         android:layout_marginTop="20dp"
         android:textSize="20sp"/>
</RelativeLayout>
```

（2）修改 MainActivity.java，代码如下：

```
...
import android.view.View.OnClickListener;   //导入 View.OnClickListener 类
public class MainActivity extends AppCompatActivity {
    private TextView mytxt=null;
    private Button mybut=null;
    private EditText myedit=null;
@Override
```

```
protected void onCreate(Bundle savedInstanceState) {
      super.onCreate(savedInstanceState);
      setContentView(R.layout.activity_main);
      this.mytxt=(TextView)super.findViewById(R.id.mytxt);        //获得组件
      this.myedit=(EditText)super.findViewById(R.id.myedit);
      this.mybut=(Button)super.findViewById(R.id.mybut);
      mybut.setOnClickListener(new ShowListener()) ;}             //注册监听
      private class ShowListener implements OnClickListener {     //定义监听处理
                                                                  //程序
         public void onClick(View v) {                    //回调方法，执行具体的操作
           String info = myedit.getText().toString() ;
           mytxt.setText("您的手机号码是: " + info) ; }}}
```

保存文件，程序运行后会显示一个文本编辑框、一个按钮。在文本编辑框中输入内容，然后单击按钮，会在下面显示相应的信息，如图 4.2 所示。

图 4.2　Exam4_1 运行结果

如果事件监听器只是临时使用一次，则建议使用匿名内部类形式的事件监听器，上面的代码可以更改如下。

```
...
this.mybut=(Button)super.findViewById(R.id.mybut);      //获得组件
mybut.setOnClickListener(new OnClickListener(){         //定义监听
public void onClick(View v) {                           //实现方法
   String info = myedit.getText().toString() ;
   mytxt.setText("您的手机号码是: " + info) ; }});}}
```

4.3　长按事件 OnLongClickListener

4.3.1　长按事件基础

Android 中提供了长按事件的处理操作，长按事件只有在触发 2 秒之后才会有反应，长按事件使用 View.OnLongClickListener 接口进行事件的处理操作。

此接口定义如下：

```
public static interface View.OnLongClickListener{
     public boolean onLongClick(View v) ; }
```

当长按事件触发后，会自动使用该接口中的 public boolean onLongClick(View v)方法进行事件处理。

说明：需要实现 onLongClick 方法，该方法的返回值为一个 boolean 类型的变量，当返回 true 时，表示已经完整地处理了这个事件，并不希望其他的回调方法再次进行处理；当返回 false 时，表示并没有完全处理完该事件，希望其他方法继续对其进行处理。

要在 Android 程序中使用长按事件，必须在程序中使用下面的语句：

```
import android.view.View.OnLongClickListener; //导入View.OnLongClickListener类
```

长按事件的实现步骤如下。

（1）通过组件 ID 获取组件实例。

例如：

```
bgimg=(ImageView)findViewById(R.id.bgimg);
```

（2）为该组件注册 OnLongClickListener 监听。

例如：

```
bgimg.setOnLongClickListener(new OnLongClickListener(){…}
```

（3）实现 onLongClick 方法。

例如：

```
public boolean onLongClick(View v) {
   …}
```

4.3.2 长按事件实例

实例 4-2：长按事件举例

新建一个项目，项目命名为 Exam4_2，编程实现：长按一张图片，将这张图片设置为背景。将准备好的图片“test1.jpg”复制到“res\drawable”文件夹下。

（1）修改 activity_main.xml 文件，代码如下：

```
…
    <TextView
        android:id="@+id/mytxt"
        android:layout_width="fill_parent"
        android:layout_height="wrap_content"
        android:text="长按图片将其设为屏幕背景" />
    <ImageView
        android:id="@+id/bgimg"
        android:layout_width="wrap_content"
        android:layout_height="wrap_content"
        android:layout_gravity="center"
        android:src="@drawable/test1" />
</LinearLayout>
```

（2）修改 MainActivity.java，代码如下：

```
…
public class MainActivity extends AppCompatActivity {
  private ImageView bgimg=null;
  private TextView showInfo=null;
  @Override
  protected void onCreate(Bundle savedInstanceState) {
      super.onCreate(savedInstanceState);
      setContentView(R.layout.activity_main);
      showInfo=(TextView)findViewById(R.id.mytxt);
      bgimg=(ImageView)findViewById(R.id.bgimg);      //获得组件
     bgimg.setOnLongClickListener(new OnLongClickListener(){//注册监听
         @Override
         public boolean onLongClick(View v) {          //实现方法
             try {
                 clearWallpaper();
setWallpaper(bgimg.getResources().openRawResource(R.drawable.test1));/*设
为桌面背景*/
                showInfo.setText("图片已经设置为手机桌面");
```

```
            } catch (IOException e) {
                //TODO Auto-generated catch block
                e.printStackTrace();
                showInfo.setText("手机桌面背景设置没有成功！");}
            return true;    }});}}      //表示已经完整地处理了这个事件
```

（3）设置手机桌面背景的操作属于手机的支持服务，必须得到相关的授权后才可以执行，双击打开 AndroidManifest.xml 文件，在</manifest>之前增加以下语句进行授权：

```
<uses-permission android:name="android.permission.SET_WALLPAPER">
</uses-permission>
```

保存所有文件并运行程序，结果如图 4.3(a)所示；长按图片的结果如图 4.3(b)所示。单击右侧的“返回”按钮，可以看出已经将这张图片设置成了手机桌面背景。

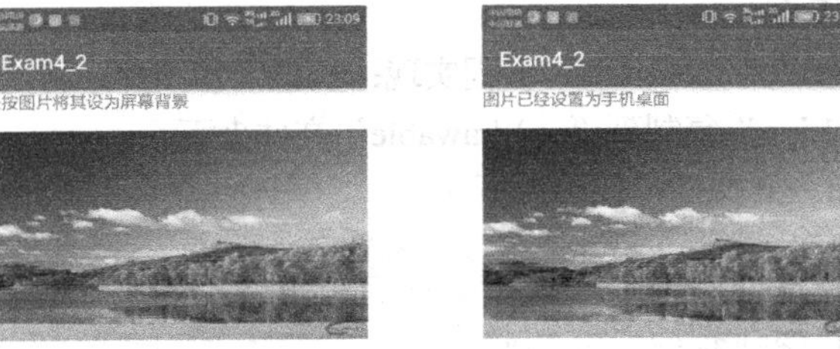

（a）运行结果　　（b）变换背景图片

图 4.3　Exam4_2 运行结果

4.4　焦点改变事件 OnFocusChangeListener

4.4.1　焦点改变事件基础

焦点改变事件是指对一个组件状态的监听，是指在组件获得或失去焦点时进行处理操作，所有的组件都存在监听焦点变化的方法，利用 OnFocusChangeListener 接口来监听焦点改变事件。焦点改变事件则是指只要该组件获得或失去焦点，不用单击任何组件即可触发相应的事件。

此接口定义如下：

```
public void setOnFocusChangeListener(View.OnFocusChangeListener l)
```

当焦点改变事件触发之后，会自动使用该接口中的 public void onFocusChange(View v, Boolean hasFocus)方法进行事件处理。

说明：需要实现 onFocusChange 方法；参数 v 表示触发该事件的事件源；参数 hasFocus 表示 v 的新状态，即 v 是否获得焦点。

要在 Android 程序中使用焦点改变事件，必须在程序中使用下面的语句导入相应类：

```
import android.view.View.OnFocusChangeListener;//导入View.OnFocusChangeListener类
```

焦点改变事件的实现步骤如下。

（1）通过组件 ID 获取组件实例。

例如：

```
this.edit = (EditText) super.findViewById(R.id.edit1);
```

（2）为该组件注册 OnFocusChangeListener 监听。

例如：

```
this.edit.setOnFocusChangeListener(new OnFocusChangeListenerImpl());
```

（3）实现 onFocusChange 方法。

例如：

```
public void onFocusChange(View v, boolean hasFocus) {…}
```

下面列出一些与焦点有关的常用方法。

setFocusable 方法：设置 View 类是否可以拥有焦点。

isFocusable 方法：监测此 View 类是否可以拥有焦点。

setNextFocusDownId 方法：设置 View 类的焦点向下移动后获得焦点 View 的 ID。

hasFocus 方法：返回了 View 类的父控件是否获得了焦点。

requestFocus 方法：尝试让此 View 类获得焦点。

isFocusableTouchMode 方法：设置 View 类是否可以在触摸模式下获得焦点，在默认情况下是不可用获得的。

4.4.2 焦点改变事件实例

实例 4-3：焦点改变事件举例

新建一个项目，项目命名为 Exam4_3，编程实现：输入信息时会要求对输入的信息的合法性做判断，例如，输入的邮箱地址要符合相应的规范，类似这样的操作都可以用焦点改变事件来实现。

（1）修改 activity_main.xml 文件，代码如下：

```
…
<TextView
    android:id="@+id/txt1"
    android:layout_width="fill_parent"
    android:layout_height="wrap_content"
    android:textSize="20dp"
    android:text="请输入您的邮箱" />
<EditText
    android:id="@+id/edit1"
    android:layout_width="fill_parent"
    android:layout_height="wrap_content"
    android:focusable="true"
      android:selectAllOnFocus="true"
      android:text="第一个文本输入框组件"/>
<EditText
    android:id="@+id/edit2"
    android:layout_width="fill_parent"
    android:layout_height="wrap_content"
    android:text="第二个文本输入框组件" />
<TextView
    android:id="@+id/txt2"
    android:layout_width="fill_parent"
    android:layout_height="wrap_content"
    android:textSize="15dp"/>
```

```
</LinearLayout>
```

（2）修改 MainActivity.java，代码如下：

```
…
public class MainActivity extends AppCompatActivity {
private EditText edit = null;
private TextView txt = null;
@Override
public void onCreate(Bundle savedInstanceState) {
    super.onCreate(savedInstanceState);
    super.setContentView(R.layout.activity_main);
    this.edit = (EditText) super.findViewById(R.id.edit1);  //取得组件
    this.txt = (TextView) super.findViewById(R.id.txt2);    //取得组件
    //注册监听焦点事件
    this.edit.setOnFocusChangeListener(new OnFocusChangeListenerImpl());}
    //设置焦点事件
private class OnFocusChangeListenerImpl implements OnFocusChangeListener {
    @Override
    public void onFocusChange(View v, boolean hasFocus) {  //具体要实现的部分
      String inputmsg=MainActivity.this.edit.getText().toString();
        if (v.getId() == MainActivity.this.edit.getId()) {
            if (hasFocus) {        //判断是否获得了焦点
                MainActivity.this.txt.setText("第一个文本输入框组件获得了焦点。");
            } else {
                if(inputmsg.matches("\\w+@\\w+\\.\\w+")) {//正则表达式判断
            MainActivity.this.txt.setText("您输入的邮箱是："+inputmsg+"符合要求！");
                } else {
            MainActivity.this.txt.setText("您输入的邮箱是："+inputmsg+"不符合要求！");
}}}}}}
```

保存文件并运行程序，结果如图 4.4 所示，可以对输入的信息进行判断并输出。

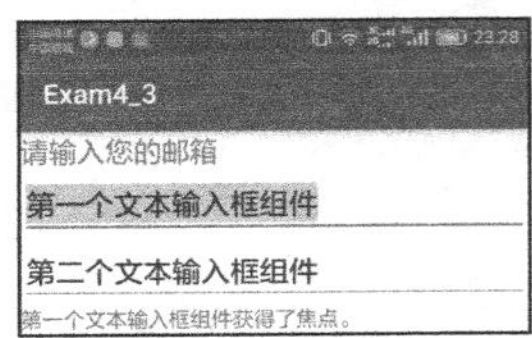

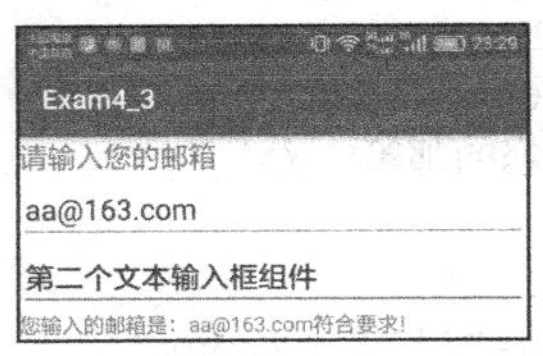

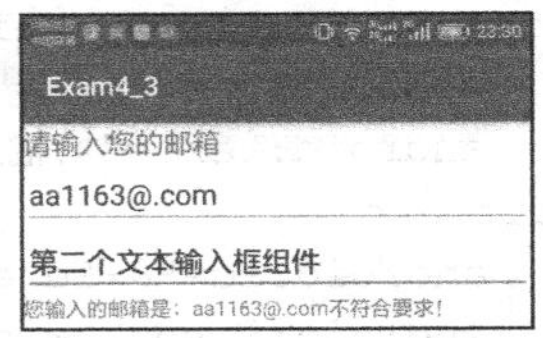

图 4.4　Exam4_3 运行结果

4.5　键盘事件 OnKeyListener

4.5.1　键盘事件基础

键盘事件是用户在利用键盘输入数据时所触发的操作，主要用于键盘的监听处理操作，键盘事件使用 OnKeyListener 接口进行事件的处理。OnKeyListener 接口定义如下：

```
public static interface View.OnKeyListener{
public boolean onKey(View v, int keyCode, KeyEvent event) ;}
```

当键盘事件触发之后，会自动使用该接口中的回调方法 public boolean onKey(View v, int keyCode,

KeyEvent event)进行事件处理。

说明：需要实现 onKey 方法；参数 v 表示事件的事件源控件；参数 keyCode 表示手机键盘的键盘码；参数 event 为键盘事件封装类的对象。其中包含了事件的详细信息，如发生的事件、事件的类型等。

要在 Android 程序中使用键盘事件，必须在程序中使用下面的语句：

```
import android.view.View.OnKeyListener;   //导入 View.OnKeyListener 类
```

还存在一个与键盘事件联系很密切的类——view.KeyEvent，可使用下面的语句来调用：

```
import android.view.KeyEvent;                 //导入 view.KeyEvent 类
```

键盘事件的实现步骤如下。

（1）通过组件 ID 获取组件实例。

例如：

```
this.edit = (EditText) super.findViewById(R.id.edit1);
```

（2）为该组件注册 OnKeyListener 监听。

例如：

```
this.edit.setOnKeyListener(new OnKeyListenerImpl());
```

（3）实现 onKey 方法。

例如：

```
public boolean onKey(View v, int keyCode, KeyEvent event) {…}
```

4.5.2 键盘事件实例

实例 4-4：键盘事件举例

新建一个项目，项目命名为 Exam4_4，编程实现：对输入信息做合法性判断，对输入的年龄做判断，小于 200 的认为合法，否则认为是不合法的年龄。

（1）修改 activity_main.xml 文件，代码如下：

```
…
<EditText
    android:id="@+id/edit1"
    android:layout_width="wrap_content"
    android:layout_height="wrap_content"
    android:selectAllOnFocus="true"
    android:numeric="integer"
    android:hint="请输入您的年龄："/>
<TextView
    android:id="@+id/txt1"
    android:layout_width="wrap_content"
    android:layout_height="wrap_content" />
</LinearLayout>
```

（2）修改 MainActivity.java 文件，代码如下：

```
…
import android.view.KeyEvent;                //导入 view.KeyEvent 类
```

```
import android.view.View.OnKeyListener;  //导入 View.OnKeyListener 类
public class MainActivity extends AppCompatActivity {
   private EditText edit = null;
   private TextView txt=null;
@Override
public void onCreate(Bundle savedInstanceState) {
    super.onCreate(savedInstanceState);
    super.setContentView(R.layout.activity_main);
    this.edit = (EditText) super.findViewById(R.id.edit1);//获得组件
    this.txt = (TextView) super.findViewById(R.id.txt1);
    this.edit.setOnKeyListener(new OnKeyListenerImpl()); } /*设置键盘事件监听*/
    private class OnKeyListenerImpl implements OnKeyListener {
    @Override
    public boolean onKey(View v, int keyCode, KeyEvent event) { //实现 onKey 方法
        switch(event.getAction()) {
        case KeyEvent.ACTION_UP:                              /*按键松开后触发*/
         String inputmsg = MainActivity.this.edit.getText().toString();
         //取出已输入内容
         int a=Integer.parseInt(inputmsg);  //将取出的信息转换为数字型数据
         if (a>0 && a<200) {                //判断年龄是否大于 0 且小于 200
           MainActivity.this.txt.setText("您输入的年龄是:"+inputmsg+", 符合
                                      要求! ");}
         else {
           MainActivity.this.txt.setText("您输入的年龄是:"+inputmsg+", 不符
                                      合要求! ");}
        case KeyEvent.ACTION_DOWN:                      //按键被按下时触发
        default:
            break ; }
        return false;    }}}                            //继续事件应有的流程
```

保存文件，程序运行结果如图 4.5 所示，可以对输入的信息进行判断并输出。

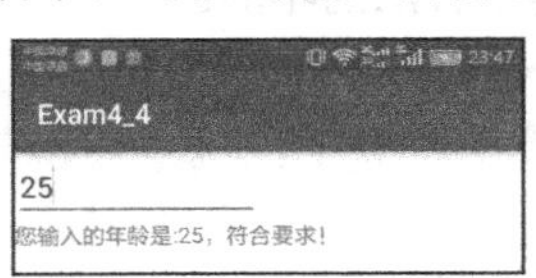

图 4.5 Exam4_4 运行结果

玩网游时经常会使用方向键来进行控制，如何知道按下了这些键呢？项目 Exam4_5 通过编写程序来提取了用户键盘输入的方向键信息，感兴趣的读者可自己下载进行学习。

4.6 触摸事件 onTouchEvent

4.6.1 触摸事件基础

触摸事件指的是当用户接触到屏幕之后所产生的事件，当用户在屏幕上划过时，可以使用触摸事件取得用户当前的坐标，OnTouchListener 接口定义如下：

```
public interface View.OnTouchListener {
    public abstract boolean onTouch (View v, MotionEvent event) ;}
```

当触摸事件触发之后，会自动使用该接口中的 public boolean onTouch(View v, MotionEvent event)方法进行事件处理。

说明：需要实现 onTouch 方法；参数 v 为事件源对象；参数 event 为事件封装类的对象。其中，封装了触发事件的详细信息，同样包括事件的类型、触发时间等信息。

返回值：该方法的返回值机理与键盘响应事件的机理相同，同样的，当已经完整地处理了该事件且不希望其他回调方法再次处理时返回 true，否则返回 false。

该方法并不像之前介绍过的方法一样只处理一种事件，一般情况下，以下 3 种情况的事件全部由 onTouchEvent 方法处理，只是 3 种情况中的动作值不同。

屏幕被按下：当屏幕被按下时，会自动调用该方法来处理事件，此时 MotionEvent.getAction()的值为 MotionEvent.ACTION_DOWN，如果在应用程序中需要处理屏幕被按下的事件，则只需重新调用该回调方法，然后在方法中进行动作的判断即可。

屏幕被抬起：当触控笔离开屏幕时触发的事件，该事件同样需要 onTouchEvent 方法来捕捉，然后在方法中进行动作判断。当 MotionEvent.getAction()的值为 MotionEvent.ACTION_UP 时，表示是屏幕被抬起的事件。

在屏幕中拖动：此方法还负责处理触控笔在屏幕上的滑动事件，调用 MotionEvent.getAction()方法来判断动作值是否为 MotionEvent.ACTION_MOVE 再进行处理。

要在 Android 程序中使用触摸事件，必须在程序中使用下面的语句：

```
import android.view.View.OnTouchListener;    //导入 View.OnTouchListener 类
```

存在一个与键盘事件联系很密切的类——view. MotionEvent，可使用下面的语句进行调用：

```
import android.view.MotionEvent;                     //导入 view. MotionEvent 类
```

触摸事件的实现步骤如下。

（1）通过组件 ID 获取组件实例。

例如：

```
this.edit = (EditText) super.findViewById(R.id.edit1);
```

（2）为该组件注册 OnTouchEventListener 监听。

例如：

```
this.locate.setOnTouchListener(new OnTouchListenerImpl());
```

（3）实现 onTouchEvent 方法。

例如：

```
public boolean onTouch (View v, MotionEvent event) {…}
```

4.6.2 触摸事件实例

实例 4-5：触摸事件举例

新建一个项目，项目命名为 Exam4_6，编程实现：在屏幕任意空白处单击、触摸，会显示出触摸位置的坐标。

（1）修改 activity_main.xml 文件，代码如下：

```
…
<TextView
    android:id="@+id/msg"
```

```
        android:layout_width="fill_parent"
        android:layout_height="fill_parent" />
</LinearLayout>
```

（2）修改 MainActivity.java，代码如下：

```
…
import android.view.MotionEvent;                    //导入 view. MotionEvent 类
import android.view.View.OnTouchListener;           //导入 View.OnTouchListener 类
public class MainActivity extends AppCompatActivity {
private TextView locate = null;
@Override
public void onCreate(Bundle savedInstanceState) {
    super.onCreate(savedInstanceState);
    super.setContentView(R.layout.activity_main);
    this.locate = (TextView) super.findViewById(R.id.msg);    //取得组件
    this.locate.setOnTouchListener(new OnTouchListenerImpl());} //设置事件监听
    private class OnTouchListenerImpl implements OnTouchListener {
     @Override
     public boolean onTouch(View v, MotionEvent event) {//实现 onTouchEvent 方法
     MainActivity.this.locate.setText("您当前的位置是: "+"X = " + event.getX()
                                        + ", Y = "+ event.getY());   //设置文本

        return false; }}}
```

保存文件并运行程序，在屏幕任意空白处单击、触摸，会显示出触摸位置的坐标，结果如图 4.6 所示。

图 4.6　触摸屏幕后显示的坐标

4.7　选择改变事件 OnCheckedChange

4.7.1　选择改变事件基础

在 RadioGroup、RadioButton（单选按钮）、CheckBox 等组件上也可以进行事件处理操作，当用户选中了某选项后也将触发相应的监听器进行相应的操作。在 Android 平台的组件中提供了选择改变事件的处理操作方法，使用 View.OnCheckedChangeListener 接口可选择并改变事件的处理操作方法。View 类指向 RadioGroup 组件或 CheckBox 组件。此接口定义如下：

```
View.setOnCheckedChangeListener(new view.OnCheckedChangeListener() {
    public void onCheckedChanged(View view, int checkedId) {…}}
```

当选择改变事件触发之后，自动使用该接口中的 public void onCheckedChanged(View view, int checkedId)方法进行事件处理。

要在 Android 程序中使用选择改变事件，必须在程序中使用下面的语句：

```
import android.widget.RadioGroup.OnCheckedChangeListener;
import android.widget.CompoundButton.OnCheckedChangeListener;
```

选择改变事件的实现步骤如下。

（1）通过组件 ID 获取组件实例。

例如：

```
group = (RadioGroup)findViewById(R.id.radiogroup1);
```

（2）为该组件注册 OnCheckedChangeListener 监听。

例如：

```
group.setOnCheckedChangeListener(new RadioGroup.OnCheckedChangeListener();
```

（3）实现 onCheckedChanged 方法。

例如：

```
public void onCheckedChanged(RadioGroup group, int checkedId) { … }
```

4.7.2 RadioGroup 选择改变事件实例

实例 4-6：选择改变事件举例 1

新建一个项目，项目命名为 Exam4_7，对选择结果进行正确或错误的判断。

（1）修改“activity_main.xml”文件，设计如图 4.7 所示界面，代码可参考 Exam3_9，这里省略代码。

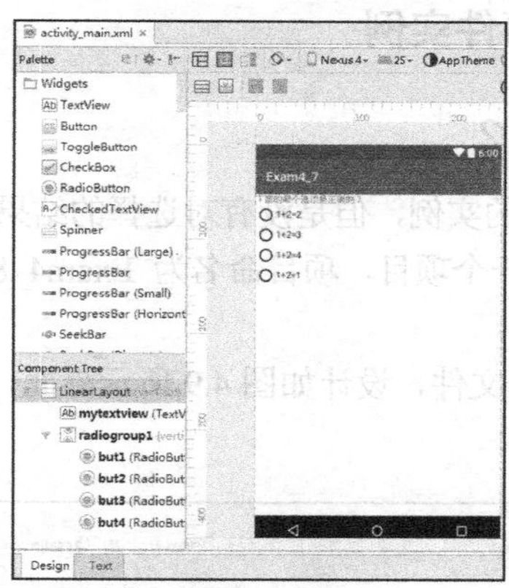

图 4.7 Exam4_7 界面

（2）修改 MainActivity.java，代码如下：

```
…
public class MainActivity extends AppCompatActivity {
    private RadioGroup group;
    private RadioButton radio1,radio2,radio3,radio4;
    @Override
    public void onCreate(Bundle savedInstanceState) {
        super.onCreate(savedInstanceState);
        setContentView(R.layout.activity_main);
        group = (RadioGroup)findViewById(R.id.radiogroup1);  //获得RadioGroup组件
        radio1 = (RadioButton)findViewById(R.id.but1);
        radio2 = (RadioButton)findViewById(R.id.but2);
        radio3 = (RadioButton)findViewById(R.id.but3);
        radio4 = (RadioButton)findViewById(R.id.but4);
        group.setOnCheckedChangeListener(new RadioGroup. OnCheckedChangeListener()
```

```
        {    //定义监听
        //实现 onCheckedChanged 方法
        public void onCheckedChanged(RadioGroup group, int checkedId) {
          if (checkedId == radio2.getId())  {
                showMessage("正确答案: " + radio2.getText()+",恭喜你, 答对了"); }
        else {
                showMessage("对不起, 您的选择错误, 再仔细思考下。"); } } });}
    public void showMessage(String str)
    {   Toast toast = Toast.makeText(this, str, Toast.LENGTH SHORT);
        toast.setGravity(Gravity.TOP, 0, 220);
        toast.show();     }  }
```

保存文件并运行程序，分别选择正确和错误的选项，结果如图 4.8 所示。

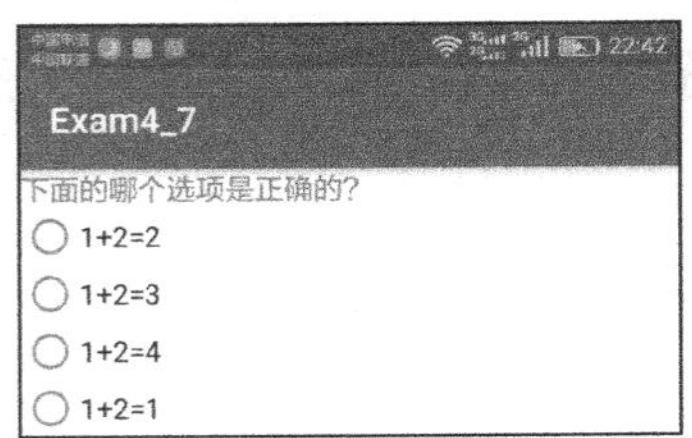

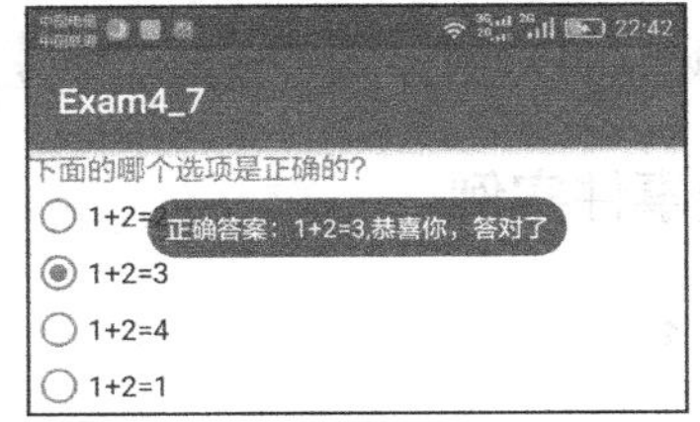

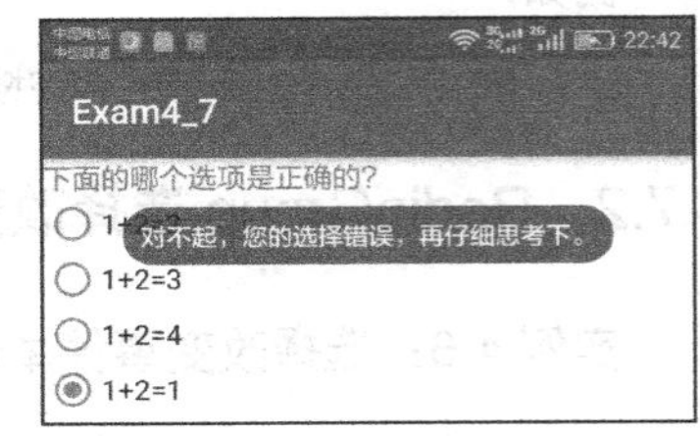

图 4.8　Exam4_7 运行结果

4.7.3　CheckBox 选择改变事件实例

实例 4-7：选择改变事件举例 2

前面章节中介绍了一个复选框的实例，但是没有对选择的结果进行判断或操作，下面就来介绍对这些选择的结果进行的操作。新建一个项目，项目命名为 Exam4_8，编程实现：对选择的结果进行提示操作。

（1）修改“activity_main.xml”文件，设计如图 4.9 所示界面，代码可参考 Exam3_10，这里省略代码。

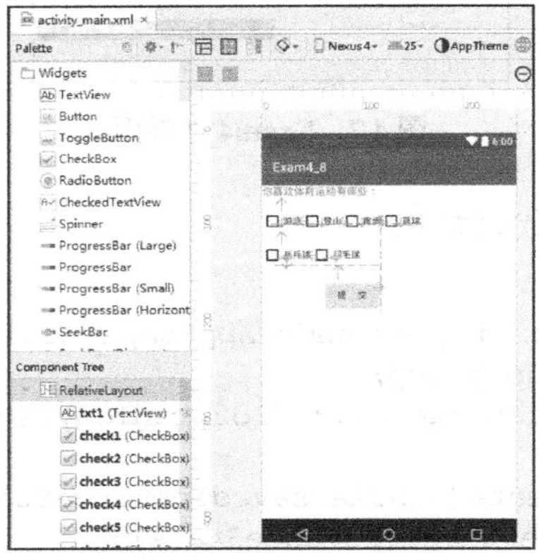

图 4.9　Exam4_8 界面

（2）修改 MainActivity.java，代码如下：

```
...
public class MainActivity extends AppCompatActivity {
```

```
    private CheckBox chk1,chk2,chk3,chk4,chk5,chk6;
    private Button mybutton;
    @Override
    public void onCreate(Bundle savedInstanceState) {
        super.onCreate(savedInstanceState);
        setContentView(R.layout.activity_main);
        mybutton = (Button)findViewById(R.id.mybut);
        chk1 = (CheckBox)findViewById(R.id.check1);
        chk2 = (CheckBox)findViewById(R.id.check2);
        chk3 = (CheckBox)findViewById(R.id.check3);
        chk4 = (CheckBox)findViewById(R.id.check4);
        chk5 = (CheckBox)findViewById(R.id.check5);
        chk6 = (CheckBox)findViewById(R.id.check6);
        chk1.setOnCheckedChangeListener(new CheckBox. OnCheckedChangeListener() {
            public void onCheckedChanged(CompoundButton arg0, boolean arg1) {
                if(chk1.isChecked())
                { showMessage("你刚才选择了"+chk1.getText());   } } });
        chk2.setOnCheckedChangeListener(new CheckBox. OnCheckedChangeListener()  {
            public void onCheckedChanged(CompoundButton arg0, boolean arg1) {
                if(chk3.isChecked())
                {  showMessage("你刚才选择了"+chk2.getText());  }  }  });
        chk3.setOnCheckedChangeListener(new CheckBox. OnCheckedChangeListener()  {
            public void onCheckedChanged(CompoundButton arg0, boolean arg1) {
                if(chk3.isChecked())
                { showMessage("你刚才选择了"+chk3.getText()); } } });
        chk4.setOnCheckedChangeListener(new CheckBox. OnCheckedChangeListener()  {
            public void onCheckedChanged(CompoundButton arg0, boolean arg1) {
                if(chk4.isChecked())
                {  showMessage("你刚才选择了"+chk4.getText()); } } });
        chk5.setOnCheckedChangeListener(new CheckBox. OnCheckedChangeListener() {
            public void onCheckedChanged(CompoundButton arg0, boolean arg1) {
                if(chk5.isChecked())
   {showMessage("你刚才选择了"+chk5.getText());}}});
        chk6.setOnCheckedChangeListener(new CheckBox. OnCheckedChangeListener()  {
            public void onCheckedChanged(CompoundButton arg0, boolean arg1) {
                if(chk6.isChecked())
                {   showMessage("你刚才选择了"+chk6.getText());}}});
            mybutton.setOnClickListener(new Button.OnClickListener(){
            public void onClick(View arg0) {
                int num = 0;
                String str="";
                if(chk1.isChecked())
                {  str=chk1.getText()+","+str;
                   num++; }
                if(chk2.isChecked())
                {  str=chk2.getText()+","+str;
                   num++; }
                if(chk3.isChecked())
                {  str=chk3.getText()+","+str;
                   num++; }
                if(chk4.isChecked())
                {  str=chk4.getText()+","+str;
```

```
                num++; }
            if(chk5.isChecked())
            {  str=chk5.getText()+","+str;
                num++; }
            if(chk6.isChecked())
            {  str=chk6.getText()+","+str;
                num++; }
            showMessage("您选择了"+str+"共"+num+"项");   }}); }
    void showMessage(String str)
    {  Toast toast = Toast.makeText(this, str, Toast.LENGTH_SHORT);
       toast.setGravity(Gravity.TOP, 0, 220);
       toast.show();   } }
```

保存文件并运行程序，选择相应的选项会相关有提示，最后单击“提交”按钮，会弹出所有选择的情况提示，如图 4.10 所示。

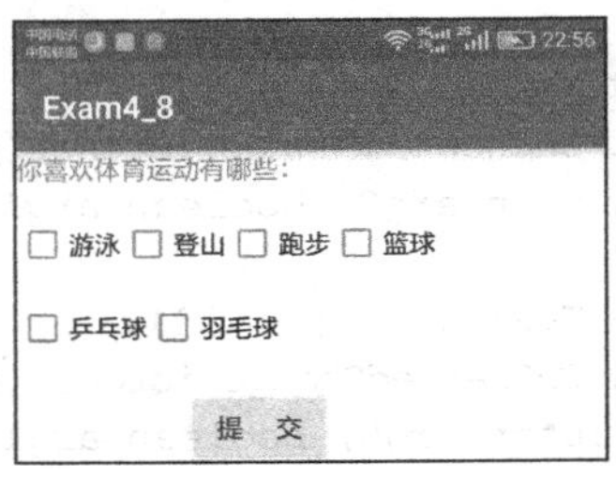

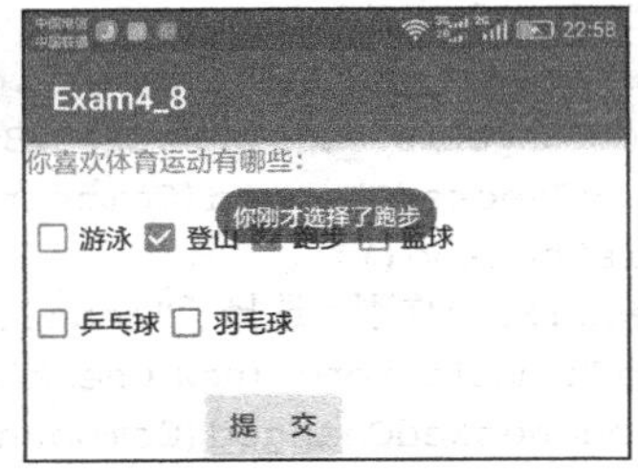

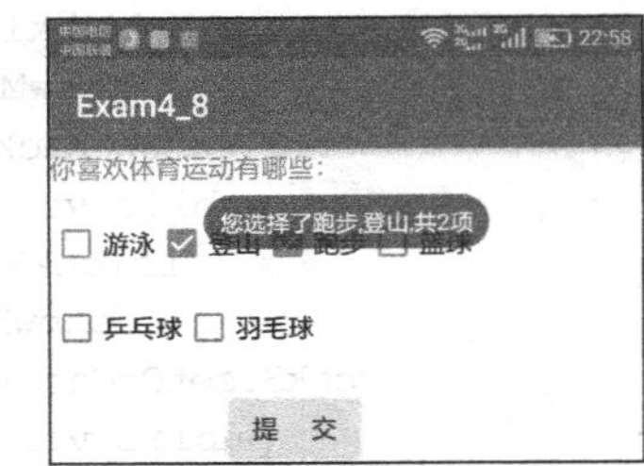

图 4.10　Exam4_8 运行结果

4.8　选项选中事件 OnItemSelected

4.8.1　选项选中事件基础

Spinner 组件的主要功能是进行下拉列表显示，当用户选择下拉列表中的某个选项后，可以使用 Spinner 类中提供的下面 3 个接口进行相应的处理操作。

（1）当列表项被选中或者被单击时触发的事件如下：

```
setOnItemClickListener(AdapterView.OnItemClickListener listener);
```

（2）当列表项改变时所触发的事件如下：

```
setOnItemSelectedListener(AdapterView.OnItemSelectedListener  listener)
```

（3）当列表项被长时间按住时所触发的事件如下：

```
setOnItemLongClickListener(AdapterView.OnItemLongClickListener listener)
```

当事件触发后，会自动使用该接口中的 public void onItemSelected()方法进行事件处理。

选项选中事件的实现步骤如下。

（1）通过组件 ID 获取组件实例。

例如：

```
spin=(Spinner)findViewById(R.id.spin);
```

（2）为该组件注册监听。

例如：

```
spin.setOnItemSelectedListener(new Spinner.OnItemSelectedListener(){…}
```

（3）实现 onItemSelected 方法。

例如：

```
public void onItemSelected(AdapterView<?> arg0, View arg1, int arg2, long arg3)
{…}
```

4.8.2 OnItemSelected 选项选中事件实例

实例 4-8：选项选中事件举例

前面介绍了 Spinner 组件的基本属性，没有对选择的结果进行判断或操作，下面介绍如何对这些选择的结果进行显示。新建一个项目，项目命名为 Exam4_9，实现对 Spinner 组件选择结果的显示。

（1）修改 activity_main.xml 文件，代码如下：

```
…
<TextView
    android:id="@+id/txt"
    android:layout_width="wrap_content"
    android:layout_height="wrap_content"
    android:text="您心中的理想专业是：" />
<Spinner
    android:id="@+id/spin"
    android:layout_width="wrap_content"
    android:layout_height="wrap_content"
    android:layout_centerHorizontal="true"
android:spinnerMode="dialog" />
</LinearLayout>
```

（2）修改 MainActivity.java，代码如下：

```
…
public class MainActivity extends AppCompatActivity{
 private static final String[] subjects={"计算机","数学","文学","英语","哲学"};
 private TextView txt;
 private Spinner spin;
 private ArrayAdapter<String> adapter;
 @Override
 public void onCreate(Bundle savedInstanceState) {
        super.onCreate(savedInstanceState);
        setContentView(R.layout.activity_main);
        txt=(TextView)findViewById(R.id.txt);
        spin=(Spinner)findViewById(R.id.spin);
        this.spin.setPrompt("请选择您的理想专业：");
        //将可选内容与 ArrayAdapter 进行连接
        adapter=new ArrayAdapter<String>(this,android.R.layout.simple_spinner_
        item,subjects);
        //设置下拉列表风格
        adapter.setDropDownViewResource(android.R.layout.simple_spinner_
```

```
                                                dropdown_item);
        spin.setAdapter(adapter);                 //将 adapter 添加到 Spinner 中
        //添加 Spinner 事件监听
        spin.setOnItemSelectedListener(new Spinner.OnItemSelectedListener()
       { public void onItemSelected(AdapterView<?> arg0, View arg1, int arg2, long arg3)
            {   txt.setText("您心中的理想专业是: "+subjects[arg2]);
                arg0.setVisibility(View.VISIBLE);  }    //设置显示当前选择的项
        public void onNothingSelected(AdapterView<?> arg0) {  }     }); } }
```

保存文件并运行程序，单击下拉列表，弹出下拉列表，选择的相应选项会在屏幕上显示出来，运行结果如图 4.11 所示。

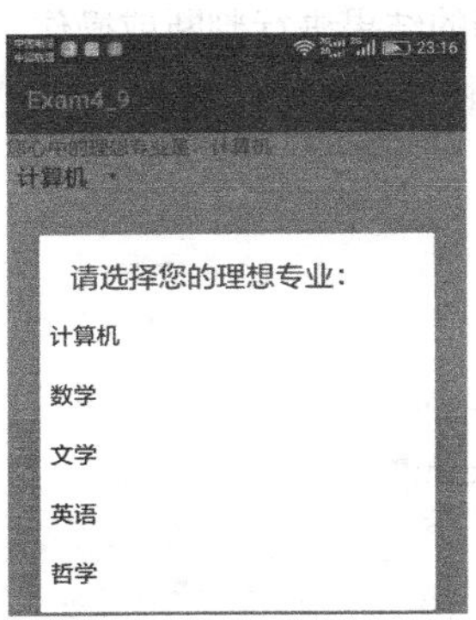

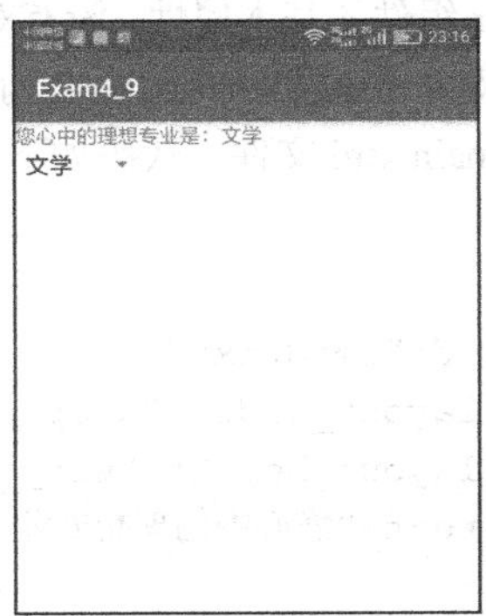

图 4.11　Exam4_9 运行结果

Spinner 组件在设计中用得较多的是二级联动应用，因为代码较长，这里不再介绍，感兴趣的读者可参照项目 Exam4_10 进行学习。Spinner 组件也可以动态地在 MainActivity 程序中添加或删除 Spinner 组件中的选项，感兴趣的读者可参照项目 Exam4_11 进行学习。

4.9　日期和时间监听事件

4.9.1　日期和时间选择器组件

日期选择器组件 DatePicker 是一个可选择年、月、日的日历布局视图，使用它可以对年、月、日进行设置，其层次关系如下：

```
java.lang.Object
     android.view.View
          android.view.ViewGroup
               android.widget.FrameLayout
                    android.widget.DatePicker
```

要在 Android 程序中使用 DatePicker 组件，必须在程序中使用下面的语句：

```
import android.widget.DatePicker;              //导入 widget.DatePicker 类
```

时间选择器组件 TimePicker 是用于选择一天中时间的视图，使用它可以进行时间的调整，其层次关系如下：

```
java.lang.Object
```

```
android.view.View
      android.view.ViewGroup
            android.widget.FrameLayout
                  android.widget.TimePicker
```

要在 Android 程序中使用 TimePicker 组件，必须在程序中使用下面的语句：

```
import android.widget.TimePicker;            //导入 widget.TimePicker 类
```

和前面介绍的组件一样，DatePicker 和 TimePicker 组件也有其属性和方法，常用的方法如表 4-1 所示。

表 4-1 DatePicker 和 TimePicker 组件常用的方法

方法	描述
public Integer getCurrentHour()	返回当前设置的小时
public Integer getCurrentMinute()	返回当前设置的分钟
public boolean is24HourView()	判断是否为 24 小时制
public void setCurrentHour(Integer currentHour)	设置当前的小时数
public void setCurrentMinute(Integer currentMinute)	设置当前的分钟
public void setEnabled(boolean enabled)	设置是否可用
public void setIs24HourView(Boolean is24HourView)	设置时间为 24 小时制
public int getYear()	取得设置的年
public void setOnTimeChangedListener (TimePicker.OnTimeChangedListener onTimeChangedListener)	设置时间调整事件的回调函数
public int getYear ()	取得设置的年份
public int getMonth()	取得设置的月
public int getDayOfMonth()	取得设置的日
public void setEnabled(boolean enabled)	设置组件是否可用
public void updateDate(int year, int monthOfYear, int dayOfMonth)	设置一个指定的日期
public void init (int year, int monthOfYear, int dayOfMonth,DatePicker. OnDateChangedListener onDateChangedListener)	初始化状态（初始化年、月、日）

4.9.2 日期和时间的设置

DatePicker 和 TimePicker 可以显示当前的系统日期和时间，与前面的基本组件类似，应在布局文件中定义组件，然后在 MainActivity 程序中调用，代码参照 Exam4_12。当要对日期和时间进行设置时，Android 平台提供了日期对话框 DatePickerDialog 及时间对话框 TimePickerDialog，其层次关系如下：

```
java.lang.Object
      android.app.Dialog
            android.app.AlertDialog
                  android.app.DatePickerDialog、android.app.TimePickerDialog
```

类似的，要在 Android 程序中使用 DatePickerDialog 和 TimePickerDialog 组件，必须在程序中使用下面的语句：

```
import android.app.DatePickerDialog; //导入 widget.DatePickerDialog 类
import android.app.TimePickerDialog; //导入 widget.TimePickerDialog 类
```

在 DatePickerDialog 组件内有一个 OnDateSetListener，可以指定监听操作；在 TimePickerDialog 组件内有一个 OnTimeSetListener，可以指定监听操作，具体方法如表 4-2 所示。

表 4-2　DatePickerDialog、TimePickerDialog 的常用方法

方法	描述
public DatePickerDialog (Context context, DatePickerDialog.OnDateSetListener callBack, int year, int monthOfYear, int dayOfMonth)	创建 DatePickerDialog 对象，同时指定监听操作及要设置的年、月、日等信息
public void updateDate (int year, int monthOfYear, int dayOfMonth)	更新显示组件上的年、月、日信息
public TimePickerDialog (Context context, TimePickerDialog.OnTimeSetListener callBack, int hourOfDay, int minute, boolean is24HourView)	创建时间对话框，同时设置时间改变的事件操作、小时、分以及是否为 24 小时制
public void updateTime (int hourOfDay, int minuteOfHour)	更新时、分

实例 4-9：DatePickerDialog 和 TimePickerDialog 组件使用举例

新建一个项目，项目命名为 Exam4_13，利用 DatePickerDialog 和 TimePickerDialog 组件设置系统的日期和时间。

（1）修改 activity_main.xml 文件，代码如下：

```
…
<TextView
    android:id="@+id/txt"
    android:layout_width="wrap_content"
    android:layout_height="wrap_content"/>
<Button
    android:id="@+id/datebut"
    android:text="设置日期"
    android:layout_width="wrap_content"
    android:layout_height="wrap_content"/>
<Button
    android:id="@+id/timebut"
    android:text="设置时间"
    android:layout_width="wrap_content"
    android:layout_height="wrap_content"/>
</LinearLayout>
```

（2）修改 MainActivity.java，代码如下：

```
…
public class MainActivity extends AppCompatActivity {
private Button datebut = null ;                      //定义按钮组件
private Button timebut = null ;                      //定义按钮组件
    @Override
public void onCreate(Bundle savedInstanceState) {
    super.onCreate(savedInstanceState);
    super.setContentView(R.layout.activity_main);
    this.datebut = (Button) super.findViewById(R.id.datebut); //取得按钮组件
    this.timebut = (Button) super.findViewById(R.id.timebut) ;  //取得按钮组件
    this.datebut.setOnClickListener(new OnClickListenerDate()) ;//设置日期单击事件
    this.timebut.setOnClickListener(new OnClickListenerTime()) ; }
 //设置日期单击事件
private class OnClickListenerDate implements OnClickListener {  //注册单击事件监听
    @Override
    public void onClick(View v) {             //实现 onClick()方法
        Dialog dialog = new DatePickerDialog(MainActivity.this,new
       DatePickerDialog.OnDateSetListener() {       //日期事件监听
```

```
            public  void onDateSet(DatePicker view, int year, int monthOfYear,
                                  int dayOfMonth)  {    //日期改变时触发
                TextView text = (TextView)findViewById(R.id.txt);
                text.setText("更新的日期为: " + year + "-" + (monthOfYear+1) + "-"+
                          dayOfMonth);        //设置文本内容
                }}, 2017, 3, 10);     //默认为2017年 3月10日
        dialog.show() ; }}              //显示对话框
  private class OnClickListenerTime implements OnClickListener {
      @Override
      public void onClick(View v) {
        Dialog dialog1= new TimePickerDialog(MainActivity.this,new
                    TimePickerDialog.OnTimeSetListener() {//时间事件监听
      public void onTimeSet(TimePicker view, int hourOfDay, int minute) {
            //时间改变时触发
            TextView text = (TextView)findViewById(R.id.txt);
            text.setText("更新的时间为: " +hourOfDay + ": " + minute);/*设置
                                                          文本内容*/

            }}, 7, 32,true);//默认为7:32
        dialog1.show() ; }}}//显示对话框
```

保存文件并运行程序，单击“设置日期”按钮，进入日期设置界面，如图 4.12 所示，设置完毕后，单击“确定”按钮会显示设置的日期。类似的，可以设置时间，如图 4.13 所示。

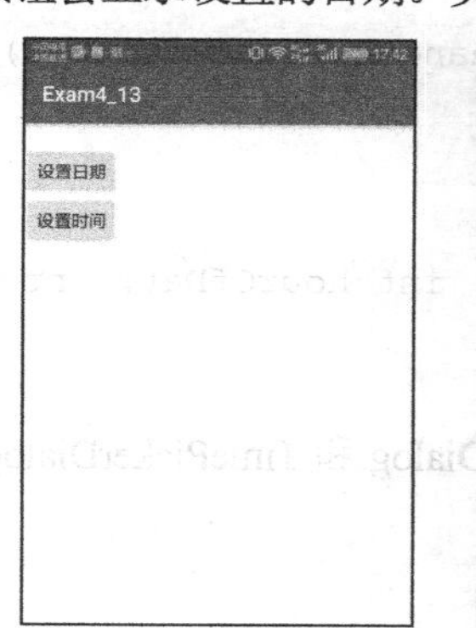

图 4.12 Exam4_13 设置日期

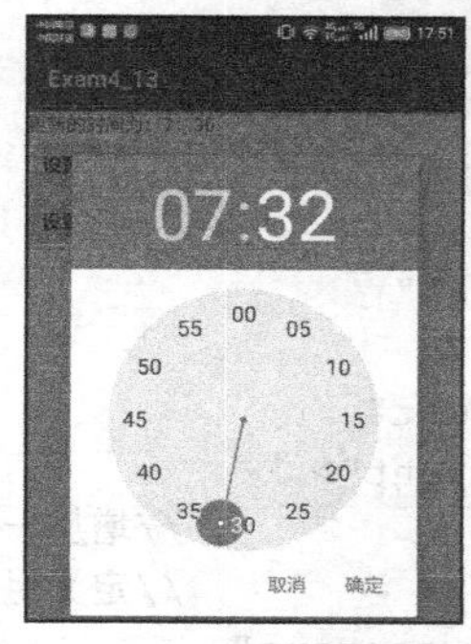

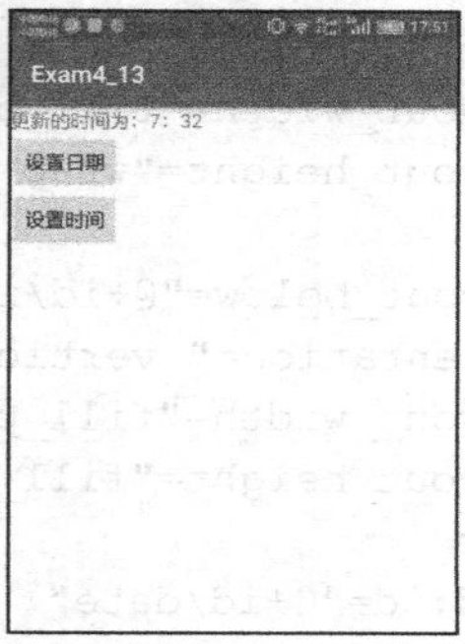

图 4.13 Exam4_13 设置时间

4.9.3 日期和时间监听事件

日期和时间选择器可以对日期和事件进行调整，当日期和时间发生变化时可以触发相应的事件，

Android 平台提供了日期、时间监听器接口来实现日期和时间监听事件。

日期监听器接口定义如下：

```
View.setOnDateChangedListener(new OnDateChangedListenerImpl());
```

事件触发之后，自动使用该接口中的 public void onDateChanged()方法进行事件处理。

时间监听器接口定义如下：

```
View.setOnTimeChangedListener(new OnTimeChangedListenerImpl());
```

事件触发后，自动使用该接口中的 public void onTimeChanged()方法进行事件处理。

要在 Android 的 Java 程序中使用日期和时间监听事件，必须在程序中使用下面的语句：

```
import android.widget.DatePicker.OnDateChangedListener;    //导入日期监听需要类
import android.widget.TimePicker.OnTimeChangedListener;    //导入时间监听需要类
```

日期和时间监听事件的实现步骤如下（以时间监听事件为例进行介绍）。

（1）通过组件 ID 获取组件实例。

例如：

```
time = (DatePicker) super.findViewById(R.id.time);
```

（2）为该组件注册监听。

例如：

```
time.setOnTimeChangedListener(new OnTimeChangedListenerImpl());…
```

（3）实现 onTimeChanged 方法。

例如：

```
public void onTimeChanged(TimePicker view, int hourOfDay, int minute) {…}
```

实例 4-10：日期和时间监听事件举例

新建一个项目，项目命名为 Exam4_14，利用 DatePickerDialog 和 TimePickerDialog 组件设置系统的日期和时间，当日期和时间发生变化时触发事件。

（1）修改 activity_main.xml 文件，代码如下：

```
…
<EditText
     android:id="@+id/input"
     android:layout_width="fill_parent"
     android:layout_height="wrap_content"/>
<LinearLayout
     android:layout_below="@+id/input"
     android:orientation=" vertical "
     android:layout_width="fill_parent"
     android:layout_height="fill_parent">
     <DatePicker                                          //增加一个日期选择器组件
         android:id="@+id/date"                           //定义组件 ID，在程序中使用
         android:layout_width="wrap_content"
         android:layout_height="wrap_content" />
     <TimePicker                                          //增加一个时间选择器组件
         android:id="@+id/time"                           //定义组件 ID，在程序中使用
         android:layout_width="wrap_content"
         android:layout_height="wrap_content" />
```

```
</LinearLayout>
</LinearLayout>
```

（2）修改 MainActivity.java，代码如下：

```
…
import android.widget.DatePicker.OnDateChangedListener;
import android.widget.TimePicker.OnTimeChangedListener;
public class MainActivity extends AppCompatActivity {
private EditText input = null;
private DatePicker date = null;
private TimePicker time = null;
@Override
public void onCreate(Bundle savedInstanceState) {
    super.onCreate(savedInstanceState);
    super.setContentView(R.layout.activity_main);
    this.input = (EditText) super.findViewById(R.id.input);
    this.date = (DatePicker) super.findViewById(R.id.date);
    this.time = (TimePicker) super.findViewById(R.id.time);
    this.time.setIs24HourView(true);                    //设置为 24 小时制
    //设置时间改变监听
    this.time.setOnTimeChangedListener(new OnTimeChangedListenerImpl());
    this.date.init(this.date.getYear(), this.date.getMonth(),
    this.date.getDayOfMonth(),
    new OnDateChangedListenerImpl());               //设置日期改变监听
    this.setDateTime(); }                           //设置文本日期
public void setDateTime() {                         //定义一个方法，设置文本内容
    this.input.setText(this.date.getYear() + "-"+ (this.date.getMonth() + 1)
        + "-" + this.date.getDayOfMonth()+ " " + this.time.getCurrentHour()
        + ":"+  this.time.getCurrentMinute());}
private class OnDateChangedListenerImpl implements OnDateChangedListener {
    public void onDateChanged(DatePicker view, int year, int monthOfYear,int
                              dayOfMonth) {
        MainActivity.this.setDateTime(); }}         //日期改变时调用方法修改文本
private class OnTimeChangedListenerImpl implements OnTimeChangedListener {
    public void onTimeChanged(TimePicker view, int hourOfDay, int minute) {
        MainActivity.this.setDateTime(); }}}        //时间改变时调用方法修改文本
```

保存文件并运行程序，进入如图 4.14 所示界面，修改日期和时间上方的提示会随之改变。

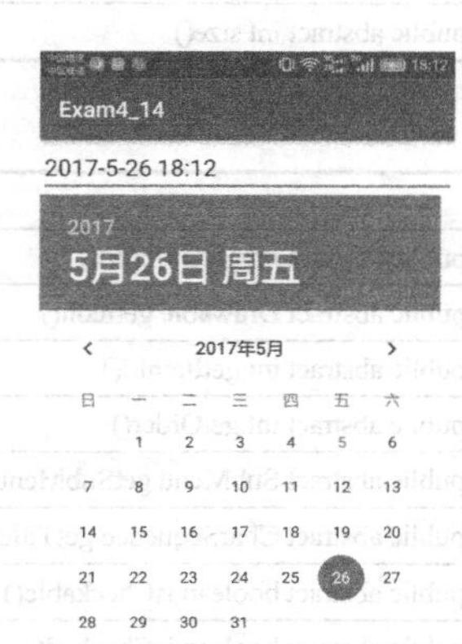

图 4.14　Exam4_14 运行结果

4.10　菜单事件

菜单是应用程序中非常重要的组成部分，能够在不占用界面空间的前提下，为应用程序提供统一的功能和设置界面，并为程序开发人员提供了易于使用的编程接口。

4.10.1　菜单事件基础

Android 系统支持 4 种菜单：选项菜单（OptionsMenu）、上下文菜单（ContextMenu）、子菜单（SubMenu）和弹出式菜单（PopupMenu）。

尽管每一种类型菜单的创建方法都不同，但是 Android 系统提供了丰富的菜单操作方法，如表 4-3、表 4-4、表 4-5 所示，因此创建菜单在 Android 系统中并不复杂。

表 4-3　常用的菜单操作方法

方法	描述
public void closeContextMenu()	关闭上下文菜单
public void closeOptionsMenu()	关闭选项菜单
public boolean onContextItemSelected(MenuItem item)	设置上下文菜单项
public void onContextMenuClosed(Menu menu)	上下文菜单关闭时触发
public void onCreateContextMenu(ContextMenu menu, View v, ContextMenu.ContextMenuInfo menuInfo)	创建上下文菜单
public boolean onCreateOptionsMenu(Menu menu)	当用户单击“Menu”按钮时调用此操作，可以生成一个选项菜单
public boolean onMenuItemSelected(int featureId, MenuItem item)	设置选项菜单项
public boolean onOptionsItemSelected(MenuItem item)	当一个选项菜单中的某个菜单项被选中时触发此操作
public void onOptionsMenuClosed(Menu menu)	当选项菜单关闭时触发此操作
public boolean onPrepareOptionsMenu(Menu menu)	在选项菜单显示之前操作，会触发此操作
public void openOptionsMenu()	打开选项菜单
public MenuInflater getMenuInflater()	取得 MenuInflater 类的对象
public void registerForContextMenu(View view)	注册上下文菜单

表 4-4　Menu 接口的常用方法及常量

方法及常量	描述
public static final int FIRST	常量，用于定义菜单项的编号
public static final int NONE	常量，表示菜单不分组
public abstract MenuItem add(int groupId, int itemId, int order, CharSequence title)	此方法用于向菜单中添加菜单项
public abstract MenuItem add(int groupId, int itemId, int order, int titleRes)	增加菜单项
public abstract SubMenu addSubMenu(int groupId, int itemId, int order, int titleRes)	增加子菜单
public abstract SubMenu addSubMenu(int groupId, int itemId, int order, CharSequence title)	增加子菜单
public abstract void removeGroup(int groupId)	删除一个菜单组
public abstract void removeItem(int id)	删除一个菜单项
public abstract void clear()	清空菜单
public abstract void close()	关闭菜单
public abstract MenuItem getItem(int index)	返回指定的菜单项
public abstract int size()	返回菜单项的个数

表 4-5　MenuItem 接口的常用方法

方法	描述
public abstract int getGroupId()	得到菜单组编号
public abstract Drawable getIcon()	得到菜单项上的图标
public abstract int getItemId()	得到菜单项上的 ID
public abstract int getOrder()	得到菜单项上的编号
public abstract SubMenu getSubMenu()	取得子菜单
public abstract CharSequence getTitle()	得到菜单项上的标题
public abstract boolean isCheckable()	判断菜单项是否可用
public abstract boolean isChecked()	判断此菜单项是否被选中
public abstract boolean isEnabled()	判断此菜单项是否可用

续表

方法	描述
public abstract boolean isVisible()	判断此菜单项是否可见
public abstract MenuItem setCheckable(boolean checkable)	设置此菜单项是否可用
public abstract MenuItem setChecked(boolean checked)	设置此菜单项是否默认选中
public abstract MenuItem setEnabled(boolean enabled)	设置此菜单项是否可用
public abstract MenuItem setIcon(Drawable icon)	设置此菜单项的图标
public abstract MenuItem setIcon(int iconRes)	设置此菜单项的图标
public abstract MenuItem setOnMenuItemClickListener(MenuItem.OnMenuItemClickListener menuItemClickListener)	设置此菜单项的监听操作
public abstract MenuItem setTitle(CharSequence title)	设置此菜单项的标题
public abstract MenuItem setVisible(boolean visible)	设置此菜单项是否可见
public abstract ContextMenu.ContextMenuInfo getMenuInfo()	得到菜单中的内容

4.10.2 选项菜单 OptionsMenu

选项菜单 OptionsMenu 在 Android 2.3.x 版本中经常用到，早期的手机上都会有一个“MENU”键，选项菜单一般通过“MENU”键打开，从 Android 3.0 开始，Android 不再要求手机设备上必须提供 MENU 键，运行时单击手机右上角的“⋮”键就相当于单击了“MENU”键。

选项菜单分为图标菜单（Icon Menu）和扩展菜单（Expanded Menu）。在 Android 2.3.x 版本中，图标菜单是能够同时显示文字和图标的菜单，不支持单选按钮和复选框，在一个菜单之中最多只会显示 6 个菜单项（MenuItem），如果菜单项超出了 6 个，则超出部分会自动隐藏，而且会自动出现一个“More”菜单项，以提示用户单击“More”按钮后会弹出第 6 项及以后的菜单项，这些菜单项被称为扩展菜单，扩展菜单是垂直的列表型菜单，不显示图标，支持单选按钮和复选框。Android 4.0 以后仍然可以正常使用图标菜单，但显示样式发生了一些变化，没有以前美观了。

要在 Android 程序中使用选项菜单，必须在程序中使用下面的语句：

```
import android.view.Menu;                        //导入 view.Menu 类
import android.view.MenuItem;                    //导入 view.MenuItem 类
```

在 OptionsMenu 中常用到如下几个操作方法。

（1）public boolean onCreateOptionsMenu(Menu menu)：设置多个菜单项（MenuItem）。

如果要实现选项菜单，则需要重载 Activity 程序的 onCreateOptionsMenu()函数，返回 true 时显示菜单，返回 false 时不显示菜单。初次使用选项菜单时，会调用 onCreateOptionsMenu()函数，用来初始化菜单子项的相关内容，设置菜单子项自身的子项 ID 和组 ID、菜单子项显示的文字和图片等。例如：

```
public boolean onCreateOptionsMenu(Menu menu) {              //显示菜单
    menu.add(Menu.NONE,                                      //菜单不分组
             Menu.FIRST + 1,                                 //菜单项 ID
             1,                                              //菜单编号
             "文件")                                          //显示标题
             .setIcon(android.R.drawable. sym_action_email); //设置显示图标
             …
             return true;} //函数的返回值，值为 true 时显示设置的菜单，否则不能够显示
```

其中，Menu.FIRST 是常量，值为 1。

sym_action_email 是 Android 系统自带图标库 android.R.drawable 中的图标，在这个库中含有大量的常用图标，用户可以查找相关资料进行了解。

使用 setIcon()函数设置显示的图标，也可以使用 setShortcut()函数添加菜单子项的快捷键。例如：

```
menu.add(Menu.NONE, Menu.FIRST + 1, 3, "设置")      //添加选项"设置"
.setShortcut('3','s');                             //添加快捷键"s"和"3"
```

setShortcut 中用两个参数来设定两个快捷键是为了应对不同的手机键盘。

第一个参数表示数字快捷键为 12 键键盘（0～9、*、#，共 12 个按键），第二个参数表示全键盘。任何键不区分大小写。

（2）public boolean onOptionsItemSelected(MenuItem item)：判断菜单项的操作。

该函数能够处理菜单选择事件，每次选择菜单子项时都会被调用，下面的代码说明了如何通过菜单子项的子项 ID 执行不同的操作。

```
public boolean onOptionsItemSelected(MenuItem item) { //选中某个菜单项
    switch (item.getItemId()) {                       //判断菜单项 ID
    case Menu.FIRST + 1:                              //ID 为 FIRST + 1 时
        MenuemailCounter=MenuemailCounter+1;          //选中该项就增加 1
        Toast.makeText(this, "您选择的是"文件菜单"项。", Toast.LENGTH
                    LONG).show();                     //显示选中的选项名
        break;                                        //退出
    case Menu.FIRST + 2:
    … }
    return false; }                                   //返回 false
```

onOptionsItemSelected ()的返回值表示是否对菜单的选择事件进行处理，如果已经处理，则返回 true，否则返回 false。

（3）public boolean onPrepareOptionsMenu(Menu menu)：在菜单显示前触发此操作。

重载 onPrepareOptionsMenu()函数，能够动态地添加、删除菜单子项，或修改菜单的标题、图标和可见性等内容，函数返回值为 true，则继续调用 onCreateOptionsMenu()方法，反之则不再调用。

下面的代码用于在用户每次打开选项菜单时，在菜单子项中显示用户打开该子项的次数：

```
static int MenuemailCounter = 0;                      //统计选项选中的计数器
public boolean onPrepareOptionsMenu(Menu menu) {      //菜单显示前调用
   MenuItem emailItem = menu.findItem(Menu.FIRST + 1); //获得 Menu.FIRST + 1 项
   emailItem.setTitle("文件选项:" +String.valueOf(MenuemailCounter)); }
      //设置标题为“文件选项”与 MenuemailCounter 的组合
```

（4）public void onOptionsMenuClosed(Menu menu)：当菜单关闭时触发此操作，例如：

```
public void onOptionsMenuClosed(Menu menu) {        //菜单退出时调用
Toast.makeText(this, "注意，选项菜单关闭！！", Toast.LENGTH_LONG).show();}
```

（5）如果希望从配置文件之中取出数据，则需要使用到下面两个常用方法：

```
public MenuInflater(Context context)              //创建 MenuInflater 类对象
public void inflater(int menuRes, Menu menu)  //将配置的资源填充到菜单之中
```

inflater 在 Android 系统中建立了从资源文件到对象的桥梁，MenuInflater 即把菜单 XML 资源转换为对象并添加到 menu 对象中，它可以通过 activity 的 getMenuInflater()得到。在 MainActivity 中重写 onCreateOptionsMenu(...)方法如下：

```
public boolean onCreateOptionsMenu(Menu menu) {
   MenuInflater inflater = getMenuInflater();
   inflater.inflate(R.menu.mainmenu, menu);
```

```
    return true;}
```

实例 4-11：选项菜单举例 1

新建一个项目，项目命名为Exam4_15，在MainActivity中编程，新建一个选项菜单OptionsMenu。

（1）修改activity_main.xml文件，代码如下：

```
…
<TextView
    android:id="@+id/mytxt"
    android:layout_width="wrap_content"
    android:layout_height="wrap_content"
    android:text="请单击右上角的…"/>
</LinearLayout>
```

（2）修改MainActivity.java，代码如下：

```
…
import android.view.Menu;
import android.view.MenuItem;
public class MainActivity extends AppCompatActivity {
   public void onCreate(Bundle savedInstanceState) {
      super.onCreate(savedInstanceState);
      super.setContentView(R.layout.activity_main);  }
   public boolean onCreateOptionsMenu(Menu menu) {          //显示菜单
      menu.add(Menu.NONE,                                   //菜单不分组
            Menu.FIRST + 1,                          //菜单项 ID
            1,                                       //菜单编号
            "文件")                                   //显示标题
            .setIcon(android.R.drawable.ic_menu_save); //设置图标
      MenuItem item1=menu.add(Menu.NONE,Menu.FIRST+2, 2, "编辑");
      item1.setIcon(android.R.drawable.ic_menu_edit);//设置菜单项
      MenuItem item2 =menu.add(Menu.NONE, Menu.FIRST + 3, 3, "设置");
      item2.setIcon(android.R.drawable.ic_menu_help);
      item2.setShortcut('3','s');                      //设置快捷键
      return true;     }                              //菜单显示
   public boolean onOptionsItemSelected(MenuItem item) {  //选中某个菜单项
      switch (item.getItemId()) {                        //判断菜单项 ID
         case Menu.FIRST + 1:
            MenuefileCounter=MenuefileCounter+1;
           Toast.makeText(this, "您选择的是"文件菜单"项。", Toast.LENGTH_LONG).
show();
            break;
         case Menu.FIRST + 2:
           Toast.makeText(this, "您选择的是"编辑菜单"项。", Toast.LENGTH_LONG).
show();
            break;
         case Menu.FIRST + 3:
           Toast.makeText(this, "您选择的是"设置菜单"项。", Toast.LENGTH_LONG).
show();
            break;          }
      return false;  }                               //返回 false
 public void onOptionsMenuClosed(Menu menu) {     //菜单退出时调用
    Toast.makeText(this, "注意，现在菜单关闭！！", Toast.LENGTH_LONG).show();}
```

```
    static int MenuefileCounter = 0;                              //统计选项选中的计数器
    public boolean onPrepareOptionsMenu(Menu menu) {             //菜单显示前调用
    MenuItem fileItem = menu.findItem(Menu.FIRST + 1);   //获得 Menu.FIRST +1 项
    fileItem.setTitle("文件:" +String.valueOf(MenuefileCounter)); //设置标题
    Toast.makeText(this, "在菜单显示之前会先执行此操作，您可以在这里进行一些预处理操作。
",Toast.LENGTH_LONG).show();
  return true;   }}          //调用 onCreateOptionsMenu()
```

保存文件，按手机右上角的“⋮”键，弹出选项菜单，反复选择“文件：0”选项，程序能够自动统计选择了多少次该选项，程序运行结果如图 4.15 所示。

在 Android 系统中，菜单不仅能够在代码中定义，还可以像界面布局一样在 XML 文件中进行定义，使用 XML 文件定义界面菜单，对代码与界面设计进行分类，有助于简化代码的复杂程度，并且更有利于界面的可视化。下面来利用 XML 中的数据建立上下文菜单中的菜单选项。

图 4.15　Exam4_15 运行结果

实例 4-12：选项菜单举例 2

新建一个项目，项目命名为 Exam4_16，新建一个选项菜单 OptionsMenu，选项数据从配置文件之中获得。

（1）右击“res”文件夹，选择“New→Folder→Res Folder”选项，如图 4.16 所示。在弹出的窗口中输入“src/main/res/menu”，新建 menu 文件夹，如图 4.17 所示，然后单击“Finish”按钮。

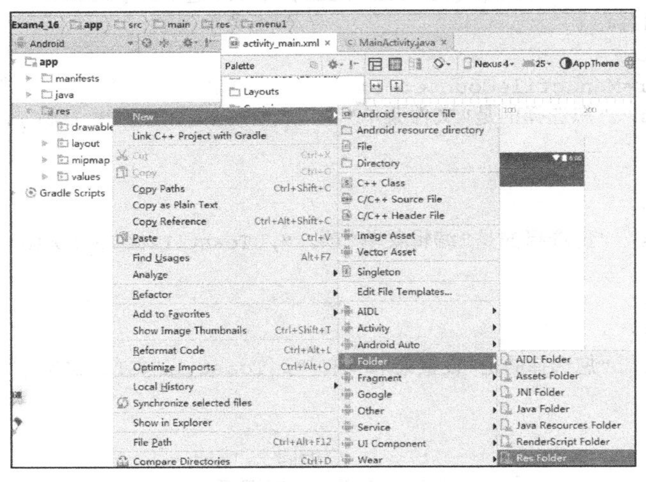

图 4.16　新建文件夹

图 4.17　输入文件夹名称

（2）右击刚才建立的“menu”文件夹，选择“New→Menu resource file”选项，如图 4.18 所示。在弹出的对话框中，在“File name”中输入“menu”，新建 menu.xml 文件，如图 4.19 所示，然后单击“OK”按钮。

（3）双击刚才新建的“menu.xml”文件，输入如下代码：

```
<?xml version="1.0" encoding="utf-8"?>
<menu xmlns:android="http://schemas.android.com/apk/res/android">
  xmlns:app="http://schemas.android.com/apk/res-auto"
  xmlns:tools="http://schemas.android.com/tools"
  tools:context=".MainActivity">
        <item android:id="@+id/file" android:title="文件" />
        <item android:id="@+id/edit" android:title="编辑" />
        <item android:id="@+id/config" android:title="设置" />
</menu>
```

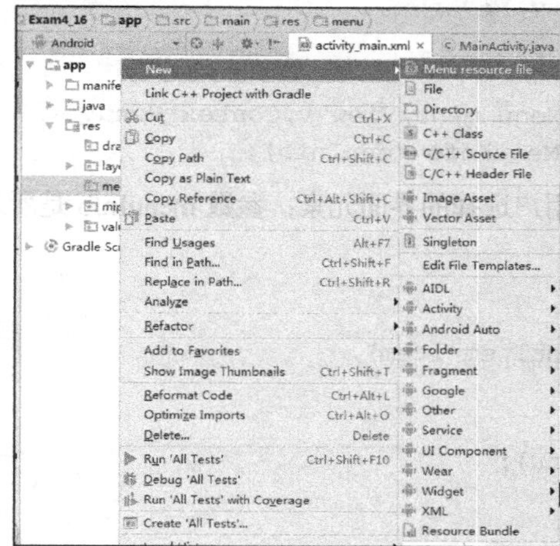

图 4.18　新建文件

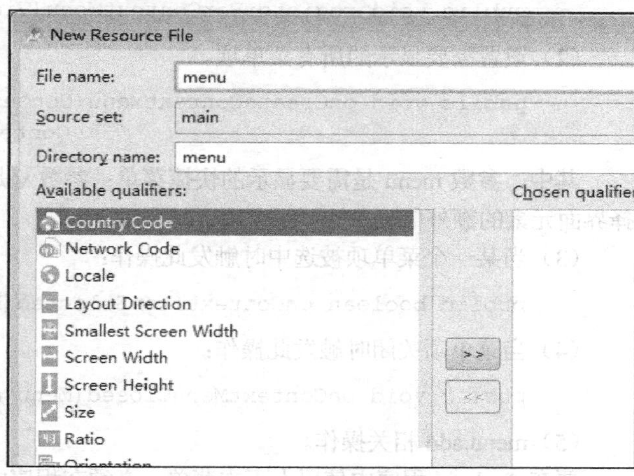

图 4.19　输入文件名称

（4）修改“MainActivity”文件，输入如下代码：

```
…
public class MainActivity extends AppCompatActivity {
   @Override
   protected void onCreate(Bundle savedInstanceState) {
      super.onCreate(savedInstanceState);
      setContentView(R.layout.activity_main);    }
   public boolean onCreateOptionsMenu(Menu menu) {
      getMenuInflater().inflate(R.menu.menu, menu);
      return true;    }
   public boolean onOptionsItemSelected(MenuItem item) {
      switch(item.getItemId()){
         case R.id.file:
            Toast.makeText(this,"选择了文件操作", Toast.LENGTH_SHORT).show();
            break;
         case R.id.edit:
            Toast.makeText(this,"选择了编辑操作",Toast.LENGTH_LONG).show();
            break;
         case R.id.config:
            Toast.makeText(this,"选择了设置操作",Toast.LENGTH_SHORT).show();
```

```
            break;
        default:}
    return true;    }}
```

保存文件，按手机右上角的“▮”键，弹出选项菜单，程序运行结果如图 4.15 所示。

4.10.3 上下文菜单 ContextMenu

上下文菜单（ContextMenu）也称为快捷菜单，类似于 Windows 系统中的右键快捷菜单，但 Android 系统是通过长按（按住不动约 2 秒钟）某个组件来弹出上下文菜单的，上下文菜单项不支持图标或快捷键。

进行上下文菜单的操作有以下几个常用的方法。

（1）将快捷菜单注册到界面控件上，注册后长按就会弹出上下文菜单。

```
public void registerForContextMenu(View v,);
```

（2）设置需要显示的所有菜单项。

```
public void onCreateContextMenu(ContextMenu menu,View v,ContextMenu.
                                        ContextMenuInfo menuInfo);
```

其中，参数 menu 是需要显示的快捷菜单；参数 v 是用户选择的界面元素；参数 menuInfo 是所选择界面元素的额外信息。

（3）当某一个菜单项被选中时触发此操作：

```
public boolean onContextItemSelected(MenuItem item);
```

（4）当菜单项关闭时触发此操作：

```
public void onContextMenuClosed(Menu menu);
```

（5）menu.add 相关操作。

要在 Android 程序中使用上下文菜单，必须在程序中使用下面的语句：

```
import android.view.ContextMenu;          //导入 view.ContextMenu 类
```

存在一个与 ContextMenu 联系密切的类——ContextMenuInfo，使用下面的语句定义：

```
import android.view.ContextMenu.ContextMenuInfo;
```

上下文菜单实现步骤如下。

（1）通过组件 ID 获取组件实例，并进行注册。

例如：

```
TextView txt = (TextView) this.findViewById(R.id.txt1);
this.registerForContextMenu(txt) ;                    //将快捷菜单注册到 txt 组件上
```

（2）为该组件注册 onCreateContextMenuListener 监听。

```
public void onCreateContextMenu(ContextMenu menu, View txt, ContextMenu.
          ContextMenuInfo menuInfo) {  //显示菜单
      super.onCreateContextMenu(menu, txt, menuInfo) ;
      menu.setHeaderTitle("信息操作") ;                        //设置显示信息头
      menu.add(Menu.NONE, Menu.FIRST + 1, 1, "添加信息");     //设置菜单项
      menu.add(Menu.NONE, Menu.FIRST + 2, 2, "查看信息");     //设置菜单项
...
```

（3）有列表操作，就应有相应的事件。上面的事件一般会与下面的方法结合使用来实现

onCreateContext Menu 方法。

```
public boolean onContextItemSelected(MenuItem item) {        //选中某个菜单项
    switch (item.getItemId()) {                              //判断菜单项 ID
    case Menu.FIRST + 1:
        Toast.makeText(this,"您选择的是"添加信息"。", Toast.LENGTH_LONG).show();
        break;
    case Menu.FIRST + 2:
        Toast.makeText(this, "您选择的是"查看信息"。", Toast.LENGTH_LONG).show();
        break;…
```

实例 4-13：上下文菜单举例 1

新建一个项目，项目命名为 exam4_17，修改 MainActivity 文件，新建一个上下文菜单。

（1）修改 activity_main.xml 文件，代码如下：

```
…
<TextView
    android:id="@+id/txt1"
    android:layout_width="fill_parent"
    android:layout_height="wrap_content"
    android:text="请长按触发快捷菜单"
    android:textSize="20dp" />
</LinearLayout>
```

（2）修改 MainActivity.java，代码如下：

```
…
public class MainActivity extends AppCompatActivity {
private TextView txt=null;
    public void onCreate(Bundle savedInstanceState) {
        super.onCreate(savedInstanceState);
        setContentView(R.layout.activity_main);
        TextView txt = (TextView) this.findViewById(R.id.txt1);
        this.registerForContextMenu(txt) ;  }        //将快捷菜单注册到 txt 组件上
    public void onCreateContextMenu(ContextMenu menu, View txt, ContextMenu.
            ContextMenuInfo menuInfo) {                         //显示菜单
        super.onCreateContextMenu(menu, txt, menuInfo) ;
        menu.setHeaderTitle("信息操作") ;                         //设置显示信息头
        menu.add(Menu.NONE, Menu.FIRST + 1, 1, "添加信息");      //设置菜单项
        menu.add(Menu.NONE, Menu.FIRST + 2, 2, "查看信息");      //设置菜单项
        menu.add(Menu.NONE, Menu.FIRST + 3, 3, "删除信息");      //设置菜单项
        menu.add(Menu.NONE, Menu.FIRST + 4, 4, "信息另存");      //设置菜单项
        menu.add(Menu.NONE, Menu.FIRST + 5, 5, "编辑信息");}     //设置菜单项
    public boolean onContextItemSelected(MenuItem item) {      //选中某个菜单项
        switch (item.getItemId()) {                            //判断菜单项 ID
        case Menu.FIRST + 1:
            Toast.makeText(this, "您选择的是"添加信息"。", Toast.LENGTH_LONG).show();
            break;
        case Menu.FIRST + 2:
            Toast.makeText(this,"您选择的是"查看信息"。", Toast.LENGTH_LONG).show();
            break;
        case Menu.FIRST + 3:
            Toast.makeText(this,"您选择的是"删除信息"。", Toast.LENGTH_LONG).show();
```

```
            break;
        case Menu.FIRST + 4:
            Toast.makeText(this, "您选择的是"另存信息"。", Toast.LENGTH_LONG).show();
            break;
        case Menu.FIRST + 5:
            Toast.makeText(this, "您选择的是"编辑信息"。", Toast.LENGTH_LONG).show();
            break;        }
        return false;     }
    public void onContextMenuClosed(Menu menu) {             //菜单退出时调用
        Toast.makeText(this, "上下文菜单关闭了", Toast.LENGTH_LONG).show(); }}
```

保存文件并运行程序，长按屏幕上的文字，结果如图 4.20 所示。

上下文菜单中的菜单选项也可以从 XML 中的数据获得，如果希望从配置文件中取出数据，需要使用下面两个常用方法：

```
public MenuInflater(Context context)              //创建 MenuInflater 类对象
public void inflate(int menuRes, Menu menu)   //将配置的资源填充到菜单之中
```

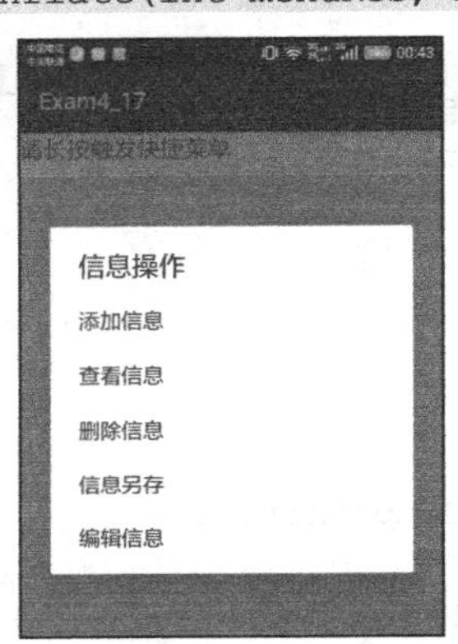

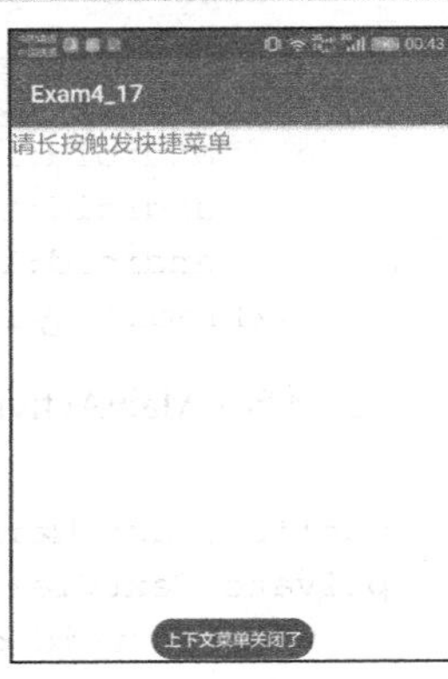

图 4.20　Exam4_17 运行结果

实例 4-14：上下文菜单举例 2

新建一个项目，项目命名为 Exam4_18，利用 XML 中的数据，新建一个上下文菜单。

（1）在 res\menu\文件夹下建立一个 menuadd.xml 文件，代码如下：

```
<?xml version="1.0" encoding="utf-8"?>
<menu xmlns:android="http://schemas.android.com/apk/res/android">
 <item    android:id="@+id/item01"     android:title="添加信息"/>
 <item    android:id="@+id/item02"     android:title="查看信息"/>
 <item    android:id="@+id/item03"     android:title="删除信息"/>
 <item    android:id="@+id/item04"     android:title="另存信息"/>
 <item    android:id="@+id/item05"     android:title="编辑信息"/>
</menu>
```

（2）修改 activity_main.xml 文件，代码与 Exam4_17 的 activity_main.xml 一致。

（3）修改 MainActivity.java，将 Exam4_17 的 MainActivity.java 的第 21 行开始的 5 行代码：

```
menu.add(Menu.NONE, Menu.FIRST + 1, 1, "添加信息");          //设置菜单项
menu.add(Menu.NONE, Menu.FIRST + 2, 2, "查看信息");          //设置菜单项
menu.add(Menu.NONE, Menu.FIRST + 3, 3, "删除信息");          //设置菜单项
menu.add(Menu.NONE, Menu.FIRST + 4, 4, "信息另存");          //设置菜单项
menu.add(Menu.NONE, Menu.FIRST + 5, 5, "编辑信息");          //设置菜单项
```

换成

```
    super.getMenuInflater().inflate(R.menu.menuadd, menu);    //设置菜单项
```

将“case Menu.FIRST + 1:”换成“case R.id.item01:”，“case Menu.FIRST + 2:”换成“case R.id.item02:”……其他代码不变，保存文件，程序运行结果和 Exam4_17 一致。

4.10.4 弹出式菜单 PopupMenu

弹出式菜单 PopupMenu 在 API 11 和更高版本中才有效。弹出式菜单会在指定组件上弹出菜单，默认情况下，PopupMenu 会显示在指定组件的下方或上方。PopupMenu 可增加多个菜单项，并可为菜单项增加子菜单。

使用 PopupMenu 创建菜单的步骤非常简单，只要按如下步骤操作即可。

（1）通过 PopupMenu 的构造函数实例化一个 PopupMenu 对象，需要传递一个当前上下文对象以及绑定的 View。

（2）调用 PopupMenu.setOnMenuItemClickListener()，来设置一个 PopupMenu 选项的选中事件。

（3）使用 MenuInflater.inflate()方法加载一个 XML 文件到 PopupMenu.getMenu()中。

（4）在需要的时候调用 PopupMenu.show()方法进行显示。

实例 4-15：弹出菜单举例

新建一个项目，项目命名为 Exam4_19，新建一个弹出式菜单。

（1）在 res 文件夹下建立 menu 文件夹，在 menu 下建立一个 mainmenu.xml 文件，代码如下：

```
<?xml version="1.0" encoding="utf-8"?>
<menu xmlns:android="http://schemas.android.com/apk/res/android">
    <item android:id="@+id/file"     android:title="文件"/>
    <item android:id="@+id/edit"     android:title="编辑"/>
    <item android:id="@+id/config"   android:title="设置"/>
</menu>
```

（2）修改 activity_main.xml 文件，代码如下：

```
...
    <Button
        android:layout_width="wrap_content"
        android:layout_height="wrap_content"
        android:layout_gravity="center"
        android:text="单击弹出菜单"
        android:id="@+id/but"/>
</LinearLayout>
```

（3）修改 MainActivity.java，代码如下：

```
...
import android.widget.PopupMenu;
import android.widget.PopupMenu.OnMenuItemClickListener;
public class MainActivity extends AppCompatActivity implements OnClickListener,
OnMenuItemClickListener{
    private Button button1;
    @Override
    protected void onCreate(Bundle savedInstanceState) {
        super.onCreate(savedInstanceState);
```

```
        setContentView(R.layout.activity_main);
        button1 = (Button)findViewById(R.id.but);
        button1.setOnClickListener(this);  } //单击按钮后，加载弹出式菜单
        public void onClick(View v) {
        //创建弹出式菜单对象
        PopupMenu popup = new PopupMenu(this, v);//第二个参数是绑定的那个 view
        MenuInflater inflater = popup.getMenuInflater();//获取菜单填充器
        inflater.inflate(R.menu.menumain, popup.getMenu());//menumain 填充菜单
        popup.setOnMenuItemClickListener(this); //绑定菜单项的单击事件
        popup.show(); } //显示(这一行代码不要忘记了)
    //弹出式菜单的单击事件处理
    @Override
    public boolean onMenuItemClick(MenuItem item) {
        // TODO Auto-generated method stub
        switch (item.getItemId()) {
            case R.id.file:
                Toast.makeText(this, "文件操作", Toast.LENGTH_LONG).show();
                break;
            case R.id.edit:
                Toast.makeText(this, "编辑操作", Toast.LENGTH_LONG).show();
                break;
            case R.id.config:
                Toast.makeText(this, "设置操作", Toast.LENGTH_LONG).show();
                break;
            default:
                break; }
        return false;       }  }
```

保存文件并运行程序，单击按钮会弹出菜单，选择某个选项会有提示，结果如图 4.21 所示。

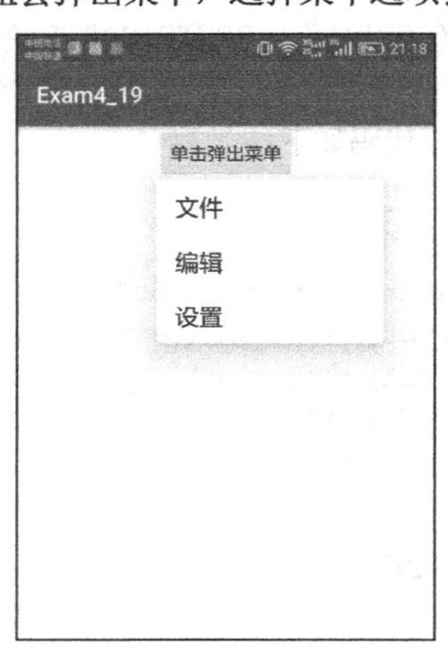

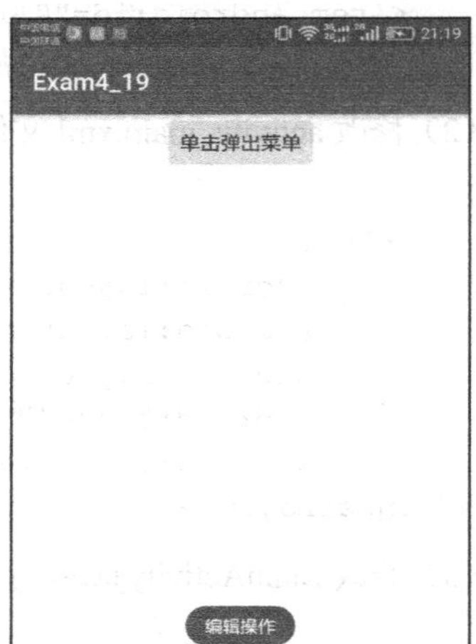

图 4.21　Exam4_19 运行结果

4.10.5　子菜单 SubMenu

Android 系统的子菜单使用非常灵活，可以在选项菜单、上下文菜单或弹出式菜单中使用子菜单。子菜单中能够显示更加详细信息的菜单子项，菜单子项使用了浮动窗体的显示形式，能够更好地适应小屏幕的显示方式，子菜单不支持嵌套、不支持显示图标。子菜单的一些使用方法如表 4-6 所示。

子菜单一般会在选项菜单或上下文菜单建立好的基础上，通过 addSubMenu(int groupId, int itemId, int order, int titleRes)方法创建和响应子菜单。

例如：

```
public boolean onCreateOptionsMenu(Menu menu) {        //建立选项菜单
  int base = Menu.FIRST;                               //设置 base 初值
  /*在 onCreateOptionsMenu()函数传递的 menu 对象上调用 addSubMenu()函数，在选项菜单
     中添加两个菜单子项，用户单击后可以打开子菜单*/
SubMenu subMenu1= menu.addSubMenu(base, base+1, Menu.NONE, "系统设置");
SubMenu subMenu2 = menu.addSubMenu(base, base+2, Menu.NONE, "文件操作");
  //子菜单中可以包括多个菜单项
   MenuItem menuitem1 = subMenu1.add(base, base+1, base+1, "显示设置");
   subMenu1.add(base, base+2, base+2, "网络设置");
   subMenu1.add(base, base+3, base+3, "高级设置");
   subMenu1.add(base, base+4, base+4, "安全设置");
   …
```

表 4-6　SubMenu 接口的常用方法

方法	描述
public abstract MenuItem getItem()	得到一个子菜单所属的父菜单对象
public abstract SubMenu setHeaderIcon(int iconRes)	设置菜单的显示图标
public abstract SubMenu setHeaderTitle(int titleRes)	设置子菜单的显示标题
public abstract SubMenu setHeaderTitle (CharSequence title)	设置子菜单的显示标题
public abstract SubMenu setIcon(int iconRes)	设置每个子菜单项的图标

实例 4-16：子菜单举例

新建一个项目，项目命名为 Exam4_20，新建一个带有子菜单的上下文菜单。

（1）修改 activity_main.xml 文件，代码如下：

```
…
<Button
    android:id="@+id/but1"
    android:layout_width="fill_parent"
    android:layout_height="wrap_content"
    android:text="请长按触发快捷菜单"     />
</LinearLayout>
```

（2）修改 MainActivity.java，代码如下：

```
…
public class MainActivity extends AppCompatActivity {
    public void onCreate(Bundle savedInstanceState) {
        super.onCreate(savedInstanceState);
        setContentView(R.layout.activity_main);
        Button but1 = (Button) this.findViewById(R.id.but1);
        this.registerForContextMenu(but1);   }
    //  重写 onCreateContextMenu 用以创建上下文菜单
    public void onCreateContextMenu(ContextMenu menu, View v,ContextMenuInfo menuInfo) {
    super.onCreateContextMenu(menu, v, menuInfo); //创建 R.id.txt1 的上下文菜单
       int base = Menu.FIRST;                                   //设置 base 初值
    //建立上下文菜单的菜单项
    SubMenu subMenu1 = menu.addSubMenu(base, base+1, Menu.NONE, "系统设置");
```

```
        SubMenu subMenu2 = menu.addSubMenu(base, base+2, Menu.NONE, "文件操作");
        //子菜单 1 可以包括多个菜单项
        MenuItem menuitem1 = subMenu1.add(base, base+1, base+1, "显示设置");
            subMenu1.add(base, base+2, base+2, "网络设置");      //添加菜单项
            subMenu1.add(base, base+3, base+3, "高级设置");
            subMenu1.add(base, base+4, base+4, "安全设置");
        //子菜单 2 可以包括多个菜单项
        MenuItem menuitem2 = subMenu2.add(base, base+1, base+1, "打开文件");
            subMenu2.add(base, base+2, base+2, "保存文件");
            subMenu2.add(base, base+3, base+3, "关闭文件");          } }
```

保存文件并运行程序，长按屏幕上的按钮，会弹出上下文菜单，选择某菜单项，会弹出子菜单，如图 4.22 所示。

子菜单的选项不仅能够在代码中定义，也能在 XML 文件中进行定义，项目 Exam4_21 就利用了 XML 中的数据建立了上下文菜单中的菜单选项，读者可下载学习。

限于篇幅，这里并没有在选择某个子菜单时弹出相应的提示，可以参照项目 Exam4_22 学习。

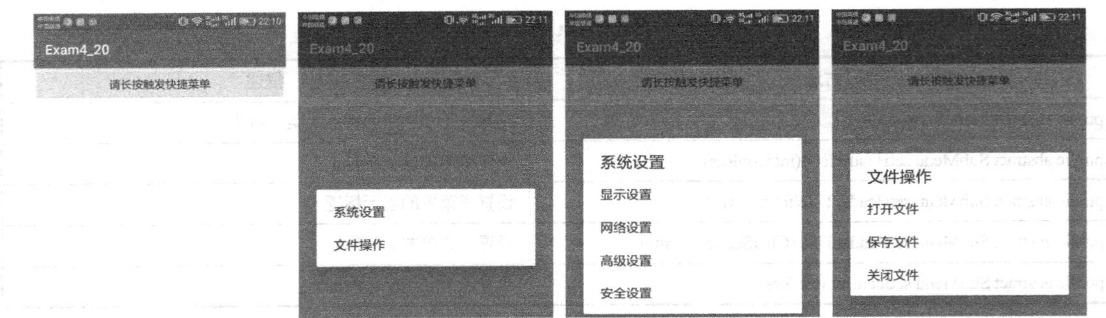

图 4.22　Exam4_20 运行结果

本章小结

本章着重分析了 6 种常见事件——单击事件、长按事件、焦点改变事件、键盘事件、触摸事件、菜单事件的基本操作，通过实例介绍加深了对各个事件处理操作步骤的理解，对选择改变事件、下拉列表选择事件、日期和时间改变触发的事件也做了简要介绍。

习题

（1）一个按钮如何实现双击事件，给出具体步骤和关键代码。

（2）一个按钮和一个文本显示组件都要实现单击事件，其主要区别在哪里？

（3）焦点改变事件的操作步骤有哪些？

（4）在键盘事件中如何判断按下了方向键的右键？写出关键代码。

（5）单选按钮组中如何知道某个选项被选中了？

（6）结合实例 4-9，哪些是判断下拉列表选择事件中某个选项被选中的代码？

（7）编写代码，将当前系统时间修改为 2017 年 10 月 20 日，12:30。

（8）当按下“▤”键时，弹出菜单是什么菜单？长按某组件后弹出的菜单是什么菜单？

（9）子菜单是通过什么方法添加的？

（10）编程实现：长按某个文本框，弹出一个菜单，该菜单有 3 个选项，其中第一个选项是含有 4 个选项的子菜单。

第 5 章　Android 常用高级组件

学习目标：

- 掌握 Android 平台中的 ListView 组件的使用。
- 了解 Android 平台中的 ExpandableListView 组件的使用。
- 掌握 Android 平台中的 ProgressBar、SeekBar、RatingBar 组件的使用。
- 掌握 Android 平台中的 ImageSwitcher、Gallery 组件的使用。
- 了解 Android 平台中的 AutoCompleteTextView 组件的使用。
- 掌握 Android 平台中的 Dialog 组件的使用。
- 掌握 Android 平台中的 TabHost 组件的使用。

在第 4 章中介绍了 Android 平台的基本组件，事实上，Android 平台提供的组件还有很多，本章会介绍一些 Android 平台常用的高级组件，例如，ListView、TabHost、Dialog 等，掌握这些常用组件的使用，可在 Android 程序设计中起到事半功倍的效果。

5.1　列表显示组件 ListView

手机屏幕的高度有限，当需要显示多组信息时，ScrollView 组件和 ListView 组件可以合理安排这些组件，浏览时可以滚屏显示组件。ScrollView 组件实现滚屏时只要将要滚动的组件添加到 ScrollView 中即可，可参照项目 Exam5_1 学习。也可以将多个组件加入到 ListView 组件中以达到组件的滚动显示，ListView 组件本身也有对应的 ListView 类支持，可以通过操作 ListView 类以完成对此组组件的操作，ListView 类的层次关系如下所示：

```
java.lang.Object
      android.view.View
           android.view.ViewGroup
                android.widget.AdapterView<T extends android.widget.Adapter>
                     android.widget.AbsListView
                          android.widget.ListView
```

要在 Android 程序中使用 ListView 组件，必须在程序中使用下面的语句：

```
import android.widget.ListView;                    //导入 widget.ListView 类
```

5.1.1　ListView 组件常见的属性和方法

ListView 组件是 View 组件的间接子类，所有第 3 章介绍的 View 类的属性它都具备。除此之外，它还具有表 5-1 列出的属性，以及表 5-2 列出的常见方法。

表 5-1　ListView 组件属性

属性	描述
android:choiceMode	规定此 ListView 所使用的选择模式
android:divider	规定 List 项目之间用某个图形或颜色来分隔
android:dividerHeight	分隔符的高度

续表

属性	描述
android:entries	引用一个将使用在此 ListView 中的数组
android:footerDividersEnabled	ListView 是否会在页脚视图前画分隔符，默认值为 true
android:headerDividersEnabled	ListView 是否会在页眉视图后画分隔符，默认值为 true

表 5-2　ListView 组件常用方法

方法	描述
public ListView(Context context)	创建 ListView 类的实例化对象
public void setAdapter(ListAdapter adapter)	设置显示的数据
public ListAdapter getAdapter()	返回 ListAdapter
public void clearChoices ()	取消之前设置的任何选择
public void addFooterView (View v)	加一个固定显示于 list 底部的视图
public void addHeaderView (View v)	加一个固定显示于 list 顶部的视图
public void addHeaderView (View v, Object data, boolean isSelectable)	加一个固定显示于 list 顶部的视图
public boolean onTouchEvent (MotionEvent ev)	用于处理触摸屏的动作事件
public void setOnItemSelectedListener(AdapterView.OnItemSelectedListener listener)	选项选中时触发
public void setOnItemClickListener(AdapterView.OnItemClickListener listener)	选项单击时触发
public void setOnItemLongClickListener(AdapterView.OnItemLongClickListener listener)	选项长按时触发

ListView 是一个经常用到的组件，ListView 中的每个子项 Item 可以是一个字符串，也可以是一个组合组件。

要实现 ListView 组件，可按如下步骤操作。

（1）准备 ListView 要显示的数据。

（2）构建适配器，适配器就是 Item 数组，动态数组有多少元素就会生成多少 Item。

适配器类型常见的有 3 种：ArrayAdapter、SimpleAdapter 和 SimpleCursorAdapter。ArrayAdapter 最简单，只能展示一行字，SimpleAdapter 有最好的扩充性，可以自定义出各种效果。SimpleCursorAdapter 是 SimpleAdapter 对数据库的简单结合，可以方便地把数据库记录以列表的形式展示出来。SimpleAdapter 继承自 AdapterView，可以通过一些方法给 ListView 添加监听器，当用户单击某一个列表项时执行相应的操作。

（3）为适配器添加 ListView，并显示出来。

实例 5-1：ListView 组件使用举例

新建一个项目，项目命名为 Exam5_2，利用 ListView 组件实现多个文本组件的滚屏显示。

（1）修改 activity_main.xml 文件，添加一个 ListView 组件。

```
...
        <ListView
            android:id="@+id/MyListView"
            android:layout_width="match_parent"
            android:layout_height="wrap_content" />
</LinearLayout>
```

（2）修改 MainActivity.java 文件：

```
...
public class MainActivity extends AppCompatActivity {
private String listdata[] = { "信息学院", "机械学院", "计算机学院","新闻学院",
    "化工学院", "美术学院","计算机学院","新闻学院", "化工学院", "美术学院",
```

```
        "体育学院", "音乐学院", "经济管理学院","南湖学院", "物理与电子学院",
        "机电工程学院", "法律学院","外语学院", "旅游学院", "科技处",
        "图书馆", "教务处","网络中心", "学工处", "财务处"};    //定义显示的数据
    private ListView listview;                              //定义 ListView 组件
    public void onCreate(Bundle savedInstanceState) {
        super.onCreate(savedInstanceState);
        super.setContentView(R.layout.activity_main);
        this.listview = (ListView)super.findViewById(R.id.MyListView);
                                                            //获得 ListView 组件
        this.listview.setAdapter(new ArrayAdapter<String>(this, //将数据包装
                android.R.layout.simple_expandable_list_item_1, //每行显示一条数据
                this.listdata)); }}                         //设置组件内容
```

保存文件并运行程序，结果如图 5.1 所示，能实现滚屏效果。

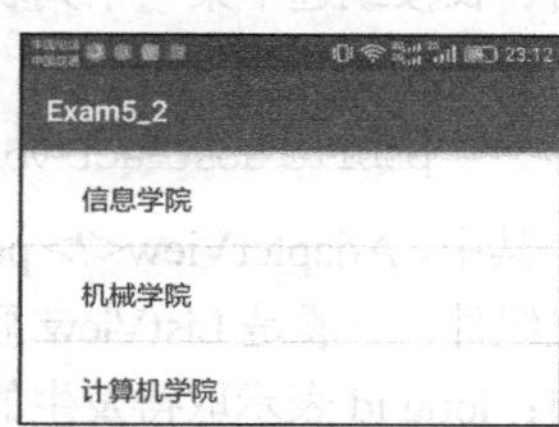

图 5.1 Exam5_2 运行结果

5.1.2 SimpleAdapter 类

上面的 ListView 实例的显示效果显得有些单一，如果希望在一行显示更多信息，就需要使用 SimpleAdapter。SimpleAdapter 类的主要功能是将 List 集合的数据转换为 ListView 可以支持的数据，定义出各种显示效果。其层次关系如下：

```
java.lang.Object
    android.widget.BaseAdapter
        android.widget.SimpleAdapter
```

要在 Android 程序中使用 SimpleAdapter 组件，必须在程序中使用下面的语句：

```
import android.widget.SimpleAdapter;          //导入 widget.SimpleAdapter 类
```

SimpleAdapter 是一个简单的适配器，可以指定一个用于显示行的布局 XML 文件，通过关键字映射到指定的布局文件，其常用方法如表 5-3 所示。

表 5-3 SimpleAdapter 类的常用方法

方法	描述
public SimpleAdapter (Context context, List<? extends Map<String, ?>> data, int resource, String[] from, int[] to)	创建 SimpleAdapter 对象，需要传递 Context 对象、封装的 List 集合、要使用的布局文件 ID、需要显示的 key（对应 Map）、组件的 ID。
public int getCount()	得到保存集合的个数
public Object getItem(int position)	取得指定位置的对象
public long getItemId(int position)	取得指定位置对象的 ID
public void notifyDataSetChanged()	当列表项发生改变时，通知更新显示 ListView

这里给出一个比较重要的构造函数，如下所示。

```
public SimpleAdapter (Context context,List<? extends Map<String, ?>> data, int
resource, String[] from, int[] to)
```

其中，参数 context 表示关联 SimpleAdapter 运行着的视图的上下文；参数 data 表示一个 Map 的列表。在列表中，每个条目对应列表中的一行，应该包含所有在 from 中指定的条目；参数 resource 表示一个定义列表项目的视图布局的资源唯一标识。布局文件将至少应包含那些在 to 中定义了的名称；参数 from 表示一个将被添加到 Map 上关联每一个项目的列名称的列表；参数 to 表示应该在参数 from 中显示列的视图。

SimlpleAdapter 是这样工作的：假设将 SimpleAdapter 用于 ListView，那么 ListView 的每一个列表项就是 resource 参数值指定的布局，而 data 参数就是要加载到 ListView 中的数据。例如：

```
//定义 ArrayList 对象名为 list
ArrayList<HashMap<String,String>>list =new ArrayList<HashMap<String,String>>();
SimpleAdapter simpleAdapter=new SimpleAdapter(this,   //定义 SimpleAdapter
list,                                                 //要显示的数据
R.layout.user,                                        //按照 user.xml 布局摆放数据
new String[]{"userId","userName","userTel"},          //每一个记录的列名称的列表
new int[]{R.id.userId,R.id.userName,R.id.userTel});//对应 user.xml 的 ID
```

意思就是将 Map 对象中 key 为 userId 的 value 绑定到 R.id.userId 上，userName 和 userTel 也类似。

SimpleAdapter 继承自 AdapterView，可以通过表 5-2 所示的方法给 ListView 添加监听器，当用户单击、长按或选中某一个列表项时执行相应的操作。

以单击事件为例：

```
public abstract void onItemClick(AdapterView<?> parent,View view,int position,long id)
```

其中，AdapterView<?> parent 表示操作的 AdapterView 对象；View view 表示取得操作 AdapterView 的父组件，一般是 ListView 显示时所使用的布局管理器；int position 表示取得 Adapter 的操作数据项的索引；long id 表示取得发生的 ListView 显示行的编号。

例如，在 Exam5_2 中添加如下语句可以在单击某个选项时将指定信息通过 Toast 显示出来。

```
listview.setOnItemClickListener(new ItemClickEvent());//注册监听
//实现监听
private final class ItemClickEvent implements AdapterView.OnItemClickListener
{   @Override
   //这里需要注意的是第三个参数 arg2，这是代表单击第几个选项
   public void onItemClick(AdapterView<?> arg0, View arg1, int arg2, long arg3){
    String text = (String) listview.getItemAtPosition(arg2);//获得选项的内容
    Toast.makeText(getApplicationContext(), text, Toast.LENGTH_LONG).show();} }
```

实例 5-2：SimpleAdapter 类使用举例

新建一个项目，项目命名为 Exam5_3，利用 SimpleAdapter 类显示数据。

（1）修改 activity_main.xml 文件：

```
...
   <ImageView android:id="@+id/picture"
      android:layout_width="100dp"
      android:layout_height="80dp"
      android:layout_margin="5dip"/>
   <LinearLayout android:orientation="vertical"
      android:layout_width="wrap_content"
      android:layout_height="wrap_content">
      <TextView android:id="@+id/title"
         android:layout_width="wrap_content"
         android:layout_height="wrap_content"
         android:textColor="#16CCDD"
         android:textSize="22sp" />
      <TextView android:id="@+id/info"
         android:layout_width="wrap_content"
         android:layout_height="wrap_content"
         android:textColor="#666666"
         android:textSize="16sp" />
   </LinearLayout>
</LinearLayout>
```

（2）修改 MainActivity.java 文件：

```
//ListActivity 本质上仍然是一个 Activity，对于 ListView 操作更方便
public class MainActivity extends ListActivity {
    public void onCreate(Bundle savedInstanceState) {
    super.onCreate(savedInstanceState);
    //注意 SimpleAdapter 适配器的用法
    SimpleAdapter adapter = new SimpleAdapter(this,    //定义 SimpleAdapter
            getData(),                                 //定义要显示的数据
            R.layout.activity_main,                    //按照指定布局摆放数据
            new String[]{"title","info","picture"},    //每一个记录的列名称的列表
            new int[]{R.id.title,R.id.info,R.id.picture});//列对应的 ID 值
    setListAdapter(adapter);    }//把数据映射到 adapter 容器
private List<Map<String, Object>> getData() {//定义 getData()方法
    List<Map<String, Object>> list = new ArrayList<Map<String, Object>>();
    Map<String, Object> map = new HashMap<String, Object>();
    map.put("title", "蓝天白云");
    map.put("info", "蓝天白云，青山绿水，让人感觉神清气爽……");
    map.put("picture", R.drawable.test1);
    list.add(map);
    map = new HashMap<String, Object>();
    map.put("title", "湖上彩虹");
    map.put("info", "湖上彩虹，令人如入仙境，如痴如醉……");
    map.put("picture", R.drawable.test2);
    list.add(map);
    map = new HashMap<String, Object>();
    map.put("title", "山川河流");
    map.put("info", "山川河流，震撼场景，壮丽景色，一览无遗……");
    map.put("picture", R.drawable.test3);
    list.add(map);
    map = new HashMap<String, Object>();
    map.put("title", "青草和路");
    map.put("info", "青草和路，风车小屋，给你无限的遐思……");
    map.put("picture", R.drawable.test4);
    list.add(map);
 return list;    }}
```

保存文件并运行该项目，结果如图 5.2 所示，能实现滚屏效果。

图 5.2　Exam5_3 运行结果图

实例 5-3：SimpleAdapter 类与 ListView 组件使用举例

新建一个项目，项目命名为 Exam5_4，利用 SimpleAdapter 类显示数据到 ListView 组件中。

（1）新建一个 user.xml 布局文件用于显示每条记录的布局，做如下的修改：

```
...
```

```
    <TextView
        android:id="@+id/userId"
        android:layout_width="100dp"
        android:layout_height="wrap_content"
        android:textSize="16dp"  />
    <TextView
        android:id="@+id/userName"
        android:layout_width="100dp"
        android:layout_height="wrap_content"
        android:textSize="16dp"  />
     <TextView
        android:id="@+id/userTel"
        android:layout_width="100dp"
        android:layout_height="wrap_content"
        android:textSize="16dp"  />
</LinearLayout>
```

（2）修改 activity_main.xml 文件，其代码如下：

```
...
 <TextView
        android:layout_width="match_parent"
        android:layout_height="wrap_content"
        android:textSize="20dp"
        android:gravity="center"
        android:text="学生信息表"   />
 <TableLayout
      android:layout_width="match_parent"
      android:layout_height="wrap_content"
      android:orientation="horizontal"    >
    <TableRow>
      <TextView
        android:text="学生编号"
        android:layout_width="100dp"
        android:layout_height="wrap_content"
        android:textSize="16dp"  />
       <TextView
        android:text="学生姓名"
        android:layout_width="100dp"
        android:layout_height="wrap_content"
        android:textSize="16dp"  />
       <TextView
        android:text="电话号码"
        android:layout_width="100dp"
        android:layout_height="wrap_content"
        android:textSize="16dp"  />
       </TableRow>
   </TableLayout>
   <ListView
     android:id="@+id/listView"
     android:layout_width="match_parent"
     android:layout_height="wrap_content" />
</LinearLayout>
```

（3）修改 MainActivity.java 文件，代码如下：

```
…
import java.util.ArrayList;
import java.util.HashMap;
import android.widget.AdapterView.OnItemClickListener;
import android.widget.ListView;
import android.widget.SimpleAdapter;
public class MainActivity extends AppCompatActivity {
    public void onCreate(Bundle savedInstanceState) {
        super.onCreate(savedInstanceState);
        setContentView(R.layout.activity_main);
        ArrayList<HashMap<String,String>>list =new ArrayList<HashMap<String,
                String>>();
        HashMap<String,String> map1=new HashMap<String,String>();//定义map1集合
        HashMap<String,String> map2=new HashMap<String,String>();
        HashMap<String,String> map3=new HashMap<String,String>();
        ListView listView=(ListView)findViewById(R.id.listView);
        map1.put("userId", "000001");        //设置userId组件显示的数据
        map1.put("userName", "陈飞翔");      //设置userName组件显示的数据
        map1.put("userTel", "8648871");      //设置userTel组件显示的数据
        list.add(map1);                      //增加数据到list中
        map2.put("userId", "000002");
        map2.put("userName", "张晓飞");
        map2.put("userTel", "8648872");
        list.add(map2);
        map3.put("userId", "000003");
        map3.put("userName", "王成功");
        map3.put("userTel", "8648873");
        list.add(map3);
        //定义一个SimpleAdapter,每一行有三个要显示的选项
        SimpleAdapter simpleAdapter=new SimpleAdapter(this,list,R.layout.
          user,new String[]{"userId","userName","userTel"},new int[]{R.id.
          userId,R.id.userName,R.id.userTel});
        //为ListView添加适配器
        listView.setAdapter(simpleAdapter);//设置listView数据为simpleAdapter
        listView.setOnItemClickListener(new OnItemClickListener() {//设置事件监听
          public void onItemClick(AdapterView<?> arg0, View arg1, int arg2,long
                              arg3) {
            ListView listView=(ListView)arg0;
            Toast.makeText(MainActivity.this, listView.getItemAtPosition
              (arg2).toString(), Toast.LENGTH_SHORT).show();  } }); } }
```

保存文件并运行该项目，可以实现滚屏，单击某条记录会有提示信息，如图 5.3 所示。

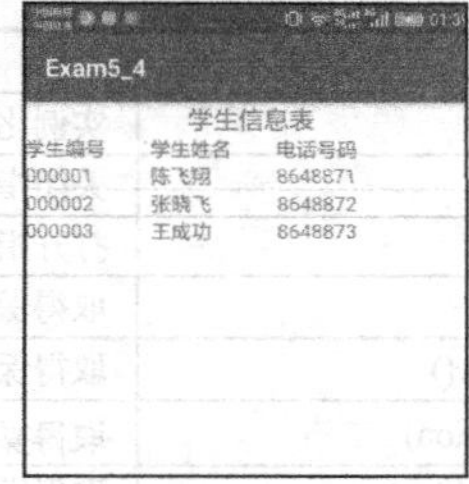

图 5.3　Exam5_4 运行结果

在 ListView 中，每行显示的不仅仅是文字，还可以是 ImageView、Button、CheckBox 等组件，读者可以按照实例 5-2 自行编写程序实现。

5.2 可展开的列表组件 ExpandableListView

在 Android 系统中，ListView 组件可以为用户提供列表的显示功能，但在某些情况下，如果希望对列表项进行分组管理并实现收缩列表效果，例如，使用 QQ 的时候，有“我的好友”“同学”“家人”组，单击其中一项会扩展此组，再次单击又会收缩回去。此时就要用 Android 的 ExpandableListView 组件来实现，ExpandableListView 类的层次关系如下：

```
java.lang.Object
 android.view.View
     android.view.ViewGroup
         android.widget.AdapterView<T extends android.widget.Adapter>
             android.widget.AbsListView
                 android.widget.ListView
                     android.widget.ExpandableListView
```

要在 Android 程序中使用 ExpandableListView 组件，应在程序中使用下面的语句：

```
import android.widget.ExpandableListView; //导入 widget.ExpandableListView 类
```

5.2.1 ExpandableListView 组件基础

1. ExpandableListView 组件常见的属性和方法

从上面的层次关系可以看出 ExpandableListView 组件是 ListView 组件的子类，所以 ListView 的属性和方法它都具备，它还具有表 5-4 列出的常用属性，以及表 5-5 列出的常用操作方法。

表 5-4 ExpandableListView 类的常用属性

属性	描述
android:childDivider	分离子列表项的图片或者颜色
android:childIndicator	在子列表项旁边显示的指示符
android:childIndicatorLeft	子列表项指示符的左边约束位置
android:childIndicatorRight	子列表项指示符的右边约束位置
android:groupIndicator	在组列表项旁边显示的指示符
android:indicatorLeft	组列表项指示器的左边约束位置
android:indicatorRight	组列表项指示器的右边约束位置

表 5-5 ExpandableListView 类的常用操作方法

方法	描述
public ExpandableListView (Context context)	实例化 ExpandableListView 类的对象
public boolean collapseGroup(int groupPos)	关闭指定的分组
public boolean expandGroup(int groupPos)	打开指定的分组
public ListAdapter getAdapter()	取得保存数据的 ListAdapter 对象
public ExpandableListAdapter getExpandableListAdapter()	取得保存数据的 ExpandableListAdapter 对象
public static int getPackedPositionType(long packedPosition)	取得操作的菜单项的类型（判断是菜单组还是菜单项）
public static int getPackedPositionGroup(long packedPosition)	取得操作所在的菜单组编号
public static int getPackedPositionChild(long packedPosition)	取得操作所在的菜单项编号

续表

方法	描述
public long getSelectedId()	取得当前所操作的菜单 ID，如果没有则返回-1
public void setAdapter(ExpandableListAdapter adapter)	设置适配器数据对象
public void setAdapter(ListAdapter adapter)	设置适配器数据对象
public boolean setSelectedChild(int groupPosition, int childPosition, boolean shouldExpandGroup)	设置选中的菜单项
public void setSelectedGroup(int groupPosition)	设置选中的菜单组
public void setOnChildClickListener (ExpandableListView.OnChildClickListener onChildClickListener)	设置菜单组的单击事件处理
public void setOnGroupClickListener (ExpandableListView.OnGroupClickListener onGroupClickListener)	设置菜单组的单击事件处理
public void setOnGroupCollapseListener (ExpandableListView.OnGroupCollapseListener onGroupCollapseListener)	设置菜单组关闭的事件处理
public void setOnGroupExpandListener (ExpandableListView.OnGroupExpandListener onGroupExpandListener)	设置菜单组打开的事件处理
public void setOnItemClickListener (AdapterView.OnItemClickListener l)	设置选项单击的事件处理

每一个可以扩展的列表项的旁边都有一个指示符（箭头）用来说明该列表项目前的状态，可以使用方法 setChildIndicator(Drawable)和 setGroupIndicator(Drawable)（或者相应的 XML 文件的属性）来设置这些指示符的样式。当然，也可以使用默认的指示符，如下所示：

```
android.R.layout.simple_expandable_list_item_1,
android.R.layout.simple_expandable_list_item_2
```

注意：在 XML 布局文件中，一般不对 ExpandableListView 的 android:layout_height 属性使用 wrap_content 值，否则可能会报错。

2. ExpandableListView 组件的适配器

与 ListView 一样，ExpandableListView 也需要一个适配器作为桥梁来取得数据，与 ListView 不同的是，ExpandableListView 是一个垂直滚动显示两级列表项的视图，它可以有两层，每一层都能够独立地展开并显示其子项。这些子项来自于与该视图关联的适配器。

BaseExpandableListAdapter 就是一个用在 ExpandableListView 组件上的适配器，此类继承 android.widget.BaseExpandableListAdapter 类。其常用覆写方法如表 5-6 所示。

表 5-6 BaseExpandableListAdapter 的子类所需要覆写的方法

方法	描述
public Object getChild(int groupPosition, int childPosition)	获得指定组中的指定索引的子项数据
public long getChildId(int groupPosition, int childPosition)	获得指定子项数据的 ID
public View getChildView(int groupPosition, int childPosition,boolean isLastChild, View convertView, ViewGroup parent)	获得指定子项的 View 组件
public int getChildrenCount(int groupPosition)	取得指定组中所有子项的个数
public Object getGroup(int groupPosition)	取得指定组数据
public int getGroupCount()	取得所有组的个数
public long getGroupId(int groupPosition)	取得指定索引的组 ID
public View getGroupView(int groupPosition, boolean isExpanded,View convertView, ViewGroup parent)	获得指定组的 View 组件
public boolean hasStableIds()	如果返回 true，则表示子项和组的 ID 始终表示一个固定的组件对象
public boolean isChildSelectable(int groupPosition, int childPosition)	判断指定的子选项是否被选中

一般适用于 ExpandableListView 的 Adapter 都要继承 BaseExpandableListAdapter 类，并且必须重载 getGroupView 和 getChildView 方法。

```
public abstract View getGroupView (int groupPosition, boolean isExpanded, View
    convertView, ViewGroup parent)
```

功能：取得用于显示给定分组的视图。这个方法仅返回分组的视图对象，要想获取子元素的视图对象，就需要调用 getChildView(int, int, boolean, View, ViewGroup)。

参数：groupPosition 决定返回视图的组位置；isExpanded 表示该组是展开状态还是收起状态；convertView 为视图对象；parent 表示该视图最终从属的父视图。

返回值：指定位置相应的组视图。

```
public abstract View getChildView (int groupPosition, intchildPosition,
    boolean isLastChild, View convertView, ViewGroup parent)
```

功能：取得显示给定分组给定子位置的数据用的视图。

参数：groupPosition 表示包含要取得子视图的分组位置；childPosition 表示分组中子视图（要返回的视图）的位置；isLastChild 表示该视图是否为组中的最后一个视图；convertView 表示视图对象；parent 表示该视图最终从属的父视图。

返回值：指定位置相应的子视图。

3．ExpandableListView 组件的事件处理

ExpandableListView 组件的监听接口如表 5-7 所示。

表 5-7　ExpandableListView 提供的监听接口

监听接口名称	作用	定义方法
ExpandableListView.OnChildClickListener	单击子选项	public boolean onChildClick (ExpandableListView parent, View v,int groupPosition, int childPosition, long id)
ExpandableListView.OnGroupClickListener	单击分组项	public boolean onGroupClick(ExpandableListView parent, View v, int groupPosition, long id)
ExpandableListView.OnGroupCollapseListener	分组关闭	public void onGroupCollapse(int groupPosition)
ExpandableListView.OnGroupExpandListener	分组打开	public void onGroupExpand(int groupPosition)

5.2.2　ExpandableListView 组件实例

ExpandableListView 组件在编程时的关键是掌握适配器 BaseExpandableListAdapter 类提供的方法。

实例 5-4：ExpandableListView 组件举例

新建一个项目，项目命名为 Exam5_5，利用 ExpandableListView 组件显示各班人员信息。

（1）修改 activity_main.xml 文件，代码如下：

```
...
<TextView
    android:layout_width="match_parent"
    android:layout_height="wrap_content"
    android:textSize="18dp"
    android:gravity="center"
    android:text="12 级信工专业学生花名册" />
<ExpandableListView
    android:id="@+id/list"
    android:layout_width="match_parent"
    android:layout_height="match_parent"
```

```
    android:background="#ffffff" />
</LinearLayout>
```

（2）修改 MainActivity.java 文件，代码如下：

```
...
import android.widget.AbsListView;
import android.widget.BaseExpandableListAdapter;//导入 BaseExpandableListAdapter
import android.widget.ExpandableListAdapter; //导入 ExpandableListAdapter
import android.widget.ExpandableListView;    //导入 ExpandableListView
import android.widget.ExpandableListView.OnChildClickListener;
public class MainActivity extends AppCompatActivity {
 protected void onCreate(Bundle savedInstanceState) {
     super.onCreate(savedInstanceState);
     setContentView(R.layout.activity_main);
     //定义 BaseExpandableListAdapter 对象
     final ExpandableListAdapter adapter = new BaseExpandableListAdapter() {
         //设置各分组的图片
         int[] logos = new int[] { R.drawable.c11, R.drawable.c12,R.drawable.c13};
         //设置各分组的显示文字
         private String[] generalsTypes = new String[] { "1 班", "2 班", "3 班" };
         //子视图显示文字
         private String[][] generals = new String[][] {
                 { "陈飞翔", "张晓", "陈俊珊", "郭小嘉", "司马德", "杨大志" },
                 { "陈康健", "王俊华", "刘致力", "田晓菲", "黄小英", "赵科健" },
                 { "王小蒙", "王俊飞", "孙启智", "姜大志", "李丽华" }};
         //子视图图片,对应每个人的图片
         public int[][] generallogos = new int[][] {
                 { R.drawable.a11, R.drawable.a12,R.drawable.a13, R.drawable.
                  a14,R.drawable.a15, R.drawable.a16 },
                 { R.drawable.a21, R.drawable.a22,R.drawable.a23, R.drawable.
                  a24,R.drawable.a25, R.drawable.a26 },
                 { R.drawable.a31, R.drawable.a32, R.drawable.a33,R.drawable.
                  a34, R.drawable.a36} };
         //自己定义一个获得文字信息的方法
         TextView getTextView() {
         //定义布局参数
             AbsListView.LayoutParams lp = new AbsListView.LayoutParams(
                 ViewGroup.LayoutParams.MATCH_PARENT, 64);
             TextView textView = new TextView(ExpandableList.this); //创建 TextView
             textView.setLayoutParams(lp);                        //设置布局参数
             textView.setGravity(Gravity.CENTER_VERTICAL);        //居中显示
             textView.setPadding(36, 0, 0, 0);                    //与边界的距离
             textView.setTextSize(16);                            //设置文字大小
             textView.setTextColor(Color.BLACK);                  //设置文字颜色
             return textView;        }                       //返回 textView 组件
         //重写 ExpandableListAdapter 中的各个方法
         public int getGroupCount() {                       //取得分组的个数
             return generalsTypes.length;}
         public Object getGroup(int groupPosition) {  //取得指定分组
             return generalsTypes[groupPosition]; }
```

```
        public long getGroupId(int groupPosition) {  //取得指定分组的 ID
            return groupPosition;    }
        public int getChildrenCount(int groupPosition) { //取得子项的个数
            return generals[groupPosition].length;    }
        public Object getChild(int groupPosition, int childPosition) {
            //取得指定子项
            return generals[groupPosition][childPosition];   }
        public long getChildId(int groupPosition, int childPosition) {
        //取得指定子项 ID
            return childPosition;               }
        public boolean hasStableIds() {  //子项和分组的 ID 始终表示一个固定的组件对象
            return true;       }
        //覆写 getGroupView 方法
        public View getGroupView(int groupPosition, boolean isExpanded,
            View convertView, ViewGroup parent) {      //取得分组显示组件
            LinearLayout ll = new LinearLayout(ExpandableList.this);
                                        //定义布局管理器
            ll.setOrientation(0);           //设置布局管理器的参数
            ImageView logo = new ImageView(ExpandableList.this);
            //定义 ImageView
            logo.setImageResource(logos[groupPosition]);//获得显示组图片
            logo.setPadding(20, 0, 0, 0);                  //设置边界
            ll.addView(logo);                              //添加 ImageView
            TextView textView = getTextView();             //定义 TextView 组件
            textView.setTextColor(Color.BLACK);            //文字颜色
            textView.setText(getGroup(groupPosition).toString());//设置显示文字
            ll.addView(textView);                          //添加 TextView 组件
            return ll;     }
        //覆写 getChildView 方法
        public View getChildView(int groupPosition, int childPosition,
            boolean isLastChild, View convertView, ViewGroup parent)
            {      //取得子项组件
            LinearLayout ll = new LinearLayout(ExpandableList.this);
        //定义布局管理器
            ll.setOrientation(0);                  //设置布局管理器的参数
        ImageView generallogo=new ImageView(ExpandableList.this);//定义 ImageView
            //获得显示子项图片
            generallogo.setImageResource(generallogos[groupPosition]
                [childPosition]);
            ll.addView(generallogo);               //添加 ImageView
            TextView textView = getTextView();     //定义 textView 组件
            textView.setText(getChild(groupPosition, childPosition).toString());
            //显示文字
            ll.addView(textView);                  //添加 TextView 组件
            return ll;                 }
        public boolean isChildSelectable(int groupPosition,int childPosition) {
            return true }};                        //判断指定的子选项是否被选中
    ExpandableListView expandableListView = (ExpandableListView) findViewById
        (R.id.list);        //获得 ExpandableListView 组件
    expandableListView.setAdapter(adapter); //获得数据
```

```
//设置 item 单击事件的监听器
expandableListView.setOnChildClickListener(new OnChildClickListener() {
    public boolean onChildClick(ExpandableListView parent, View v,int
      groupPosition, int childPosition, long id) {   //单击该子项后的操作
        Toast.makeText(ExpandableList.this,"您刚才单击了" +
            adapter.getChild(groupPosition, childPosition),Toast.LENGTH_
                    SHORT).show();
        return false;}   }); }}
```

保存文件并运行该项目，单击可将列表项展开或收缩，单击某个选项会触发事件，如图 5-4 所示。

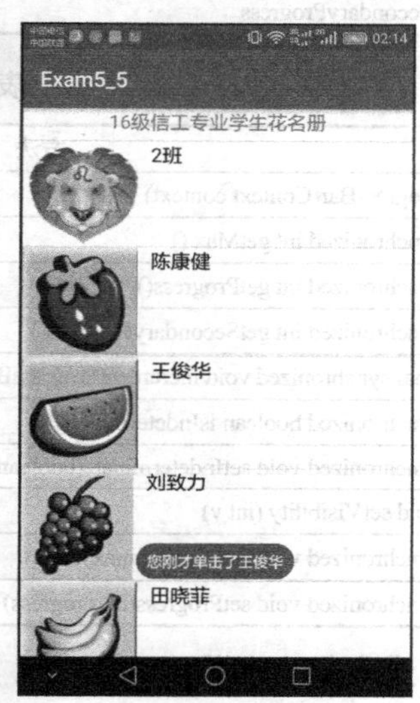

图 5.4 Exam5_5 运行结果

5.3 进度条组件 ProgressBar

进度条组件 ProgressBar 是某些操作的进度发展情况指示器，为用户呈现操作的进度，操作完成时，进度条会被填满。进度条能直观地帮助用户了解等待一定时间的操作所需的时间。ProgressBar 的层次关系如下：

```
java.lang.Object
    android.view.View
        android.widget.ProgressBar
```

要在 Android 程序中使用 ProgressBar 组件，必须在程序中使用下面的语句：

```
import android.widget.ProgressBar;          //导入 widget.ProgressBar 类
```

5.3.1 ProgressBar 组件基础知识

1. ProgressBar 组件常见的属性和方法

ProgressBar 组件是 View 组件的子类，所以第 3 章介绍的 View 类的属性和方法它都具备。除此之外，它还具有表 5-8 列出的属性，以及表 5-9 列出的常见方法。

表 5-8　ProgressBar 组件属性

属性	描述
android:progressBarStyle	默认进度条样式，不确定模式
android:progressBarStyleHorizontal	水平进度条样式
android:progressBarStyleLarge	大号进度条样式，也是不确定进度模式
android:progressBarStyleSmall	小号进度条样式，也是不确定进度模式
android:progress	定义默认进度条的范围
android:max	设置最大进度值
android:secondaryProgress	设置次要进度条

表 5-9　ProgressBar 组件常用方法

方法	描述
public ProgressBar(Context context)	创建一个 ProgressBar 实例
public synchronized int getMax ()	返回这个进度条的范围的上限
public synchronized int getProgress()	返回进度
public synchronized int getSecondaryProgress()	返回次要进度
public final synchronized void incrementProgressBy(int diff)	指定增加的进度
public synchronized boolean isIndeterminate()	指示进度条是否为不确定模式
public synchronized void setIndeterminate(boolean indeterminate)	设置不确定模式
public void setVisibility (int v)	设置该进度条是否可视
public synchronized void setMax(int max)	设置这个进度条的范围的上限
public synchronized void setProgress(int progress)	设置进度

2. 进度条的分类

Android 系统提供的进度条有对话框进度条、标题进度条和水平进度条。其中对话框进度条在本书后面介绍。进度条的样式也分为有条状的水平进度条和圆形转动的圆形进度条。

圆形进度条一般表示一个运转的过程，例如，发送短信、连接网络等，表示一个过程正在执行中。一般只要在 XML 布局中定义即可：

```
<ProgressBar
  android:id="@+id/PBar"
  android:layout_width="wrap_content"
  android:layout_height="wrap_content"
  android:layout_gravity="center_vertical"/>
```

此时，若没有设置它的风格，那么它就是圆形的、一直会旋转的进度条。如果希望得到一个超大号圆形 ProgressBar，只需要设置 style 属性，即 style="?android:attr/progressBarStyleLarge"；小号 ProgressBar 对应的是 style= "?android:attr/progressBarStyleSmall"；标题型 ProgressBar 对应的是 style="?android:attr/progressBarStyleSmallTitle"；水平进度条对应的是 style="?android:attr/progressBarStyleHorizontal"。

3. 标题栏进度条创建

（1）调用 Activity 程序的 requestWindowFeatures()方法，获得进度条。

例如：

```
requestWindowFeature(Window.FEATURE_PROGRESS);    //请求一个窗口进度条特性风格
```

（2）调用 Activity 程序的 setProgressBarIndeterminateVisibility()方法，显示进度条对话框。

例如：

```
setProgressBarVisibility(true);                   //设置进度条可视
```

（3）设置进度值。

例如：

```
setProgress(myProgressBar.getProgress() * 100);  /*设置标题栏中前一个进度条的进度值*/
setSecondaryProgress(myProgressBar.getSecondaryProgress() * 100); /*设置后一个进度条的进度值*/
```

4．水平进度条创建

（1）在布局文件中声明 ProgressBar：

```
<ProgressBar
…
  android:id="@+id/PBar"
  style="?android:attr/progressBarStyleHorizontal"     //设置为水平进度条
  android:max="100"                                     //最大进度值为 100
  android:progress="50"                                 //初始化的进度值
  android:secondaryProgress="70" />                     //初始化次要进度值
```

（2）在 Activity 中获得 ProgressBar 实例：

```
private ProgressBar myProgressBar;                      //定义 ProgressBar
myProgressBar = (ProgressBar) findViewById(R.id.PBar);  //获得 ProgressBar 实例
```

（3）调用 ProgressBar 的 incrementProgressBy()方法增加和减少进度：

```
myProgressBar.incrementProgressBy(5);//ProgressBar      //进度值增加 5
myProgressBar.incrementSecondaryProgressBy(5);//ProgressBar//次要进度条进度值增加 5
```

5．事件监听

onSizeChanged(int w, int h, int oldw, int oldh)：当进度值改变时引发此事件。

Exam5_6 是一个关于 ProgressBar 组件的实例，读者可以自行学习，一般应用中单独采用进度条的情不多，一般会结合对话框来进行，这些会在后面的对话框组件 ProgressDialog 中进行介绍。

5.4 拖动条组件 SeekBar

拖动条（SeekBar）组件与 ProgressBar 组件水平形式的显示进度条类似，最大的区别在于，拖动条可以由用户自己进行手工调节，例如，当用户需要调整音乐的播放速度或者要调整音量时都会使用到拖动条 SeekBar 类。其层次关系如下：

```
java.lang.Object
    android.view.View
        android.widget.ProgressBar
            android.widget.AbsSeekBar
                android.widget.SeekBar
```

要在 Android 程序中使用 SeekBar 组件，必须在程序中使用下面的语句：

```
import android.widget.SeekBar;                          //导入 widget.SeekBar 类
```

5.4.1 SeekBar 组件基础知识

1．SeekBar 组件常见的属性和方法

因为 SeekBar 组件是 ProgressBar 组件的子类，所以 ProgressBar 组件所有的属性和方法 SeekBar 组件都继承了。此外，该组件还有如表 5-10 所示的方法。

表 5-10　SeekBar 组件常用方法

方法	描述
public SeekBar(Context context)	创建 SeekBar 类的对象
public void setOnSeekBarChangeListener (SeekBar.OnSeekBarChangeListener l)	设置改变监听操作
public synchronized void setMax(int max)	设置增长的最大值
public synchronized void setProgress(int progress)	设置进度值
public synchronized void setSeconddaryProgress (int progress)	设置第二拖动条的数值
public synchronized int getProgress()	返回进度

2. 拖动条的事件

实现 SeekBar.OnSeekBarChangeListener 接口需要监听 3 个事件：数值改变（onProgressChanged）、开始拖动（onStartTrackingTouch）、停止拖动（onStopTrackingTouch）。

onStartTrackingTouch 开始拖动时触发，它与 onProgressChanged 的区别是停止拖动前只触发一次，而 onProgressChanged 只要在拖动，就会重复触发。

5.4.2　SeekBar 组件实例

实例 5-5：ProgressBar 组件举例

新建一个项目，项目命名为 Exam5_7，建立拖动条，当拖动时会触发相应的事件。

（1）修改 activity_main.xml 文件，代码如下：

```
…
    <TextView
        android:id="@+id/myTextView"
        android:layout_width="wrap_content"
        android:layout_height="wrap_content"
        android:textSize="16sp" />
    <SeekBar
        android:id="@+id/mySeekBar"
        android:layout_width="match_parent"
        android:layout_height="wrap_content" />
</LinearLayout>
```

（2）修改 MainActivity.java 文件，代码如下：

```
import android.widget.SeekBar;
import android.widget.SeekBar.OnSeekBarChangeListener;
public class MainActivity extends AppCompatActivity {
    private SeekBar seek;
    private TextView myTextView;
    protected void onCreate(Bundle savedInstanceState) {
        super.onCreate(savedInstanceState);
        setContentView(R.layout.activity_main);
        myTextView = (TextView) findViewById(R.id.myTextView);
        seek = (SeekBar) findViewById(R.id.mySeekBar);
        seek.setProgress(60);        //初始化
        seek.setOnSeekBarChangeListener(seekListener);
```

```
        myTextView.setText("当前值为: -" + 60);     }
        //设置监听
    private OnSeekBarChangeListener seekListener = new OnSeekBarChangeListener(){
        //数值改变时触发操作
        public void onProgressChanged(SeekBar seekBar, int progress,boolean
         fromUser) {
  MainActivity.this.myTextView.append("正在拖动,当前值:"+seekBar.get
        Progress()+"\n");}
        //开始拖动时候触发操作
        public void onStartTrackingTouch(SeekBar seekBar) {
    MainActivity.this.myTextView.append("开始拖动,当前值:"+seekBar.getProgress()
        +"\n"); }
        //停止拖动时，触发操作
        public void onStopTrackingTouch(SeekBar seekBar) {
MainActivity.this.myTextView.append("停止拖动,当前值:"+seekBar.getProgress()
     +"\n");}};}
```

保存文件并运行该项目，结果如图 5.5 所示，拖动 SeekBar，会触发相应的事件。

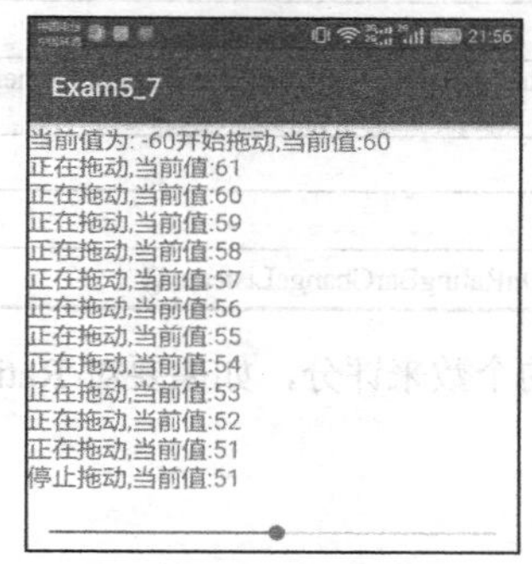

图 5.5 Exam5_7 运行结果

5.5 星级评分条组件 RatingBar

星级评分条组件（RatingBar）一般是用来做评分用的，用星形来显示等级评定，如，网站的满意度调查，有时也可以把它当做一个水平的进度条。RatingBar 的层次关系如下：

```
java.lang.Object
    android.view.View
        android.widget.ProgressBar
            android.widget.AbsSeekBar
                android.widget.RatingBar
```

要在 Android 程序中使用 RatingBar 组件，必须在程序中使用下面的语句：

```
import android.widget.RatingBar;                    //导入 widget.RatingBar 类
```

5.5.1 RatingBar 组件基础

由上面的层次关系可发现 RatingBar 是 ProgressBar 的子类，所以 ProgressBar 所有的属性和方法 RatingBar 都继承了。此外，它还有如表 5-11 所示的属性和表 5-12 所示的方法。

表 5-11　RatingBar 组件常见的属性

属性	描述
android:isIndicator	RatingBar 是否为一个指示器（用户无法进行更改）
android:numStars	显示的星形数量，必须是一个整型值，如“100”
android:rating	默认的评分，必须是浮点类型，如“1.2”
android:stepSize	评分的步长，必须是浮点类型，如“1.2”

表 5-12　RatingBar 组件常用的方法

方法	描述
public RatingBar(Context context)	创建 RatingBar 对象
public int getNumStars()	取得评分数量
public float getRating()	取得当前评分值
public float getStepSize()	取得设置的步长
public boolean isIndicator()	判断是否可以操作
public void setIsIndicator(boolean isIndicator)	是否可以操作
public synchronized void setMax(int max)	设置评分等级的范围
public void setNumStars(int numStars)	设置评分星的个数
public void setOnRatingBarChangeListener(RatingBar. OnRatingBarChangeListener listener)	设置操作监听
public void setRating(float rating)	设置当前的评分值
public void setStepSize(float stepSize)	设置每次增长的步长
public RatingBar.OnRatingBarChangeListener getOnRatingBarChangeListener ()	监听器（可能为空）监听评分改变事件

RatingBar 默认情况下使用星形的个数来评分，如果要对 RatingBar 进行美化，可以用自定义的图片替换系统默认的图片。

5.5.2　RatingBar 组件实例

实例 5-6：RatingBar 组件举例

新建一个项目，项目命名为 Exam5_8，建立 RatingBar，当改变 RatingBar 的值时会触发相应的事件。

（1）修改 activity_main.xml 文件，代码如下：

```
…
<RatingBar
    android:id="@+id/myratingBar"
    android:layout_height="wrap_content"
    android:layout_width="wrap_content" />
</LinearLayout>
```

（2）修改 MainActivity.java 文件，代码如下：

```
…
import android.widget.RatingBar;
public class MainActivity extends AppCompatActivity {
    public void onCreate(Bundle savedInstanceState) {
        super.onCreate(savedInstanceState);
        setContentView(R.layout.activity_main);
        RatingBar ratingBar = (RatingBar) findViewById(R.id.myratingBar);
```

```
        ratingBar.setNumStars(5);            //定义星级
        ratingBar.setRating(4);               //设置默认星级
        //监听星级改变并显示改变值
        ratingBar.setOnRatingBarChangeListener(new RatingBar.OnRating
          BarChangeListener() {
            public void onRatingChanged(RatingBar ratingBar,
                    float rating, boolean fromUser) {
                    float pp=(int)((rating*20/100)*100);
                    Toast.makeText(MainActivity.this, "您的满意度为: " + String.
                    valueOf(pp)+"%", Toast.LENGTH_LONG).show();}});}}
```

保存文件并运行该项目，结果如图 5.6 所示，手动修改 RatingBar 时会触发相应的事件。

图 5.6 Exam5_8 运行结果

5.6 自动完成文本框 AutoCompleteTextView

在上网查询资料的时候常常会遇到这样的情况：在输入框中输入几个字后，后面会自动出现一些文字信息供用户选择，Android 系统中的自动完成文本框（AutoCompleteTextView）就可以实现这样的功能。

自动完成文本框能够对用户键入的文本进行有效扩充提示，不需要用户输入整个内容。这个功能的实现要依靠 android.widget.AutoCompleteTextView 类来完成，其层次关系如下：

```
    java.lang.Object
        android.view.View
                android.widget.TextView
            android.widget.EditText
                android.widget.AutoCompleteTextView
```

要在 Android 程序中使用 AutoCompleteTextView 组件，必须在程序开始时使用下面的语句：

```
    import android.widget.AutoCompleteTextView; //导入widget.AutoCompleteTextView类
```

5.6.1 AutoCompleteTextView 组件基础

由上面的层次关系可发现 AutoCompleteTextView 是 EditText 的子类，所以 EditText 所有的属性和方法 AutoCompleteTextView 都继承了。此外，它还有如表 5-13 所示的属性和表 5-14 所示的方法。

表 5-13　AutoCompleteTextView 组件常见的属性

属性	描述
android:completionHint	设置显示下拉列表的提示题目
android:completionThreshold	至少输入几个字符，它才会具有自动提示的功能
android:dropDownAnchor	后接一个 View 的 ID，会在这个 View 下弹出自动提示
android:dropDownHeight	设置下拉列表的高度
android:dropDownWidth	设置下拉列表的宽度
android:popupBackground	设置下拉列表的背景

表 5-14　AutoCompleteTextView 组件常用的方法

方法	描述
public void clearListSelection()	清除所有的下拉列表项
public ListAdapter getAdapter()	取得数据集
public void setAdapter(T adapter)	设置数据集
public void setCompletionHint(CharSequence)	设置出现下拉列表的提示标题
public void setThreshold(int)	至少输入几个字符才会显示提示
public void setDropHeight(int)	设置下拉列表的高度
public void setDropWidth(int)	设置下拉列表的宽度
public void setDropDownbackgroundResource(int)	设置下拉列表的背景
public void setOnClickListener(View.OnClickListener listener)	设置单击事件
public void setOnItemClickListener(AdapterView. OnItemClickListener l)	在选项上设置单击事件
public void setOnItemSelectedListener(AdapterView. OnItemSelectedListener l)	选项选中时的单击事件

AutoCompleteTextView 组件可以很好地帮助用户进行文本信息的输入，它可以和一个字符串数组或 List 对象绑定，当用户输入两个及以上字符时，系统将在 AutoCompleteTextView 组件下方列出字符串数组中所有以输入字符开头的字符串。

AutoCompleteTextView 组件使用的关键是需要使用 AutoCompleteTextView 类的 setAdapter 方法指定一个 Adapter 对象，例如，“this.Auto.setAdapter(adapter);”，其中 adapter 是一个数据集，可以是字符串数组或 List 对象。

5.6.2　AutoCompleteTextView 组件实例

实例 5-7：AutoCompleteTextView 组件举例

新建一个项目，项目命名为 Exam5_9，利用 AutoCompleteTextView 组件实现文本提示功能。

（1）修改 activity_main.xml 文件，代码如下：

```
...
<AutoCompleteTextView
    android:id="@+id/auto"
    android:layout_width="match_parent"
    android:layout_height="wrap_content" />
</LinearLayout>
```

（2）修改 MainActivity.java 文件，代码如下：

```
...
import android.widget.AutoCompleteTextView;
public class MainActivity extends AppCompatActivity {
```

```
    private static final String autoString[] = new String[] { "湖南", "湖南理工",
          "湖南理工学院", "湖南省岳阳市", "湖南的辣椒" };                 //设置数据
    private AutoCompleteTextView Auto = null;                      //定义文本提示组件
    public void onCreate(Bundle savedInstanceState) {
        super.onCreate(savedInstanceState);
        super.setContentView(R.layout.activity_main);
        ArrayAdapter <String> adapter = new ArrayAdapter<String> (this,android.R.
            layout.simple_dropdown_item_1line, autoString);            //定义数据集
        this.Auto=(AutoCompleteTextView) super.findViewById(R.id.auto);/*取得文
本提示组件*/
        this.Auto.setAdapter(adapter);   }}                           //设置数据集
```

保存文件并运行该项目，输入文字“湖南”后会弹出相应的提示，如图 5.7 所示。还可以使用 MultiAutoCompleteTextView 组件来完成连续输入的功能，读者可以查阅相关 Android API 资料进行学习。

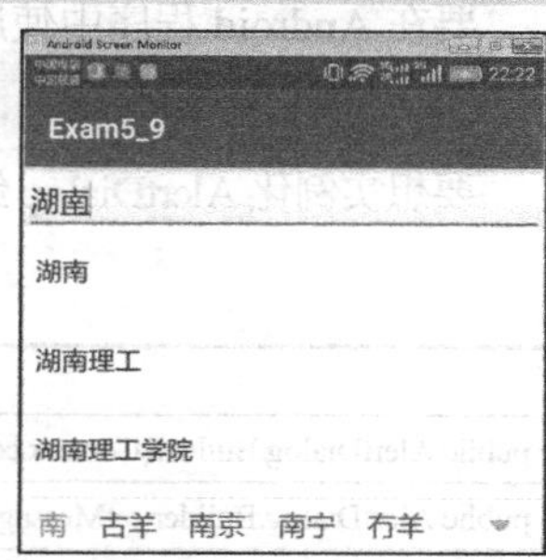

图 5.7 Exam5_9 运行结果

5.7 对话框组件 Dialog

在程序设计时经常需要在界面上弹出一些对话框，用来提示用户输入信息或者让用户做出选择而进行一些简单的交互操作，这就是对话框的功能。在 Android 平台的开发之中，所有的对话框都是使用 Dialog 类来实现的，其常用的方法如表 5-15 所示，其层次关系如下：

```
java.lang.Object
    android.app.Dialog
```

表 5-15 Dialog 类定义的常用方法

方法	描述
public void setTitle(CharSequence title)	设置对话框的显示标题
public void show()	显示对话框
public void hide()	隐藏对话框
public boolean isShowing()	判断对话框是否显示
public void setContentView(View view)	设置组件
public void setContentView(int layoutResID)	设置组件的 ID
public void dismiss()	隐藏对话框
public void closeOptionsMenu()	关闭选项菜单
public void setDismissMessage(Message msg)	设置隐藏对话框时的消息
public void setCancelable(boolean flag)	设置是否可以取消
public void setCancelMessage(Message msg)	设置对话框取消时的消息
public void cancel()	取消对话框，与 dismiss()方法类似
public Window getWindow()	取得 Window 对象
public void setOnShowListener (DialogInterface.OnShowListener listener)	对话框弹出时触发事件
public void setOnDismissListener (DialogInterface.OnDismissListener listener)	对话框隐藏时触发事件
public void setOnCancelListener (DialogInterface.OnCancelListener listener)	对话框取消时触发事件

Android 提供的常见对话框有：Alertialog——用于实现警告对话框、ProgressDialog——用于实现带进度条的对话框、DatePickerDialog——用于实现日期选择对话框和 TimePickerDialog——用于实现时

间选择对话框。其中，DatePickerDialog 和 TimePickerDialog 见前面相关章节。

5.7.1 警告对话框

警告对话框是在程序运行时，弹出一个对话框并显示一条警告信息，提示用户的后续操作，它是项目中经常出现的一种对话框，Android 系统提供了 AlertDialog 类来实现警告对话框，其层次关系如下：

```
java.lang.Object
    android.app.Dialog
        android.app.AlertDialog
```

要在 Android 程序中使用 AlertDialog 组件，必须在程序开始时使用下面的语句：

```
import android.app.AlertDialog;             //导入 app.AlertDialog 类
```

要想实例化 AlertDialog 组件，可以通过 AlertDialog.Builder 类来实现，其常用方法如表 5-16 所示。

表 5-16 AlertDialog.Builder 类定义的常用方法

方法	描述
public AlertDialog.Builder(Context context)	创建 AlertDialog.Builder 对象
public AlertDialog.Builder setMessage (int messageId)	设置显示信息的资源 ID
public AlertDialog.Builder setMessage (CharSequence message)	设置显示信息的字符串
public AlertDialog.Builder setView(View view)	设置显示的 View 组件
public AlertDialog.Builder setSingleChoiceItems (CharSequence[ ] items, int checkedItem, DialogInterface.OnClickListener listener)	设置对话框显示一个单选的 List，指定默认选项，同时设置监听处理操作
public AlertDialog.Builder setSingleChoiceItems (ListAdapter adapter, int checkedItem, DialogInterface.OnClickListener listener)	设置对话框显示一个单选的 List，指定默认选项，同时设置监听处理操作
public AlertDialog.Builder setMultiChoiceItems (CharSequence[ ] items, boolean[ ] checkedItems, DialogInterface.OnMultiChoiceClickListener listener)	设置对话框显示一个复选的 List，同时设置监听处理操作
public AlertDialog.Builder setPositiveButton (CharSequence text, DialogInterface.OnClickListener listener)	为对话框添加一个确认按钮，同时设置监听操作
public AlertDialog.Builder setPositiveButton (int textId, DialogInterface.OnClickListener listener)	为对话框添加一个确认按钮，显示内容由资源文件指定，并设置监听操作
public AlertDialog.Builder setNegativeButton (CharSequence text, DialogInterface.OnClickListener listener)	为对话框设置一个取消按钮，并设置监听操作
public AlertDialog.Builder setNegativeButton (int textId, DialogInterface.OnClickListener listener)	为对话框设置一个取消按钮，显示内容由资源文件指定，并设置监听操作
public AlertDialog.Builder setNeutralButton(CharSequence text, DialogInterface.OnClickListener listener)	设置一个按钮，并设置监听操作
public AlertDialog.Builder setNeutralButton(int textId, DialogInterface.OnClickListener listener)	设置一个按钮，显示内容由资源文件指定，并设置监听操作
public AlertDialog.Builder setItems(CharSequence[ ] items, DialogInterface.OnClickListener listener)	将信息内容设置为列表项，同时设置监听操作
public AlertDialog.Builder setItems(int itemsId, DialogInterface.OnClickListener listener)	将信息内容设为列表项，列表项内容由资源文件指定，同时设置监听操作
public AlertDialog create()	创建 AlertDialog 的实例化对象
public AlertDialog.Builder setIcon(Drawable icon)	设置显示的图标文件
public AlertDialog.Builder setIcon(int iconId)	设置要显示图标来源的 ID

比较常见的警告框有：只带提示信息的简单对话框、带按钮的警告框、列表式警告框、带复选框或单选框式警告框、带按钮和输入框的警告框。

使用 AlertDialog.Builder 类创建警告框的一般步骤如下。

（1）通过 AlertDialog.Builder(Context)获取一个构造器 Builder。

例如：

```
AlertDialog.Builder builder=new AlertDialog.Builder(this);
```

（2）使用这个 Builder 类的公共方法来定义警告对话框的所有属性。

例如：

```
AlertDialog.Builder builder=new AlertDialog.Builder(MainActivity.this);
//实例化对象
builder.setTitle("请选择您喜爱的交通工具");          //设置显示标题
   …
```

（3）通过 Builder.Create()来创建 AlertDialog 对象。

例如：

```
AlertDialog alert=builder.create();   //通过 create()来创建 AlertDialog 对象
```

（4）直接调用 Builder.Show()显示对话框。

例如：

```
alert.show();                          //调用 show()显示对话框
```

有时不调用 Builder.Create()方法，而是在设置好警告对话框的所有属性后直接调用 show()方法来显示 AlertDialog。

在警告对话框中增加按钮时，需设置 DialogInterface.OnClickListener 事件监听接口对象，此接口主要负责进行对话框中按钮的事件处理，此接口定义如下：

```
public interface DialogInterface.OnClickListener {
    public abstract void onClick (DialogInterface dialog, int which) ;}
```

对话框事件处理接口如表 5-17 所示。

表 5-17　对话框事件处理接口

接口名称	描述
DialogInterface.OnClickListener	对话框单击事件处理接口
DialogInterface.OnCancelListener	对话框取消事件处理接口
DialogInterface.OnDismissListener	对话框隐藏事件处理接口
DialogInterface.OnKeyListener	对话框键盘事件处理接口
DialogInterface.OnMultiChoiceClickListener	对话框多选事件处理接口
DialogInterface.OnShowListener	对话框显示事件处理接口

5.7.2　AlertDialog 组件实例

实例 5-8：AlertDialog 组件举例

新建一个项目，项目命名为 Exam5_10，利用 AlertDialog 组件，实现不同的警告框，并实现相应的事件处理操作。

（1）修改 activity_main.xml 文件，代码如下：

```
…
<Button
```

```
        android:id="@+id/Alert"
        android:layout_width="match_parent"
        android:layout_height="wrap_content"
        android:text="普通警告框"
        android:gravity="center"/>
    <Button
        android:id="@+id/butAlert"
        android:layout_width="match_parent"
        android:layout_height="wrap_content"
        android:text="带按钮的警告框"
        android:gravity="center"/>
    <Button
        android:id="@+id/listAlert"
        android:text="列表式的警告框"
        android:layout_width="match_parent"
        android:layout_height="wrap_content"/>
    <Button
        android:id="@+id/singleAlert"
        android:text="带单选功能的警告框"
        android:layout_width="match_parent"
        android:layout_height="wrap_content"/>
    <Button
        android:id="@+id/checkAlert"
        android:text="带复选功能的警告框"
        android:layout_width="match_parent"
        android:layout_height="wrap_content"/>
</LinearLayout>
```

（2）修改 MainActivity.java 文件，代码如下：

```
…
import android.app.AlertDialog;
import android.app.Dialog;
import android.content.DialogInterface;
import android.view.View.OnClickListener;
public class MainActivity extends AppCompatActivity {
    private Button Alert = null ;
    private Button butAlert = null ;
    private Button listAlert = null ;
    private Button singleAlert = null ;
    private Button checkAlert = null ;
public void onCreate(Bundle savedInstanceState) {
        super.onCreate(savedInstanceState);
        super.setContentView(R.layout.activity_main);
        this.Alert = (Button) super.findViewById(R.id.Alert) ;
        this.butAlert = (Button) super.findViewById(R.id.butAlert) ;
        this.listAlert = (Button) super.findViewById(R.id.listAlert) ;
        this.singleAlert = (Button) super.findViewById(R.id.singleAlert) ;
        this.checkAlert = (Button) super.findViewById(R.id.checkAlert) ;
        this.Alert.setOnClickListener(new AlertImpl());          //设置单击事件
        this.butAlert.setOnClickListener(new butAlertImpl());  //设置单击事件
```

```
        this.listAlert.setOnClickListener(new listAlertImpl());   //设置单击事件
        this.singleAlert.setOnClickListener(new singleAlertImpl());
        this.checkAlert.setOnClickListener(new editAlertImpl()); }
    private class AlertImpl implements OnClickListener {
        public void onClick(View v) {
        Dialog dialog = new AlertDialog.Builder(MainActivity.this)  // 实例化对象
            .setIcon(R.drawable.alert)                              // 设置显示图片
            .setTitle("雷电警告")                                    // 设置显示标题
            .setMessage("雷雨多发季节，注意雷电！！")                  // 设置显示内容
            .create();                                              // 创建 Dialog
        dialog.show(); }}                                           // 显示对话框
    private class butAlertImpl implements OnClickListener {
        public void onClick(View v) {
            Dialog dialog = new AlertDialog.Builder(MainActivity.this)
                                                                    //实例化对象
                .setIcon(R.drawable.alert)                          //设置显示图片
                .setTitle("确定删除？")                              //设置显示标题
                .setMessage("您确定要删除这个文件吗？")                //设置显示内容
                .setPositiveButton("确定",                           //增加一个确定按钮
                    new DialogInterface.OnClickListener() { //设置操作监听
                        public void onClick(DialogInterface dialog,
                        int whichButton) { } //单击事件
                    }).setNeutralButton("查看文件",                  //增加普通按钮
                    new DialogInterface.OnClickListener() {//设置监听操作
                        public void onClick(DialogInterface dialog,int
                        whichButton) {}}).setNegativeButton("取消",
                        //增加取消按钮
                    new DialogInterface.OnClickListener() {//设置操作监听
                        public void onClick(DialogInterface dialog,
                            int whichButton) {   //单击事件
                        }}).create();                               //创建 Dialog
        dialog.show(); }}                                           //显示对话框
    private class editAlertImpl implements OnClickListener {
        boolean[] flags=new boolean[]{false,false,false,false,false}; //初始复选情况
        public void onClick(View v) {
        final String[] items={"登山","跑步","篮球","羽毛球","乒乓球"}; //定义数组
        AlertDialog.Builder builder = new AlertDialog.Builder(MainActivity.
            this);  builder.setTitle("请选择您喜爱的体育运动"); //设置显示标题
        builder.setMultiChoiceItems(items, flags, new DialogInterface.
        OnMultiChoice ClickListener(){//设置要显示的复选项，第 2 项默认选中，触发事件
          public void onClick(DialogInterface dialog, int which, boolean
            isChecked) {
                flags[which]=isChecked;
                String result = "您选择的是：";
                for (int i = 0; i < flags.length; i++) {  //确定已选择的选项
                    if(flags[i]){
                        result=result+items[i]+"、"; }  }
            Toast.makeText(getApplicationContext(), result, Toast.LENGTH_
                SHORT).show();}});                                  //显示选择的选项
        //添加一个确定按钮
```

```
        builder.setPositiveButton(" 确定", new DialogInterface.OnClickListener(){
            public void onClick(DialogInterface dialog, int which) {  }  });
        AlertDialog alert = builder.create();                    //创建 Dialog
        alert.show(); }}                                          //显示对话框
  private class singleAlertImpl implements OnClickListener {
      public void onClick(View v) {
      final CharSequence[] items={"A.1+1=2","B.1+1=3","C.1+2=2","D.1+3=2"};
      //定义数组
      AlertDialog.Builder builder = new AlertDialog.Builder(MainActivity.this);
      //实例化对象
      builder.setTitle("请选择您认为正确的答案");                    //设置显示标题
      //设置要显示的单选项，第 2 项默认选中，触发事件
      builder.setSingleChoiceItems(items,1,new DialogInterface.OnClickListener(){
        public void onClick(DialogInterface dialog, int item){  //事件的具体操作
            Toast.makeText(getApplicationContext(), "您选择的是："+items
                [item], Toast.LENGTH_SHORT).show();}});  //显示选择的选项
            //添加一个确定按钮
            builder.setPositiveButton(" 确 定 ", new DialogInterface.
                OnClickListener(){
           public void onClick(DialogInterface dialog, int which) { } });
              AlertDialog alert = builder.create();       //创建 Dialog
              alert.show(); }}                             //显示对话框
  private class listAlertImpl implements OnClickListener {
      public void onClick(View v) {
        final CharSequence[] items = { "飞机", "轮船", "火车" };  //定义数组
        AlertDialog.Builder builder=new AlertDialog.Builder(MainActivity.this);
                                                          //实例化对象
        builder.setTitle("请选择您喜爱的交通工具");          //设置显示标题
        //设置对话框要显示的 List，触发事件
        builder.setItems(items, new DialogInterface.OnClickListener(){
          public void onClick(DialogInterface dialog, int item){
          //事件的具体操作
            Toast.makeText(getApplicationContext(), "您选择的是："+items[item],
              Toast.LENGTH_SHORT).show();} });            //显示选择的选项
            AlertDialog alert = builder.create();   //创建 Dialog
            alert.show();  }} }                      //调用 show()方法显示
```

保存文件并运行该项目，进入如图 5.8 所示的界面，单击各按钮会弹出不同的警告框，进行不同的选择会执行不同的操作，如图 5.8 所示。

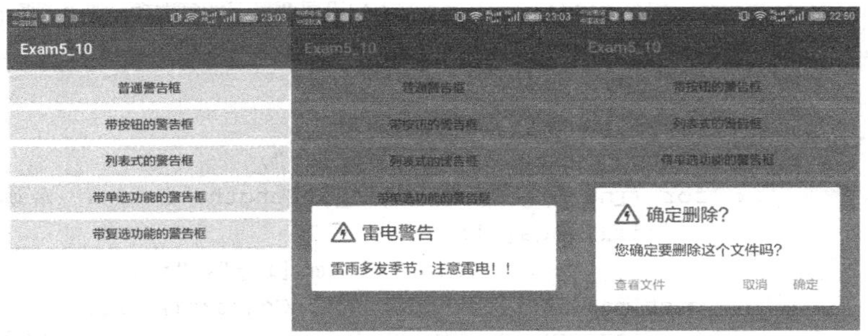

图 5.8　Exam5_10 运行结果

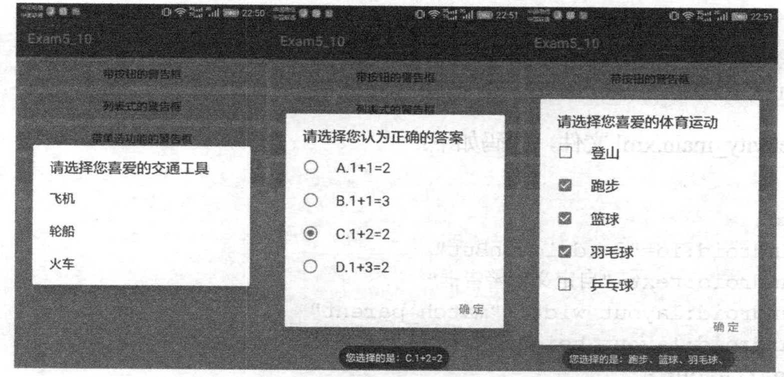

图 5.8　Exam5_10 运行结果（续）

5.7.3　自定义对话框

前面介绍的对话框界面比较简单，在程序设计的时候往往需要显示一些复杂的界面，要求定制一些对话框，如登录提示的对话框。可以这样来实现：先定义一个布局文件来定义显示需要显示的组件，再将这个布局文件显示包含到对话框之中，这个包含需要 Android 的 LayoutInflater 类的支持。LayoutInflater 类主要的方法有以下两个：

```
public static LayoutInflater from(Context context)//创建 LayoutInflater 对象
public View inflate(int resource, ViewGroup root)//将布局文件转变为 View 对象
```

实例 5-9：各种属性的运用

新建一个项目，项目命名为 Exam5_11，利用 LayoutInflater 类建立一个登录提示的对话框。

（1）在 res\layout 目录下，建立一个 TableLayout 布局的文件 login.xml，代码如下：

```
...
<TableRow>
    <TextView
        android:text="用户名："
        android:layout_marginLeft="18dip"
        android:textSize="10pt"
        android:layout_width="wrap_content"
        android:layout_height="wrap_content"/>
    <EditText
        android:width="65pt"
        android:layout_height="wrap_content"/>
</TableRow>
<TableRow>
    <TextView
        android:text="密　码："
        android:layout_marginLeft="18dip"
        android:textSize="10pt"
        android:layout_width="wrap_content"
        android:layout_height="wrap_content"/>
    <EditText
        android:password="true"
        android:width="65pt"
```

```
        android:layout_height="wrap_content"/>
    </TableRow>
</TableLayout>
```

（2）修改 activity_main.xml 文件，代码如下：

```
…
<Button
    android:id="@+id/loginBut"
    android:text="自定义的警告框"
    android:layout_width="match_parent"
    android:layout_height="wrap_content"/>
</LinearLayout>
```

（3）修改 MainActivity.java 文件，代码如下：

```
…
import android.app.AlertDialog;
import android.app.Dialog;
import android.content.DialogInterface;
import android.view.LayoutInflater;
public class MainActivity extends AppCompatActivity {
 private Button loginbut = null ;
 public void onCreate(Bundle savedInstanceState) {
     super.onCreate(savedInstanceState);
     super.setContentView(R.layout.activity_main);
     this.loginbut = (Button) super.findViewById(R.id.loginBut) ;
     this.loginbut.setOnClickListener(new OnClickListenerImpl()) ;} //设置单击事件
private class OnClickListenerImpl implements OnClickListener {
     public void onClick(View v) {
      LayoutInflater flater=LayoutInflater.from(MainActivity.this);  /*创建
        LayoutInflater 对象*/
        View dialogView = flater.inflate(R.layout.login,null); /*将布局文件
        转换为 View*/
        Dialog dialog = new AlertDialog.Builder(MainActivity.this)   /*创
        建 Dialog*/
               .setIcon(R.drawable.c12)                        //设置显示图片
               .setTitle("用户登录界面")                        //设置显示标题
               .setView(dialogView)                            //设置显示组件
               .setPositiveButton("确定",                      //设置确定按钮
                  new DialogInterface.OnClickListener() {//单击确定触发的事件
                     public void onClick(DialogInterface dialog, int
                        whichButton) {}})
               .setNegativeButton("取消",                      //设置取消按钮
                     new DialogInterface.OnClickListener() {/*单击取消触
                     发的事件*/
                        public void onClick(DialogInterface dialog, int
                           whichButton) {}}).create();  //创建对话框
        dialog.show();  }}  }                                  //显示对话框
```

保存文件并运行该项目，会弹出一个按钮，单击该按钮，进入如图 5.9 所示界面。

5.7.4 带进度条的对话框 ProgressDialog

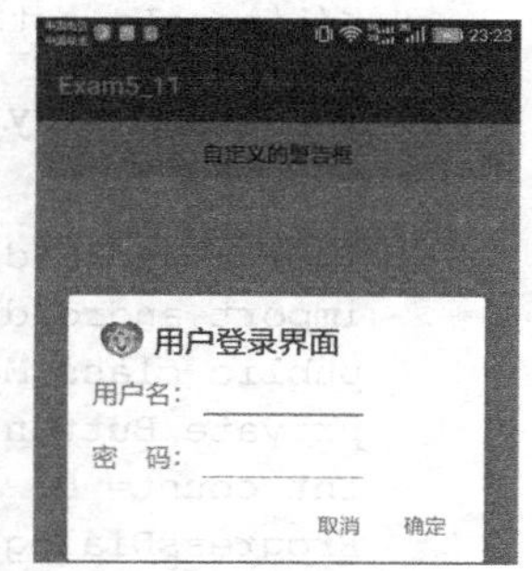

图 5.9 Exam5_11 运行结果

在进行复杂操作时往往需要有一段操作时间，这时会弹出一个等待的对话框，用户可以通过进度处理对话框显示进度的相关情况，进度处理对话框（ProgressDialog）可以实现这样的功能。其层次关系如下所示：

```
java.lang.Object
      android.app.Dialog
          android.app.AlertDialog
              android.app.ProgressDialog
```

要在 Android 程序中使用 ProgressDialog 组件，必须在程序中使用下面的语句：

```
import android.widget.ProgressDialog;    //导入 widget.ProgressDialog 类
```

从上面可以看出 ProgressDialog 是 AlertDialog 子类，所以 AlertDialog 的属性和方法 ProgressDialog 都具备。此外，它还有表 5-18 列出的常用方法。

表 5-18 ProgressDialog 常用的方法

方法	描述
public static final int STYLE_HORIZONTAL	常量，水平进度显示风格
public static final int STYLE_SPINNER	常量，圆形进度显示风格
public ProgressDialog(Context context)	创建进度对话框
public void setMessage(CharSequence message)	设置显示信息
public int getMax()	获得进度条的最大增长值
public int getProgress()	获得当前进度
public int getSecondaryProgress()	获得次要进度条
public void onStart()	启动进度框
public void setProgressStyle (int style)	设置进度条的显示风格
public static ProgressDialog show (Context context, CharSequence title, CharSequence message)	直接创建进度对话框，指定对话框的标题及信息
public void incrementProgressBy(int diff)	设置进度条每次增长的值
public void setMax(int max)	设置进度条的最大增长值
public void setProgress(int value)	设置当前进度

实例 5-10：ProgressDialog 组件举例

新建一个项目，项目命名为 Exam5_12，利用 ProgressDialog 类分别建立圆形和水平进度提示对话框。

（1）修改 activity_main.xml 文件，代码如下：

```
...
<Button
    android:id="@+id/circleBut"
    android:text="圆形进度条"
    android:layout_width="match_parent"
    android:layout_height="wrap_content"/>
<Button
    android:id="@+id/lineBut"
    android:text="水平进度条"
    android:layout_width="match_parent"
    android:layout_height="wrap_content"/>
```

```
</LinearLayout>
```

（2）修改 MainActivity.java 文件，代码如下：

```
…
import android.app.ProgressDialog;              //导入 app.ProgressDialog 类
import android.content.DialogInterface;
public class MainActivity extends AppCompatActivity {
private Button circleBut,lineBut;
int count= 0;
ProgressDialog ProDialog;                              //声明进度条对话框
   public void onCreate(Bundle savedInstanceState) {
       super.onCreate(savedInstanceState);
       setContentView(R.layout.activity_main);
       circleBut = (Button)this.findViewById(R.id.circleBut);
       lineBut = (Button)this.findViewById(R.id.lineBut);
       circleBut.setOnClickListener(new Button.OnClickListener(){ //单击事件
          public void onClick(View v) {
          ProDialog = new ProgressDialog(MainActivity.this);  //定义进度对话框
          //设置进度条风格，风格为圆形
          ProDialog.setProgressStyle(ProgressDialog.STYLE_SPINNER);
          ProDialog.setTitle("网络连接");                      //设置标题
          ProDialog.setMessage("正在连接网络，请等待......"); //设置提示信息
          ProDialog.setIndeterminate(false);              //设置进度条是否不明确
          ProDialog.setCancelable(true);                  //设置是否可按返回键取消
          ProDialog.setButton("确定", new DialogInterface.OnClickListener()
                                   {//设置按钮
          public void onClick(DialogInterface dialog, int which) {
          ProDialog.cancel();}});                      //单击"确定"按钮取消对话框
          ProDialog.show();} });
          lineBut.setOnClickListener(new Button.OnClickListener() {
          //单击事件
          public void onClick(View v) {
          count= 0;                                                //计数
          ProDialog = new ProgressDialog(MainActivity.this);//定义进度对话框
          //设置进度条风格，风格为水平的
          ProDialog.setProgressStyle(ProgressDialog.STYLE_HORIZONTAL);
          ProDialog.setTitle("程序安装");                 //设置标题
          ProDialog.setMessage("程序正在安装，请等待......");    //设置提示信息
          ProDialog.setIndeterminate(false);          //设置进度条是否不明确
          ProDialog.setCancelable(true);              //设置是否可按返回键取消
          ProDialog.show();                           //显示进度条对话框
          new Thread(){                               //定义线程对象
             public void run(){                       //线程主体方法
               while(count<=100){                     //从 1 到 100 循环
                 ProDialog.setProgress(count++);      //由线程来控制进度条
                   try {Thread.sleep(100);}           //休眠 0.1 秒
                     catch (InterruptedException e) {
                        e.printStackTrace();}}
          ProDialog.cancel();}}.start();} }); }}  //线程启动
```

保存文件并运行该项目，弹出两个按钮，分别单击弹出的这两个按钮，会弹出相应的对话框，如图 5.10 所示。代码中在水平进度条对话框中使用了一个线程，以模拟进度情况。

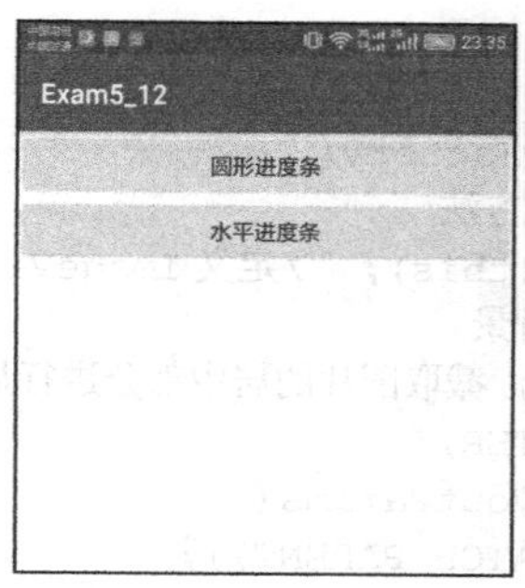

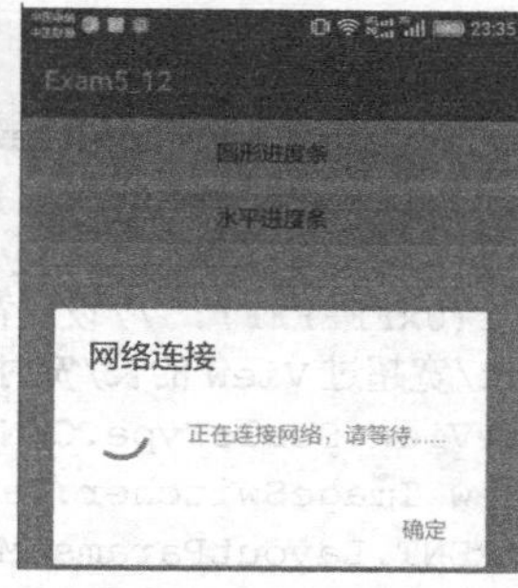

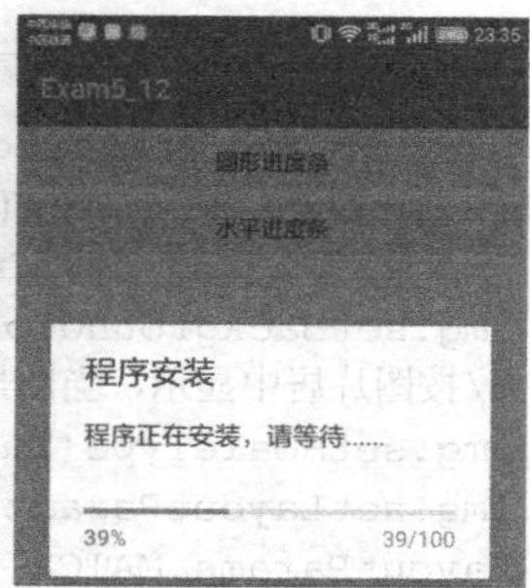

图 5.10　Exam5_12 运行结果

5.8　图片切换组件 ImageSwitcher

要在 Windows 平台上查看多张图片，可以通过 Windows 图片和传真查看器在“下一张”和“上一张”之间切换，在 Android 平台中可以通过 ImageSwitcher 组件实现效果。ImageSwitcher 组件的主要功能是完成图片的切换显示，其层次关系如下：

```
java.lang.Object
      android.view.View
            android.view.ViewGroup
                 android.widget.FrameLayout
                       android.widget.ViewAnimator
                             android.widget.ViewSwitcher
                                    android.widget.ImageSwitcher
```

要在 Android 程序中使用 ImageSwitcher 组件，必须在程序中使用下面的语句：

```
import android.widget.ImageSwitcher;    //导入widget.ImageSwitcher类
```

从上面可以看出 ImageSwitcher 是 View 的子类，所以 View 类的属性和方法 ImageSwitcher 都具备。此外，它还有表 5-19 列出的常用方法。

表 5-19　ImageSwitcher 类的常用方法

方法	描述
public ImageSwitcher(Context context)	创建 ImageSwitcher 对象
public void setFactory(ViewSwitcher.ViewFactory factory)	设置 ViewFactory 对象，用于完成两个图片切换时 ViewSwitcher 的转换操作
public void setImageResource(int resid)	设置显示的图片 ID
public void setInAnimation(Animation inAnimation)	图片读进 ImageSwitcher 时的动画效果
public void setOutAnimation(Animation outAnimation)	图片从 ImageSwitcher 消失时的动画效果
public void setImageDrawable (Drawable drawable)	绘制图片
public void setImageURI (Uri uri)	设置图片地址，可读取网络图片

利用 ImageSwitcher 浏览图片的基本步骤如下。

（1）定义需要显示的图片，例如：

```
private int[] images=new int[]{R.drawable.a11,R.drawable.a12,R.drawable.13,
     R.drawable.a14,R.drawable.a15 };  //图片数据
```

（2）调用 setFactory 方法，返回一个 View 对象，例如：

```
private class ViewFactoryImp implements ViewFactory{
  public View makeView() {   //覆写 makeView()方法
    ImageView img=new ImageView(MainActivity.this); //定义 ImageView 组件
    img.setBackgroundColor(0xFFFFFF); //设置背景
    //按图片居中显示，当图片长/宽超过 View 的长/宽时，截取图片的居中部分进行显示
    img.setScaleType(ImageView.ScaleType.CENTER);
    img.setLayoutParams(new ImageSwitcher.LayoutParams(
    LayoutParams.MATCH_PARENT,LayoutParams.MATCH_PARENT));
    return img;}}}
```

（3）利用 setImageResource 方法指定图片来源，例如：

```
this.imageswitch.setImageResource(this.images[this.count]);
```

（4）利用 setInAnimation 方法指定图片读入这个 ImageSwitcher 中时实现的动画效果，一般动画效果是从 Android.R 系统文件中读取的。

例如：

```
this.imageswitch.setInAnimation(AnimationUtils.loadAnimation(this,android
    .R.anim.fade_in)); //进入时的动画效果
this.imageswitch.setOutAnimation( AnimationUtils.loadAnimation(this,andro
    id.R.anim.fade_out));} //离开时的动画效果
```

（5）显示指定图片，例如：

```
MainActivity.this.imageswitch.setImageResource(MainActivity.this.images[-
    -MainActivity.this.count]);}//获得指定图片，并将图片计数减 1
```

实例 5-11：ImageSwitcher 组件举例

新建一个项目，项目命名为 Exam5_13，利用 ImageSwitcher 组件实现图片的浏览，先将 a11…a15 图片复制到 drawable 目录中。

（1）修改 activity_main.xml 文件，代码如下：

```
…
 <ImageSwitcher
       android:id="@+id/myImageS"
       android:layout_width="match_parent"
       android:layout_height="wrap_content"/>
 <LinearLayout
    android:layout_width="match_parent"
    android:layout_height="match_parent"
    android:orientation="horizontal" >
     <Button
        android:id="@+id/previous"
        android:layout_width="wrap_content"
        android:layout_height="wrap_content"
        android:text="上一张"/>
     <Button
        android:id="@+id/next"
        android:layout_width="wrap_content"
        android:layout_height="wrap_content"
        android:text="下一张"/>
     </LinearLayout>
</LinearLayout>
```

（2）修改 MainActivity.java 文件，代码如下：

```
…
import android.widget.ImageSwitcher;          //导入 widget.ImageSwitcher 类
import android.widget.ViewSwitcher.ViewFactory; //导入 ViewSwitcher.ViewFactory 类
public class MainActivity extends AppCompatActivity {
private ImageSwitcher imageswitch=null;      //定义 ImageSwitcher 组件
private Button previous=null;
private Button next=null;
private int count=0;                         //用于计数的变量
private int[] images=new int[]{R.drawable.a11,R.drawable.a12,R.drawable.13,
    R.drawable.a14,R.drawable.a15 };         //图片数据
protected void onCreate(Bundle savedInstanceState) {
super.onCreate(savedInstanceState);
setContentView(R.layout.activity_main);
//获得 ImageSwitcher 组件
this.imageswitch=(ImageSwitcher) super.findViewById(R.id.myImageS);
this.previous=(Button) super.findViewById(R.id.previous);
this.next=(Button) super.findViewById(R.id.next);
this.previous.setOnClickListener(new OnPreButClickImp()); //为“上一张”按钮添
加方法
this.next.setOnClickListener(new OnNextButClickImp());    //为“下一张”按钮添
加方法
this.imageswitch.setFactory(new ViewFactoryImp());     //设置转换工厂
this.imageswitch.setImageResource(this.images[this.count]);
this.imageswitch.setInAnimation(AnimationUtils.loadAnimation(this,android
    .R.anim.fade_in));                                  //进入时的动画效果
this.imageswitch.setOutAnimation( AnimationUtils.loadAnimation(this,andro
    id.R.anim.fade_out));}                   //离开时的动画效果
public class OnPreButClickImp implements OnClickListener{//“上一张”按钮触发
的事件
    public void onClick(View v) {
        if(MainActivity.this.count!=0){
            MainActivity.this.imageswitch.setImageResource(MainActivity.
                this.images[--MainActivity.this.count]);}/*获得指定图片，并
                将图片计数减 1*/
        else{Toast.makeText(MainActivity.this, "已经是第一张，不能往前了！",
            Toast.LENGTH_SHORT).show();}}}
public class OnNextButClickImp implements OnClickListener{//“下一张”按钮触发
的事件
    public void onClick(View v) {
        if(MainActivity.this.count!=MainActivity.this.images.length-1){
            MainActivity.this.imageswitch.setImageResource(MainActivity.this.
                images[++MainActivity.this.count]);} /*获得指定图片，并将图片计数加
1*/
        else{
    Toast.makeText(MainActivity.this, "已经是最后一张!不能往后了！", Toast.LENGTH_
        SHORT).show();}}}
    private class ViewFactoryImp implements ViewFactory{
        public View makeView() {    //覆写 makeView()方法
            ImageView img=new ImageView(MainActivity.this); //定义 ImageView 组件
            img.setBackgroundColor(0xFFFFFF); //设置背景
            //按图片居中显示，当图片长/宽超过 View 的长/宽时，截取图片居中部分进行显示
            img.setScaleType(ImageView.ScaleType.CENTER);
```

```
        img.setLayoutParams(new ImageSwitcher.LayoutParams(
        LayoutParams.MATCH_PARENT,LayoutParams.MATCH_PARENT));
        return img;}}}
```

保存文件并运行该项目，结果如图 5.11 所示，单击“下一张”按钮，能实现图片的浏览功能。

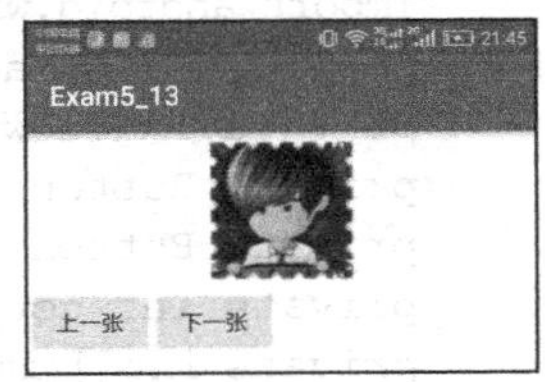

图 5.11　Exam5_13 运行结果

5.9　选项卡组件 TabHost

选项卡是界面设计时经常使用的界面组件，可以实现多个分页之间的快速切换，每个分页可以显示不同的内容。Android 平台提供了使用 TabHost 组件实现选项卡的功能，选项卡组件的主要功能是进行应用程序分类管理。每个选项卡称为一个Tab，而包含多个选项卡的容器称为 TabHost。TabHost 类的层次关系如下：

```
java.lang.Object
    android.view.View
        android.view.ViewGroup
            android.widget.FrameLayout
                android.widget.TabHost
```

TabHost 是整个 Tab 的容器，包括两部分：TabWidget 和 FrameLayout。其中 TabWidget 是每个 Tab 的选项，FrameLayout 则是 Tab 内容。

要在 Android 程序中使用 TabHost 组件，必须在程序中使用下面的语句：

```
import android.widget.TabHost;                    //导入 widget.TabHost 类
```

5.9.1　TabHost 组件基础

1．TabHost 组件常用方法

TabHost 组件有如表 5-20 所示的常用方法。

表 5-20　TabHost 组件常用方法

方法	描述
public TabHost(Context context)	创建 TabHost 类对象
public void addTab(TabHost.TabSpec tabSpec)	增加一个 Tab
public TabHost.TabSpec newTabSpec(String tag)	创建一个 TabHost.TabSpec 对象
public void clearAllTabs ()	清除所有关联到当前 TabHost 的选项卡
public View getCurrentView()	取得当前的 View 对象
public int getCurrentTab()	获取当前选项卡的 ID
public String getCurrentTabTag()	获得当前选项卡的 Tag 选项内容
public View getCurrentTabView()	获取当前选项卡的视图
public TabHost.TabSpec newTabSpec(String tag)	获取一个 TabHost.TabSpec，并关联到当前 TabHost
public void setup()	建立 TabHost 对象
public void setCurrentTab(int index)	设置当前显示的 Tab 编号
public void setCurrentTabByTag(String tag)	设置当前显示的 Tab 名称
public FrameLayout getTabContentView()	获取并保存选项卡内容
public void setOnTabChangedListener (TabHost.OnTabChangeListener l)	设置选项改变时触发
public void setup()	使用 findViewById()加载 TabHost，在新增一个选项卡之前，需要调用它

要实现选项卡的显示界面，有以下两种实现途径。

（1）直接让一个 Activity 程序继承 TabActivity 类。

（2）利用 findViewById()方法取得 TagHost 组件，要进行一些必要的配置。

如果使用 findViewById()方法取得 TagHost 组件，那么在新增一个选项卡之前，需要调用 setup()方法。在 TabActivity 里使用 getTabHost()方法获取 TabHost 组件时，不需要调用 setup()方法。

2. 直接使用一个 Activity 程序继承 TabActivity 类的方式实现选项卡功能

这种方法相对比较简单，直接用 TabActivity 类提供的方法即可实现操作，TabActivity 类提供的常用方法如表 5-21 所示。

可以直接用 getTabHost()方法获得一个TabHost类的对象，不能使用 findViewById()方法进行 TabHost 对象的实例化，要通过 LayoutInflater 类完成布局管理器中定义组件的实例化操作。

表 5-21 TabActivity 类的常用方法

方法	描述
public TabActivity()	无参构造，方便子类继承时调用
public TabHost getTabHost()	取得 TabHost 类的对象
public TabWidget getTabWidget()	取得 TabWidget 类的对象
public void setDefaultTab(String tag)	设置默认选中的选项
public void setDefaultTab(int index)	设置默认选中的选项

LayoutInflater 类的作用类似于前面介绍过的 findViewById()，不同的是 LayoutInflater 是查找 layout 下的布局管理文件，并且将它实例化，而 findViewById()是查找布局管理文件下的具体组件（如 Button、EditText 等）。LayoutInflater 类常用的方法如表 5-22 所示。

表 5-22 LayoutInflater 类的常用方法

方法	描述
LayoutInflater(Context context)	创建一个 LayoutInflater 对象
public View inflate (int resource, ViewGroup root)	设置所需要的布局管理器的资源 ID，设置组件的容器及是否包含设置组件的参数
public View inflate(int resource, ViewGroup root, boolean attachToRoot)	设置所需要的布局管理器的资源 ID，设置组件的容器及是否包含设置组件的参数
public static LayoutInflater from(Context context)	从指定的容器之中获得 LayoutInflater 对象

其中，resource 表示 View 的 layout 的 ID；root 如果返回 null，则将此 View 作为根，此时即可应用此 View 中的其他控件，如果返回非 null，则将默认的 layout 作为 View 的根；attachToRoot 如果为 true，则将此解析的 XML 作为 View 根，为 flase 则默认的 XML 为根视图 View。

Android 程序想要创建一个界面的时候，一般的做法是新建一个类，继承 Activity 基类，然后在 onCreate 中使用 setContentView 方法来载入一个 XML 中定义好的界面。

其实，在 Activity 程序中就是使用 LayoutInflater 类来载入 XML 文件指定界面的，通过 getSystemService(Context.LAYOUT_INFLATER_SERVICE)方法可以获得一个 LayoutInflater 对象，然后使用 inflate 方法来载入 layout 的 XML 文件，对于一个没有被载入或者想要动态载入的界面，需要使用 inflate 来载入。对于一个已经载入的界面，可以使用这个界面调用 findViewById 方法来获得其中的组件。

getSystemService()是 Android 中很重要的一个 API，它是 Activity 的一个方法，根据传入的 NAME 来取得对应的 Object，然后转换成相应的服务对象。其常见系统服务如表 5-23 所示。

表 5-23　getSystemService 返回的系统服务对象

传入的 Name	返回的对象	描述
WINDOW_SERVICE	WindowManager	管理打开的窗口程序
LAYOUT_INFLATER_SERVICE	LayoutInflater	取得 XML 中定义的 view
ACTIVITY_SERVICE	ActivityManager	管理应用程序的系统状态
POWER_SERVICE	PowerManger	电源的服务
ALARM_SERVICE	AlarmManager	闹钟的服务
NOTIFICATION_SERVICE	NotificationManager	状态栏的服务
KEYGUARD_SERVICE	KeyguardManager	键盘锁的服务
LOCATION_SERVICE	LocationManager	位置的服务，如 GPS
SEARCH_SERVICE	SearchManager	搜索的服务
VEBRATOR_SERVICE	Vebrator	手机震动的服务
CONNECTIVITY_SERVICE	Connectivity	网络连接的服务
WIFI_SERVICE	WifiManager	Wi-Fi 服务
TELEPHONY_SERVICE	TelephonyManager	电话服务

如果要增加一个选项使用方法 addTab(TabHost.TabSpec tabSpec)，则有多个选项就要增加多个 TabHost.TabSpec 的对象，TabHost.TabSpec 类是 TabHost 定义的内部类，如果要想取得此类的实例化对象，就要依靠 TabHost 类中的 newTabSpec()方法来完成。

每个选项卡都有一个选项卡指示符、内容和 tag 选项，TabHost.TabSpec 类的常用方法如表 5-24 所示。

表 5-24　TabHost.TabSpec 类定义的常用方法

方法	描述
public TabHost.TabSpec setContent(int viewId)	设置要显示的组件 ID
public TabHost.TabSpec setContent(Intent intent)	指定一个加载 activity 的 Intent 对象作为选项卡内容
public TabHost.TabSpec setContent(TabHost. TabContentFactory contentFactory)	指定 TabHost.TabContentFactory 用于创建选项卡的内容
public TabHost.TabSpec setIndicator(View view)	指定一个视图作为选项卡指示符
public TabHost.TabSpec setIndicator(CharSequence label)	设置一个选项
public TabHost.TabSpec setIndicator(CharSequence label, Drawable icon)	为选项卡指示符指定一个选项和图标
public String getTag ()	获取 tag 选项字符串

以继承 TabActivity 类的方式实现选项卡功能的步骤如下。

（1）设计所有的分页的界面布局。

（2）建立一个类继承 TabActivity 类。

例如：

```
public class MainActivity extends TabActivity {…}
```

（3）通过方法获得 TabHost 对象，例如：

```
TabHost th = getTabHost();
```

（4）将指定布局管理文件实例化。

```
LayoutInflater.from(this).inflate(R.layout.activity_main,th.getTabContent
    View(), true);
```

（5）设置选项卡的标题和内容。

例如：

```
th.addTab(th.newTabSpec("tab1").setIndicator("文件").setContent(R.id.file));
```

虽然这种方法很简单，但是建议不要采用这种形式实现选项卡功能，因为 TabActivity 类从 API 级别 13 开始已经废弃不用了，移植到手机上时可能会因为版本的问题出现错误。

3. 在布局管理器之中定义 TabHost 组件实现选项卡功能

如果使用布局管理器的方式实现选项卡功能，则必须先了解 android.widget.TabWidget 类。其层次关系如下：

```
java.lang.Object
    android.view.View
        android.view.ViewGroup
            android.widget.LinearLayout
                android.widget.TabWidget
```

如果想通过配置实现选项卡功能，则对配置文件的编写有以下要求。

（1）所有用于选项配置的文件，必须以“<TabHost>”为根节点。

（2）为了保证选项页和选项内容显示正常，可以采用一个布局管理器进行布局。

（3）定义一个“<TagWidget>”选项，用于表示整个选项容器。另外，在定义此组件的时候要引入“tabs”的组件，表示允许加入多个选项页。

（4）由于 TabHost 是 FrameLayout 的子类，所以想定义选项页必须使用 FrameLayout 布局，而后在此布局中定义所需要的选项页组件，且框架布局中必须引用 tabcontent 组件(android:id="@android:id/tabcontent")，这与 getTabContentView()功能类似。

TabWidget 类常用属性如表 5-25 所示，常用方法如表 5-26 所示。

表 5-25 TabWidget 类常用属性

方法	描述
android:divider	可绘制对象，被绘制在选项卡窗口间充当分割物
android:tabStripEnabled	确定是否在选项卡中绘制
android:tabStripLeft	用来绘制选项卡下面的分割线左边部分的可视化对象
android:tabStripRight	用来绘制选项卡下面的分割线右边部分的可视化对象

表 5-26 TabWidget 类常用方法

方法	描述
public TabWidget(Context context)	创建 TabWidget 实例
public void addView(View child)	向 TabWidget 增加组件
public int getTabCount()	返回选项卡的数量
public void setEnabled(boolean enabled)	配置是否启用
public void focusCurrentTab(int index)	设置当前选项卡并且使其获得焦点
public void setCurrentTab (int index)	设置当前选项卡

5.9.2 TabHost 组件实例

可以直接用一个 Activity 程序继承 TabActivity 类的方式实现选项卡功能，项目 Exam5_14 就是采用这种方式实现的，但是 TabActivity 类从 API 级别 13 开始已经废弃不用了，移植到手机上时可能会因为版本的问题出现错误。建议采用在布局管理器中定义 TabHost 组件，然后在 MainActivity 中调用组件的方法来实现选项卡功能。

实例 5-12：TabHost 组件举例

新建一个项目，项目命名为 Exam5_15，在布局管理器中定义 TabHost 组件以实现选项卡功能。

（1）修改 activity_main.xml 文件，代码如下：

```
…
<LinearLayout                    //采用一个布局管理器进行布局
    android:orientation="vertical"
    android:layout_width="match_parent"
    android:layout_height="match_parent">
<TagWidget                    //定义一个"<TagWidget>"的选项，用于表示整个选项容器
android:id="@android:id/tabs"        //引入"tabs"的组件，表示允许加入多个选项页
    android:layout_width="match_parent"
    android:layout_height="wrap_content"
    android:layout_alignParentTop="true"/>
    <FrameLayout                    // 定义选项页必须使用 FrameLayout 布局
        android:id="@android:id/tabcontent"  //框架布局上必须引用 tabcontent 组件
        android:layout_width="match_parent"
        android:layout_height="match_parent">
      <LinearLayout
       android:id="@+id/file"
       android:layout_width="match_parent"
       android:layout_height="match_parent"
       android:orientation="vertical"
       android:gravity="center_horizontal">
       <Button
          android:id="@+id/open"
          android:layout_width="150dp"
          android:layout_height="wrap_content"
          android:text="打开文件"/>
      <Button
          android:id="@+id/save"
          android:layout_width="150dp"
          android:layout_height="wrap_content"
          android:text="保存文件" />
     <Button
          android:id="@+id/saveAs"
          android:layout_width="150dp"
          android:layout_height="wrap_content"
          android:text="文件另存为" />
</LinearLayout>
 <LinearLayout
      android:id="@+id/edit"
      android:layout_width="match_parent"
      android:layout_height="match_parent"
      android:orientation="vertical"
      android:gravity="center_horizontal">
      <Button
          android:id="@+id/copy"
          android:layout_width="match_parent"
          android:layout_height="wrap_content"
```

```
                android:text="复制"/>
            <Button
                android:id="@+id/paste"
                android:layout_width="match_parent"
                android:layout_height="wrap_content"
                android:text="粘贴"/>
    </LinearLayout>
    <LinearLayout
            android:id="@+id/seek"
            android:layout_width="match_parent"
            android:layout_height="match_parent"
            android:orientation="vertical"
            android:gravity="center_horizontal">
           <EditText
                  android:id="@+id/edit1"
                  android:layout_width="wrap_content"
                  android:layout_height="wrap_content"
                  android:text="请输入检索关键字..."
                  android:textSize="18sp"/>
          <Button
                  android:id="@+id/seekbut1"
                  android:layout_width="wrap_content"
                  android:layout_height="wrap_content"
                  android:text="搜索"/>
        </LinearLayout>
        <LinearLayout
          android:id="@+id/time"
          android:layout_width="match_parent"
          android:layout_height="match_parent"
          android:orientation="vertical"
          android:gravity="center_horizontal">
          <TimePicker
                  android:id="@+id/seekbut"
                  android:layout_width="wrap_content"
                  android:layout_height="wrap_content"
                  android:text="设置时间"/>
        </LinearLayout>
      </FrameLayout>
    </LinearLayout >
    </TabHost>
```

（2）修改 MainActivity.java 文件，代码如下：

```
…
import android.widget.TabHost;
import android.widget.TabHost.TabSpec;
import android.widget.TabHost.OnTabChangeListener;
public class MainActivity extends AppCompatActivity {
 private TabHost myTabHost;                    //定义 TabHost
```

```
private int[] layRes = { R.id.file, R.id.edit, R.id.seek, R.id.time };
/*定义内嵌布局管理器 ID*/
public void onCreate(Bundle savedInstanceState) {
    super.onCreate(savedInstanceState);
    super.setContentView(R.layout. activity_main) ;  //调用默认布局管理器
    this.myTabHost = (TabHost) super.findViewById(R.id.tabhost); //取得对象
    this.myTabHost.setup() ;                          //建立 TabHost 对象
    TabSpec myTab1 = myTabHost.newTabSpec("文件");     //定义 TabSpec
    myTab1.setIndicator("文件") ;                      //设置选项文字
    myTab1.setContent(this.layRes[0]) ;               //设置显示的组件
    this.myTabHost.addTab(myTab1) ;                   //增加选项
    TabSpec myTab2 = myTabHost.newTabSpec("编辑");     //定义 TabSpec
    myTab2.setIndicator("编辑") ;                      //设置选项文字
    myTab2.setContent(this.layRes[1]) ;               //设置显示的组件
    this.myTabHost.addTab(myTab2) ;                   //增加选项
    TabSpec myTab3 = myTabHost.newTabSpec("查看");     //定义 TabSpec
    myTab3.setIndicator("查看") ;                      //设置选项文字
    myTab3.setContent(this.layRes[2]) ;               //设置显示的组件
    this.myTabHost.addTab(myTab3) ;                   //增加选项
    TabSpec myTab4 = myTabHost.newTabSpec("设置时间"); //定义 TabSpec
    myTab4.setIndicator("设置时间") ;                  //设置选项文字
    myTab4.setContent(this.layRes[3]) ;               //设置显示的组件
    this.myTabHost.addTab(myTab4) ;                   //增加选项
    this.myTabHost.setCurrentTab(0) ;                 //设置开始默认选项
    myTabHost.setOnTabChangedListener(new OnTabChangeListener() {//事件处理
        public void onTabChanged(String tabId) {
            Dialog dialog = new AlertDialog.Builder(MainActivity.this)
                .setTitle("提示")
                .setMessage("当前选中: "+tabId+"选项")
                .setPositiveButton("确定",new DialogInterface.OnClickListener(){
                    public void onClick(DialogInterface dialog, int whichButton)
                        { dialog.cancel(); }   }).create();//创建按钮
                  dialog.show(); } });}}
```

保存文件并运行该项目，选择某个选项卡，如“文件”选项卡，会触发事件，结果如图 5.12 所示。

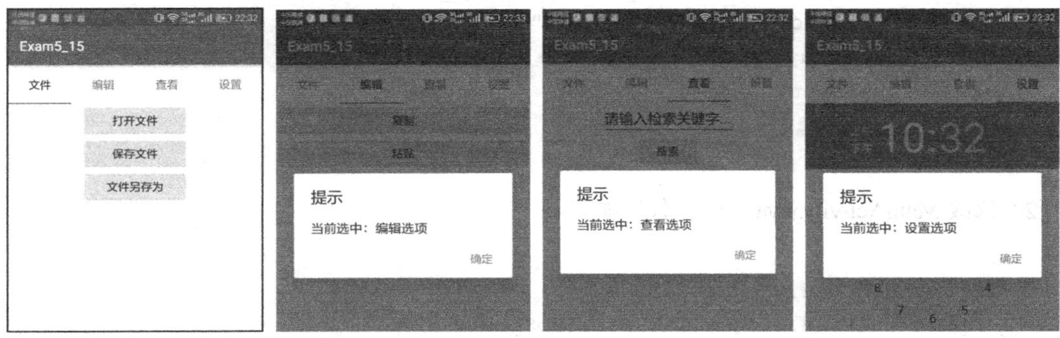

图 5.12　Exam5_15 运行结果

Android 系统提供的组件还有很多，新的组件还在不断地开发出来，限于篇幅，其他组件本书中不再赘述，感兴趣的用户可以参考 Android API 文档资料进行学习。

本章小结

本章着重介绍了 Android 系统提供的常见高级组件，如列表显示、进度条、对话框、选项卡组件等，这些组件都有自己的相应属性、方法和事件触发处理机制。

习题

（1）ListView 组件中有哪些事件？请写出其代码。
（2）ProgressBar 组件与 ProgressDialog 组件的区别与联系有哪些？
（3）利用 DatePickerDialog 和 TimePickerDialog 将当前时间设置为 2017-10-15 上午 12:30。
（4）AlertDialog 有哪些种类？
（5）AutoCompleteTextView 组件实现自动提示功能的步骤有哪些？
（6）参照实例 5-13，编写程序，利用 GridView 组件显示一组图片。
（7）假设 ImageSwitcher 组件中显示的图片比本身大，应该在哪里修改什么语句？
（8）用两种方法设置 TabHost 组件，TabHost 组件中含有 2 个选项。

第 6 章　Android 组件之间的通信

学习目标：

- 了解使用 Intent 进行组件间通信的原理。
- 掌握使用 Intent 启动 Activity 的方法。
- 掌握获取 Activity 返回值的方法。
- 掌握 Message、Handler、Looper 类的使用及消息的传递。
- 掌握 Service 的定义及使用。
- 了解系统提供的 Service 程序。
- 掌握发送和接收广播消息的方法。

前面的章节学习了一些组件和事件的处理机制，所涉及的程序都是在 Activity 程序中进行的，在一个项目中，往往会包含多个 Activity 程序，多个 Activity 程序之间控制权的转换及数据的通信，主要依据 Intent 组件来实现。

6.1　Android 四大组件

Android 四大基本组件分别是 Activity（活动）、Service（服务）、Content Provider（内容提供者）、BroadcastReceiver（广播接收者）。并不是每一个 Android 应用程序都需要这四种组件，例如，前面介绍的程序都是在一个 Activity 程序中进行的。下面对这四个组件进行一些简单的说明，让大家有一个整体的认识。

Activity：Activity 是活动的意思，一个 Activity 通常表现为一个可视化的用户界面，是 Android 程序与用户交互的窗口，也是 Android 组件中最基本、最复杂的一个组件。从外部来看，一个 Activity 占据当前的窗口，响应所有窗口事件，具有控件、菜单等界面元素。从内部逻辑来看，Activity 需要为了保持各个界面状态，而管理生命周期和一些转跳逻辑。对于开发者而言，需要派生一个 Activity 的子类，进行编码以实现各种功能方法。

Service：Service 是服务的意思，服务是运行在后台的一个组件，它就像一个没有界面的 Activity。它的很多方面与 Activity 类似，例如，封装有一个完整的功能逻辑实现，接收上层指令，完成相关的事件，定义好需要接收的 Intent 提供同步和异步的接口等。服务不提供用户界面，例如，在后台下载文件、播放音乐，在播放音乐的同时做其他事情，不会妨碍用户与其他活动的交互。另一个组件（如 Activity）可以启动一个服务，并运行或者绑定到 Service。

BroadcastReceiver：BroadcastReceiver 是广播接收者的意思，它不执行任何任务。广播是一种广泛运用在应用程序之间传输信息的机制，而 BroadcastReceiver 是对发送出来的广播进行过滤接收并响应的一类组件。

BroadcastReceiver 不包含任何用户界面。然而，它可以启动一个 Activity 以响应接收到的信息，或者通过 NotificationManager 通知用户。可以通过多种方式使用户知道有新的通知产生，如手机震动、闹钟等。

ContentProvider：ContentProvider 是内容提供者的意思，作为应用程序之间唯一的共享数据的途径，ContentProvider 的主要功能就是存储并检索数据以及向其他应用程序提供访问数据的接口。

在 Android 中，每个应用程序都是使用自己的用户 ID 并在自己的进程中运行的。这样做的好处是，可以有效地保护系统及应用程序，避免被其他应用程序所影响，每个进程都拥有独立的进程地址空间和虚拟空间。Android 的数据都属于应用程序自身，其他的应用不能直接进行操作。如果要实现不同应用之间的数据共享，就要用到 ContentProvider 组件，本书将在第 8 章中详细介绍 ContentProvider 组件的应用。

Android 中还有一个很重要的概念——Intent，Intent 是一个对动作和行为的抽象描述，负责组件之间、程序之间的消息传递。而 BroadcastReceiver 组件提供了一种把 Intent 作为一个消息广播出去，由所有对其感兴趣的程序对其做出反应的机制。

在 Android 中，Intent 作为连接组件的纽带，除了 ContentProvider 是通过 Content Resolver 来激活的之外，其他 3 种组件——Activity、Service 和 BroadcastReceiver 都是由 Intent 激活的，Intent 在不同的组件之间传递消息，将一个组件的请求意图传给另一个组件。

6.2 Intent

Intent 的中文意思是“意图”，Intent 组件在 Android 中是一个十分重要的组件，它是连接不同应用的桥梁和纽带，也是让组件复用（Activity 和 Service）成为可能的一个重要因素。Intent 组件的主要作用是在运行相同或不同应用程序的 Activity、Service，BroadcastReceiver 间进行切换以及数据的传递。Intent 组件常用的方法如表 6-1 所示。

表 6-1 Intent 常用的方法

方法	描述
public void startActivity(Intent intent)	启动一个 Activity，并通过 Intent 传送数据
public void startActivityForResult(Intent intent, int requestCode)	启动并接收另一个 Activity 程序回传数据，当 requestCode 大于 0 时才可以触发 onActivityResult()
public Intent getIntent()	返回启动当前 Activity 程序的 Intent
protected void onActivityResult(int requestCode, int resultCode, Intent data)	当需要接收 Intent 回传数据的时候覆写此方法，对回传操作进行处理
public void finish()	调用此方法会返回之前的 Activity 程序，并自动调用 onActivity Result()方法
public final Cursor managedQuery (Uri uri, String[ ] projection, String selection, String[ ] selectionArgs, String sortOrder)	处理返回的 Cursor 结果集

要在 Android 程序中使用 Intent 组件，必须在程序中使用下面的语句：

```
import android.content.Intent;      //导入 content.Intent 类
```

6.2.1 利用 Intent 启动 Activity

一个 Activity 类似于 Web 开发中的一个页面，在 Web 设计中经常会用超级链接来从一个页面跳转到另外的一个页面，在 Android 系统中，应用程序一般有多个 Activity，多个 Activity 间需要通信，Intent 组件既可以在多个 Activity 之间传递要操作的信息，也可以启动其他的 Activity 程序。

启动 Activity 的方式有两种：显式启动和隐式启动。显式启动：必须在 Intent 中指明启动的 Activity 所在的类。隐式启动：Android 系统根据 Intent 的动作和数据来决定启动哪一个 Activity，也就是说，在隐式启动时，Intent 中只包含需要执行的动作和所包含的数据，而无须指明具体启动哪一个 Activity，选择权由 Android 系统和最终用户来决定。

使用 Intent 显式启动 Activity 的基本步骤如下。

（1）创建一个 Intent。

（2）指定当前的应用程序上下文以及要启动的 Activity，例如：

```
//实例化 Intent，指定当前的应用程序以及要启动的 Activity
Intent it = new Intent(Send.this, Receive.class);
```

（3）把创建好的 Intent 作为参数传递给 startActivity()方法或者 startActivityForResult()方法，这两个方法的区别在于是否接收另一个 Activity 程序回传的数据，startActivityForResult 会有数据的回传，当 requestCode 大于 0 时，才可以触发 onActivityResult()。例如：

```
Send.this.startActivity(it);                      //启动 Activity，不会接收数据的回传
Send.this.startActivityForResult(it, 1);  //启动 Activity，会接收数据的回传
```

（4）在 AndroidManifest.xml 文件中注册两个 Activity 时，应使用<activity>标签，并嵌套在<application>标签内部。例如：

```
<activity android:name="Send" android:label="@string/app_title">
     <intent-filter>
         <action android:name="android.intent.action.MAIN" />
         <category android:name="android.intent.category.LAUNCHER" />
     </intent-filter>
</activity>
<activity android:name="Receive" android:label="@string/receive_name" />
</application>
```

<application>节点下共有两个<activity>节点，分别代表应用程序中所使用的两个 Activity：Send 和 Receive。其中，Send 是程序的主入口，Receive 是通过 Send 启动的。

6.2.2 利用 Intent 在 Activity 之间传递数据

1．利用 Intent 的 startActivity 传递数据

startActivity 可以把数据传递到指定的地方，但是不可以获得从接收方反馈的数据。

传递数据方的 Activity 中的关键代码如下：

```
Intent it = new Intent(Send.this, Receive.class); //实例化 Intent，接收方是 Receive
String info1=name.getText().toString();            //取得文本框输入的内容
it.putExtra("sendinfo", info1) ;        //将 sendinfo 赋值为字符串 info1 的值，传出
Send.this.startActivity(it);            //启动 Activity，不会接收数据的回传
```

接收数据方的 Activity 中的关键代码如下：

```
Intent it = super.getIntent() ;                    //取得启动此程序的 Intent
String info = it.getStringExtra("sendinfo") ;//取得传来的 sendinfo 值
```

2．利用 Intent 的 startActivityForResult 传递数据

startActivityForResult 可以把数据传到指定的地方，还可以把接收方的数据传过来。

传递数据方的 Activity 中的关键代码如下：

```
Intent it = new Intent(Send.this, Receive.class);  //实例化 Intent，接收方是 Receive
String info1=name.getText().toString();            //取得文本框中输入的内容
it.putExtra("sendinfo", info1) ;       //将 sendinfo 赋值为字符串 info1 的值，传出
//启动 Activity，会接收数据的回传，requestCode 的值大于 0 时才触发 onActivityResult()
Send.this.startActivityForResult(it, 1);
```

重载 onActivityResult 方法，用来接收传递过来的数据。

如果想要接收回传的数据，则需要 Activity 常量的支持。Activity 提供的操作常量有以下几类。

RESULT_OK：表示操作正常的状态码。

RESULT_CANCELED：表示操作取消的状态码。

RESULT_FIRST_USER：表示用户自定义的操作状态码。

重载 onActivityResult 的关键代码如下：

```
protected void onActivityResult(int requestCode, int resultCode, Intent data) {
    switch (resultCode) {                            //判断操作类型
    case RESULT_OK:                                  //成功操作
        msg.setText("返回的内容是：" + data.getStringExtra("returninfo"));
        break;
    case RESULT_CANCELED:                            //取消操作
        msg.setText("操作取消。");
        break ;
    default:
        break;      }   }}
```

接收数据方的 Activity 中的关键代码如下：

```
Intent it = super.getIntent() ;                        //取得启动此程序的 Intent
String info = it.getStringExtra("sendinfo") ;//取得传送来的 sendinfo 值
…
//返回信息变量名为 returninfo，它的值为 retu
Receive.this.getIntent().putExtra("returninfo",retu);
//设置返回数据的状态，RESULT_OK 与 Send.java 中的 onActivityResult()中判断的对应
Receive.this.setResult(RESULT_OK, Receive.this.getIntent()) ;
Receive.this.finish() ;                                //结束 Intent
```

6.2.3 Intent 组件传递数据实例

实例 6-1：Intent 组件传递数据举例 1

新建一个项目，项目命名为 Exam6_1，利用 Intent 实现不同 Activity 之间的跳转，程序运行结果如图 6.1 所示，代码简单但是较多，这里就不给出了，可以参照 Exam6_1 进行学习。

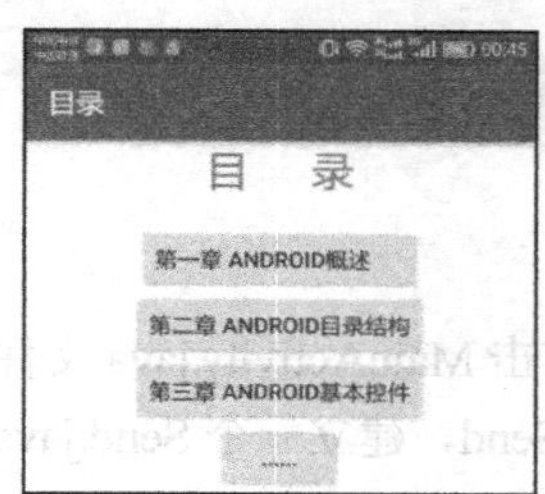

图 6.1 Exam6_1 运行结果

实例 6-2：Intent 组件传递数据举例 2

新建一个项目，项目命名为 Exam6_2，利用 Intent 的 startActivityForResult 可以在 Activity 之间进行数据的传递和回传。

（1）在 res 文件夹下选择 layout 文件，在其下建立 send.xml 文件，代码如下：

```
…
<TextView
```

```
    android:id="@+id/txt"
    android:layout_width="wrap_content"
    android:layout_height="wrap_content"
    android:text="这是发送方"/>
  <EditText
    android:id="@+id/myedit"
    android:layout_width="wrap_content"
    android:layout_height="wrap_content"
    android:selectAllOnFocus="true"
    android:text="输入您的姓名"/>
  <Button
    android:id="@+id/mybut"
    android:layout_width="wrap_content"
    android:layout_height="wrap_content"
    android:text="传送数据到 Receive"/>
  <TextView
    android:id="@+id/msg"
    android:layout_width="wrap_content"
    android:layout_height="wrap_content"/>
</LinearLayout>
```

（2）在 res 文件夹下选择 layout 文件，在其下建立 receive.xml 文件，代码如下：

```
 …
 <TextView
    android:id="@+id/txt"
    android:layout_width="wrap_content"
    android:layout_height="wrap_content"
    android:text="这是接收方"/>
  <TextView
    android:id="@+id/show"
    android:layout_width="wrap_content"
    android:layout_height="wrap_content"/>
  <Button
    android:id="@+id/retu"
    android:layout_width="wrap_content"
    android:layout_height="wrap_content"
    android:text="返回数据到 Send。"/>
</LinearLayout>
```

（3）在 src 文件夹下选择文件 org.hnist.demo，右击 MainActivity.java 文件，选择“Refactor”选项，再选择“Rename”选项，将 MainActivity 重命名为 Send，建立一个 Send.java 文件，做如下修改：

```
 …
 import android.content.Intent;                    //导入 content.Intent 类
 public class Send extends AppCompatActivity {
    private Button mybut = null ;
    private TextView msg = null ;
    private EditText name = null ;
    public void onCreate(Bundle savedInstanceState) {
       super.onCreate(savedInstanceState);
       super.setContentView(R.layout.send_main);     //默认布局管理器
       this.mybut = (Button) super.findViewById(R.id.mybut) ;
```

```
        this.msg = (TextView) super.findViewById(R.id.msg) ;
        this.name = (EditText) super.findViewById(R.id.myedit) ;
        this.mybut.setOnClickListener(new OnClickListenerImpl());}}//定义事件
    private class OnClickListenerImpl implements OnClickListener {
        public void onClick(View view) {
            Intent it = new Intent(Send.this, Receive.class);  //实例化 Intent
            String info1=name.getText().toString();     //取得文本框输入的内容
            it.putExtra("sendinfo", info1) ;            //设置附加信息
            Send.this.startActivityForResult(it, 1);}}  //启动 Activity
protected void onActivityResult(int requestCode, int resultCode, Intent data) {
        switch (resultCode) {                                   //判断操作类型
        case RESULT_OK:                                         //成功操作
            msg.setText("返回的内容是: " + data.getStringExtra("returninfo"));
            break;
        case RESULT_CANCELED:                                   //取消操作
            msg.setText("操作取消。");
            break ;
        default:
            break;     }   }}
```

（4）右击 Send.java 文件，选择“copy”选项，然后选择“Paste”选项，输入 Receive.java，单击“OK”按钮，建立一个 receive.java 文件，做如下修改：

```
…
import android.content.Intent;                          //导入 content.Intent 类
public class Send extends AppCompatActivity {
    private TextView show = null ;
    private Button rebut = null ;
    @Override
    public void onCreate(Bundle savedInstanceState) {
        super.onCreate(savedInstanceState);
        super.setContentView(R.layout.receive_main);   //调用默认布局管理器
        this.show = (TextView) super.findViewById(R.id.show) ;
        this.rebut = (Button) super.findViewById(R.id.retu) ;
        Intent it = super.getIntent() ;              //取得启动此程序的 Intent
        String info = it.getStringExtra("sendinfo") ;  //取得设置的附加信息
        this.show.setText(info) ;                      //设置文本显示信息
        this.rebut.setOnClickListener(new OnClickListenerImpl()) ; }  //设置监听
    private class OnClickListenerImpl implements OnClickListener {
        public void onClick(View view) {
            String retu="我收到你发来的信息:"+show.getText().toString();;
            Receive.this.getIntent().putExtra("returninfo", retu) ; //返回信息
            //设置返回数据的状态
            Receive.this.setResult(RESULT_OK, Receive.this.getIntent()) ;
            Receive.this.finish() ;}      }}                     //结束 Intent
```

（5）修改 AndroidManifest.xml 文件，添加一个 activity，代码如下：

```
<?xml version="1.0" encoding="utf-8"?>
<manifest xmlns:android="http://schemas.android.com/apk/res/android"
    …
        <activity android:name="Send" android:label="Send">
            <intent-filter>
```

```
                <action android:name="android.intent.action.MAIN" />
                <category android:name="android.intent.category.LAUNCHER" />
            </intent-filter>
        </activity>
        <activity android:name="Receive" android:label="Receive" />
    </application>
</manifest>
```

保存所有文件，运行该项目，进入 Send 界面，在文本框中输入内容，单击按钮，会将输入的信息在另一个 Activity（Receive）中显示出来，单击其中的返回按钮，可将指定信息回传到 Send 界面中，如图 6.2 所示。

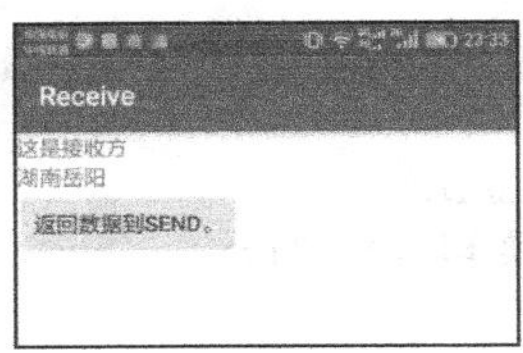

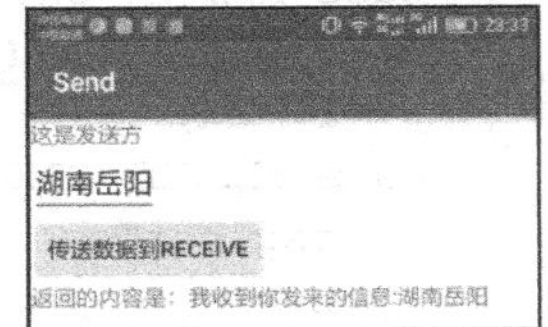

图 6.2　Exam6_2 运行结果

6.3　深入了解 Intent

Android 中提供了 Intent 组件来协助应用程序之间的交互与通信，Intent 负责对应用中一次操作的动作、动作涉及的数据、附加数据进行描述，根据此 Intent 的描述，Android 会找到对应的组件，将 Intent 传递给调用的组件，并完成组件的调用。

前面的例子说明 Intent 能够实现 Activity 之间的切换和数据的传递，事实上，这仅仅是利用 Intent 实现了对附加信息的传递，Intent 的功能还有很多，如打开网页、打电话、发短信、E-mail 等，采用 Intent 都可以轻松实现。

6.3.1　Intent 的构成

Intent 数据结构中两个最重要的部分是动作和动作对应的数据。典型的动作类型有：MAIN、VIEW、DAIL、EDIT 等，而动作对应的数据一般以 URI 的形式进行表示。

1．Intent 的动作（Action）

在 Intent 中，Action 就是希望触发的动作，当用户指明了一个 Action，执行者就会依照这个动作的指示，接收相关输入，表现对应行为，产生相应的输出。可以通过 setAction()方法进行设置，通过 getAction()方法进行读取。Android 系统中已经为用户准备好了一些表示 Action 操作的常量，如表 6-2 所示。

表 6-2　常用的 Action 常量

Action 名称	AndroidManifest.xml 配置名称	描述
ACTION_MAIN	android.intent.action.MAIN	作为一个程序的入口
ACTION_VIEW	android.intent.action.VIEW	用于数据的显示
ACTION_DIAL	android.intent.action.DIAL	调用电话拨号程序
ACTION_EDIT	android.intent.action.EDIT	用于编辑给定的数据
ACTION_PICK	android.intent.action.PICK	从特定的一组数据中进行数据选择操作

续表

Action 名称	AndroidManifest.xml 配置名称	描述
ACTION_RUN	android.intent.action.RUN	运行数据
ACTION_SEND	android.intent.action.SEND	调用发送短信程序
ACTION_GET_CONTENT	android.intent.action.GET_CONTENT	根据指定 Type 来选择打开操作内容的 Intent
ACTION_CHOOSER	android.intent.action.CHOOSER	创建文件操作选择器

除了上述的常用常量之外，更多的动作请参考 Intent 类的 API，要注意的是，使用这些常量的时候可能要在 AndroidManifest.xml 文件中做相应的配置，这样才会生效。

动作在很大程度上决定了剩下的 Intent 如何构建，特别是数据和附加字段，应该尽可能明确指定动作，并紧密关联到其他 Intent 字段。

2. Intent 动作对应的数据

Intent 动作对应的数据一共有 6 种：数据（Data）、数据类型（Type）、操作类别（Category）、附加信息（Extras）、组件（Component）、标志（Flags）。

1）数据

数据是指作用于动作的数据的 URI 和数据的 MIME 类型。不同的动作有不同的数据规格。例如，如果动作是 ACTION_VIEW，则数据字段是 http:URI，接收活动将被调用以下载和显示 URI 指向的数据。不同的动作对应不同的数据，常见的数据如表 6-3 所示。

表 6-3 常见的动作与数据对应表

操作类型	Data（URI）格式	范例
浏览网页	http://网页地址	http://www.hnist.cn
拨打电话	tel:电话号码	tel:07308648870
发送短信	smsto:短信接收人号码	smsto: 13207304568
查找 SD 卡文件	file:///sdcard/文件或目录	file:///sdcard/aa.jpg
显示地图	geo:坐标,坐标	geo:35.89, 29.6

如果要设置数据，就要借助 android.net.Uri 类完成，参照 API 了解其方法和属性。

2）数据类型

数据类型指要传送数据的 MIME 类型，可以直接通过 setType()方法进行设置，getType()用于读取类型。一般而言，Intent 的数据类型能够根据数据本身进行判定，但是可以通过 setType()方法设置属性，可以强制采用显式指定的类型而不再进行推导。常见 MIME 类型如表 6-4 所示。

表 6-4 常见的几种 MIME 类型

作用	MIME 类型
发送短信	vnd.android-dir/mms-sms
设置图片	image/png
普通文本	text/plain
设置音乐	audio/mp3

3）操作类别

有时通过 Action，配合数据或数据类别，就可以准确地表达出一个完整的意图，但加一些约束在里面才能够更精准。在 Android 中，所有应用的主 Activity 都需要一个操作类别为 CATEGORY_LAUNCHER，Action 为 ACTION_MAIN 的 Intent。

这个选项指定了将要执行的 Action 的其他额外约束，可以通过 addCategory()方法设置多个类别，removeCategory()方法可删除一个之前添加的种类，getCategories()方法可获取 Intent 对象中的所有种类。常见的 Category 如表 6-5 所示。

表 6-5　常见的几种 Category

Category 名称	AndroidManifest.xml 配置名称	描述
CATEGORY_LAUNCHER	android.intent.category.LAUNCHER	表示此程序显示在应用程序列表中
CATEGORY_HOME	android.intent.category.HOME	显示主桌面，即开机时的第一个界面
CATEGORY_PREFERENCE	android.intent.category.PREFERENCE	运行后将出现一个选择面板
CATEGORY_BROWSABLE	android.intent.category.BROWSABLE	显示一张图片、E-mail 信息
CATEGORY_DEFAULT	android.intent.category.DEFAULT	设置一个操作的默认执行
CATEGORY_OPENABLE	android.intent.category.OPENABLE	当 Action 设置为 GET_CONTEN 时用于打开指定的 URI
CATEGORY_GADGET		设置活动可以嵌入另一个活动

4）附加信息

附加信息传递的是一组键值对，可以使用 putExtra()方法进行设置，主要的功能是传递数据所需要的一些额外的操作信息。使用附加信息可以为组件提供扩展信息，例如，执行“发送电子邮件”这个动作时，可以将电子邮件的标题、正文等保存在附加信息中，再传给电子邮件发送组件。常见的附加消息如表 6-6 所示。

表 6-6　常见的几种附加消息

操作数据	附加信息	作用
短信操作	sms_body	表示要发送短信的内容
彩信操作	Intent.EXTRA_STREAM	设置发送彩信的内容
指定接收人	Intent.EXTRA_BCC	指定接收 E-mail 或信息的接收人
Email 收件人	Intent.EXTRA_EMAIL	用于指定 E-mail 的接收者，接收一个数组
Email 标题	Intent.EXTRA_SUBJECT	用于指定 E-mail 邮件的标题
Email 内容	Intent.EXTRA_TEXT	用于设置邮件内容

5）组件

组件指明了将要处理的 Activity 程序，通常 Android 会根据 Intent 中包含的其他属性的信息，如 Action、Data/Type、Category 等进行查找，最终找到一个与之匹配的目标组件。但是，如果已经指定了 component 属性，则可直接使用它指定的组件，而不再执行上述查找过程。所有的组件信息都被封装在一个 ComponentName 对象之中，这些组件都必须在 AndroidManifest.xml 文件的“<application>”中注册。

6）标志

在 android.content.Intent 中一共定义了 20 种不同的标志，标志是一个整型数，由一些列的标志位构成，这些标志是用来指明运行模式的。例如，指示 Android 系统如何启动一个活动和启动之后如何对待它。这些标志都定义在 Intent 类中，可以通过使用 addFlags()方法进行增加。

6.3.2　Intent 常用用法示例

Android 系统之中提供了多种 Intent 动作，可以在 Intent 中指定程序要执行的动作（如 view、edit、dial），以及程序执行该动作时所需要的数据，然后调用 startActivity()方法，Android 系统会自动寻找最符合指定要求的应用程序，并按要求执行该程序。

Intent 常用的一些操作如下。

（1）从 Google 中搜索内容，代码如下：

```
Intent intent = new Intent();                          //实例化 Intent
intent.setAction(Intent.ACTION_WEB_SEARCH);            //指定动作
intent.putExtra(SearchManager.QUERY,"searchString")    //设置数据
startActivity(intent);                                 //启动此应用
```

（2）浏览网页，代码如下：

```
Uri uri =Uri.parse("http://www.hnist.cn");                    //定义 URI 数据
Intent it = new Intent(Intent.ACTION_VIEW,uri);//实例化 Intent，指定动作和数据
startActivity(it);                                            //启动此应用
```

（3）显示地图，代码如下：

```
Uri uri = Uri.parse("geo:38.899533,-77.036476");
Intent it = new Intent(Intent.Action_VIEW,uri);
startActivity(it);
```

（4）路径规划，代码如下：

```
Uri uri =Uri.parse("http://maps.google.com/maps?f=dsaddr=startLat%20startLng&
    daddr=endLat%20endLng&hl=en");
Intent it = new Intent(Intent.ACTION_VIEW, uri);
startActivity(it);
```

（5）拨打电话，代码如下：

```
Uri uri =Uri.parse("tel:07308748870");
Intent it = new Intent(Intent.ACTION_DIAL,uri);
startActivity(it);
```

（6）调用发短信的程序，代码如下：

```
Intent it = new Intent(Intent.ACTION_VIEW);
it.putExtra("sms_body", "TheSMS text");
it.setType("vnd.android-dir/mms-sms");
startActivity(it);
```

（7）发送 E-mail，代码如下：

```
Uri uri =Uri.parse("mailto:xxx@abc.com");
Intent it = new Intent(Intent.ACTION_SENDTO, uri);
startActivity(it);
```

或者

```
Intent it = new Intent(Intent.ACTION_SEND);
it.putExtra(Intent.EXTRA_EMAIL,"me@abc.com");
it.putExtra(Intent.EXTRA_TEXT, "The email body text");
it.setType("text/plain");
startActivity(Intent.createChooser(it,"Choose Email Client"));
```

或者

```
Intent it=new Intent(Intent.ACTION_SEND);
String[] tos={"me@abc.com"};
String[]ccs={"you@abc.com"};
it.putExtra(Intent.EXTRA_EMAIL, tos);
it.putExtra(Intent.EXTRA_CC, ccs);
it.putExtra(Intent.EXTRA_TEXT, "Theemail body text");
it.putExtra(Intent.EXTRA_SUBJECT, "The email subject text");
it.setType("message/rfc822");
startActivity(Intent.createChooser(it,"Choose Email Client"));
```

发送附件，代码如下：

```
Intent it = new Intent(Intent.ACTION_SEND);
it.putExtra(Intent.EXTRA_SUBJECT, "Theemail subject text");
it.putExtra(Intent.EXTRA_STREAM,"file:///sdcard/mysong.mp3");
sendIntent.setType("audio/mp3");
startActivity(Intent.createChooser(it,"Choose Email Client"));
```

（8）播放多媒体，代码如下：

```
Uri uri=Uri.withAppendedPath(MediaStore.Audio.Media.INTERNAL_CONTENT_URI, "1");
Intent it = new Intent(Intent.ACTION_VIEW,uri);
startActivity(it);
```

（9）打开照相机，代码如下：

```
Intent intent = new Intent(MediaStore.ACTION_IMAGE_CAPTURE);
startActivityForResult(intent, 1);
```

（10）打开联系人列表，代码如下：

方式一：

```
Intent i = new Intent();
i.setAction(Intent.ACTION_GET_CONTENT);
i.setType("vnd.android.cursor.item/phone");
startActivityForResult(i, REQUEST_TEXT);
```

方式二：

```
Uri uri = Uri.parse("content://contacts/people");
Intent it = new Intent(Intent.ACTION_PICK, uri);
startActivityForResult(it, REQUEST_TEXT);
```

要注意的是，有些动作可能要在 AndroidManifest.xml 文件中申明其权限，Android 常用权限如表 6-7 所示。

表 6-7 常见的 AndroidManifest.xml 权限设置

设置属性（前缀 android.permission）	描述
SEND_SMS	发短信
READ_SMS	读短信
CALL_PHONE	打电话
SET_TIME_ZONE	设置时间
INTERNET	访问网络
CHANGE_NETWORK_STATE	改变网络状态
CHANGE_WIFI_STATE	改变 Wi-Fi 状态
CAMERA	照相机
SET_WALLPAPER	设置壁纸
VIBRATE	允许震动
ACCESS_FINE_LOCATION	访问位置信息
BLUETOOTH	访问蓝牙设备
READ_CALENDAR	读取日历
READ_CONTACTS	读取联系人
WRITE_CONTACTS	修改联系人
RECORD_AUDIO	录音
MOUNT_UNMOUNT_FILESYSTEMS	在 SD 卡中创建或删除文件
WRITE_EXTERNAL_STORAGE	向 SD 卡中写入数据
READ_PHONE_STATE	读取电话状态

6.3.3 Intent 操作实例

在 Android 系统中操作 Intent 的一般步骤如下：指定数据，实例化 Intent，指定 Action，设置数据，启动 Activity。

实例 6-3：Intent 组件操作举例

新建一个项目，项目命名为 Exam6_3，利用 Intent 编程实现浏览网页和拨打电话的功能。

（1）新建布局管理文件 activity_main.xml，代码如下：

```
…
    <EditText
        android:id="@+id/tel"
        android:layout_width="match_parent"
        android:layout_height="wrap_content"
        android:selectAllOnFocus="true"
        android:text="输入电话号码或网址"/>
    <Button
        android:id="@+id/telbut"
        android:layout_width="match_parent"
        android:layout_height="wrap_content"
        android:text="拨打电话"/>
    <Button
        android:id="@+id/netbut"
        android:layout_width="match_parent"
        android:layout_height="wrap_content"
        android:text="打开网页"/>
</LinearLayout>
```

（2）新建 Activity 文件 MainActivity.java，代码如下：

```
…
import android.content.Intent;
import android.net.Uri;
public class MainActivity extends AppCompatActivity{
    private Button telbut = null ;
    private Button netbut = null ;
    private EditText tel = null ;
    public void onCreate(Bundle savedInstanceState) {
        super.onCreate(savedInstanceState);
        super.setContentView(R.layout.activity_main);
        this.telbut = (Button) super.findViewById(R.id.telbut) ;
        this.netbut = (Button) super.findViewById(R.id.netbut) ;
        this.tel = (EditText) super.findViewById(R.id.tel) ;
        this.netbut.setOnClickListener(new OnClickListenernet()); //定义事件
        this.telbut.setOnClickListener(new OnClickListenertel());}//定义事件
    private class OnClickListenernet implements OnClickListener {
        public void onClick(View view) {
            String netaddress = MainActivity.this.tel.getText().toString() ;
            Uri uri = Uri.parse("http://" + netaddress) ;       //指定数据
            Intent it = new Intent() ;                          //实例化 Intent
            it.setAction(Intent.ACTION_VIEW);                   //指定 Action
            it.setData(uri) ;                                   //设置数据
```

```
            MainActivity.this.startActivity(it);    }    }    //启动 Activity
    private class OnClickListenertel implements OnClickListener {
        public void onClick(View view) {
            String tel = MainActivity.this.tel.getText().toString() ;
            Uri uri = Uri.parse("tel:" + tel) ;                  //指定数据
            Intent it = new Intent() ;                           //实例化 Intent
            it.setAction(Intent.ACTION_DIAL);                    //指定 Action
            it.setData(uri) ;                                    //设置数据
            MainActivity.this.startActivity(it); } }}            //启动 Activity
```

（3）要拨打电话，必须在修改 AndroidManifest.xml 文件后，才能生效，在</application>与</manifest>之间加入如下代码：

```
    <uses-permission  android:name="android.permission.CALL_PHONE"/>
```

保存所有文件，运行该项目，进入界面，在输入框中输入“073086848870”，单击“拨打电话”按钮，可以拨打电话，返回后在输入框中输入“www.hnist.cn”，单击“打开网页”按钮，可以打开相应网页，如图 6.3 所示。

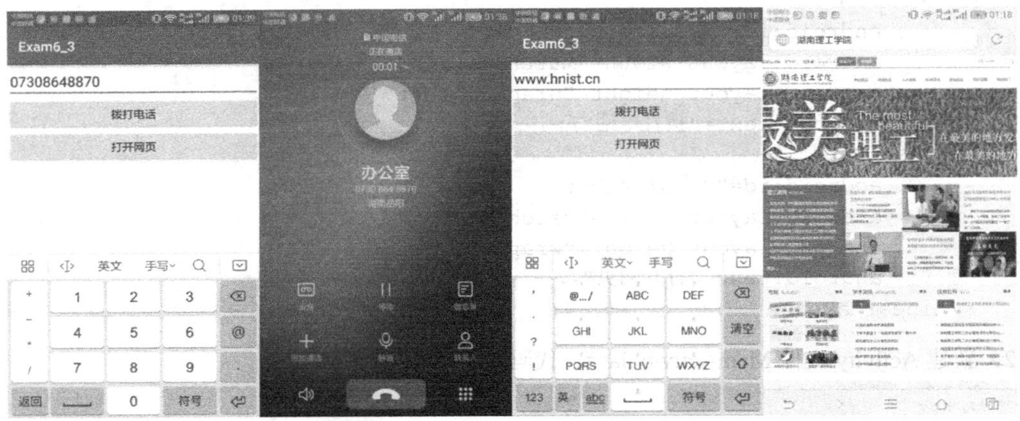

图 6.3　Exam6_3 运行结果

这里只给出了打开网页和拨打电话的代码，读者可以自己参照 6.3.2 给出的相关代码实现其他 Intent 操作。

6.4　Activity 的生命周期

Activity 是整个 Android 平台的基本组成，系统中的 Activity 被一个 Activity 栈管理，同一时刻只有最上端的那个 Activity 是处于运行状态的，当一个新的 Activity 启动时，将被放置到栈顶，处于运行状态，前一个 Activity 保留在栈中，不再放到前台，直到新的 Activity 退出为止。一个 Activity 生命周期包含以下 3 个阶段。

（1）运行状态（Running State）：在屏幕的前台（Activity 栈顶），此时 Activity 程序显示在屏幕前台，并且具有焦点，可以和用户的操作动作进行交互，例如，向用户提供信息、捕获用户单击按钮的事件并做出处理。

（2）暂停状态（Paused State）：一个 Activity 失去焦点，但是依然可见。一个暂停状态的 Activity 依然保持活力（保持所有的状态、成员信息，和窗口管理器保持连接），但是不可以与其进行交互，在系统内存不够的时候将被结束。

（3）停止状态（Stopped State）：一个 Activity 被另外的 Activity 完全覆盖。它依然保持所有状态和成员信息，但是它不再可见，当系统内存不够的时候将被结束。

如果一个 Activity 处于 Paused 或者 Stopped 状态，则系统可以将该 Activity 从内存中删除，Android 系统采用两种方式进行删除，要么要求该 Activity 结束，要么直接结束它的进程。当该 Activity 再次显示给用户时，它必须重新开始和重置前面的状态。

当 Activity 程序在不同状态之间进行切换时，可以通过 Activity 类提供的方法来进行操作，常用的方法如表 6-8 所示。

表 6-8　Activity 程序的生命周期控制方法

方法	可关闭否	描述
protected void onCreate(Bundle savedInstanceState)	不可以	当启动新的 Activity 的时候被调用
protected void onRestart()	不可以	当 Activity 对用户即将可见时调用
protected void onStart()	不可以	重新启动 Activity 时调用（此方法即重启留在缓存中的 Activity）
protected void onResume()	不可以	当 Activity 界面可与用户交互时调用
protected void onPause()	可以	当系统要启动一个其他的 Activity 时调用，用于保存当前数据
protected void onStop()	可以	该 Activity 已经不可见时调用
protected void onDestroy()	可以	当 Activity 被 finish 或手机内存不足被销毁的时候调用

在一个 Activity 正常启动的过程中，它们被调用的顺序是 onCreate→onStart→onResume，一个 Activity 被另一个 Activity 关掉时的顺序是 onPause→onStop→onDestroy，这就是一个完整的生命周期。

onCreate：在这里创建界面，做一些数据的初始化工作。

onStart：此时变成用户可见但是还不能进行交互。

onResume：此时可以和用户进行交互。

onPause：此时用户可见但不可交互，此时程序的优先级降低，有可能被系统收回。在这里保存的数据应该早在 onResume 中就读出来了。要注意的是，在这个方法里操作的时间要短，因为下一个 Activity 不会等到这个方法完成才启动。

onStop：此时变得不可见，被下一个 Activity 完全覆盖了。

onDestroy：这是 Activity 被关掉前最后一个被调用的方法，可能的原因是调用了 finish()方法或者系统为了节省空间而将它暂时关掉。

图 6.4 显示了 Activity 的重要状态转换。

整个生命周期，从 onCreate()开始到 onDestroy()结束。Activity 在 onCreate()中设置所有的“全局”状态，在 onDestory()中释放所有的资源。

可以看见的生命周期，从 onStart()开始到 onStop()结束。在这段时间内，可以看到 Activity 在屏幕上，尽管有可能不在前台，不能和用户交互。在这两个接口之间，需要保持显示给用户的 UI 数据和资源等。

前台的生命周期，从 onResume()开始到 onPause()结束。在这段时间里，该 Activity 处于所有 Activity 的最前面，和用户进行交互。Activity 可以在 Resumed 和 Paused 状态之间进行切换。

所有的 Activity 都需要实现 onCreate(Bundle)方法以初始化设置，大部分 Activity 需要实现 onPause()方法以提交更改过的数据，当前大部分的 Activity 也需要实现 onFreeze()接口，以便恢复在 onCreate(Bundle)中设置的状态。

从流程图中可以看出，无论现在是 onPause 状态还是 onStop 状态，当系统的内存不足时，都会使该 Activity 结束，发生这种情况时一些重要的数据和状态可能还来不及保存，Google 专门提供了一个回调函数——onSaveInstanceState(Bundle)，通过该函数，可以把这些数据在销毁前进行保存，然后供onCreate()或 onRestoreInstanceState()重新读出。

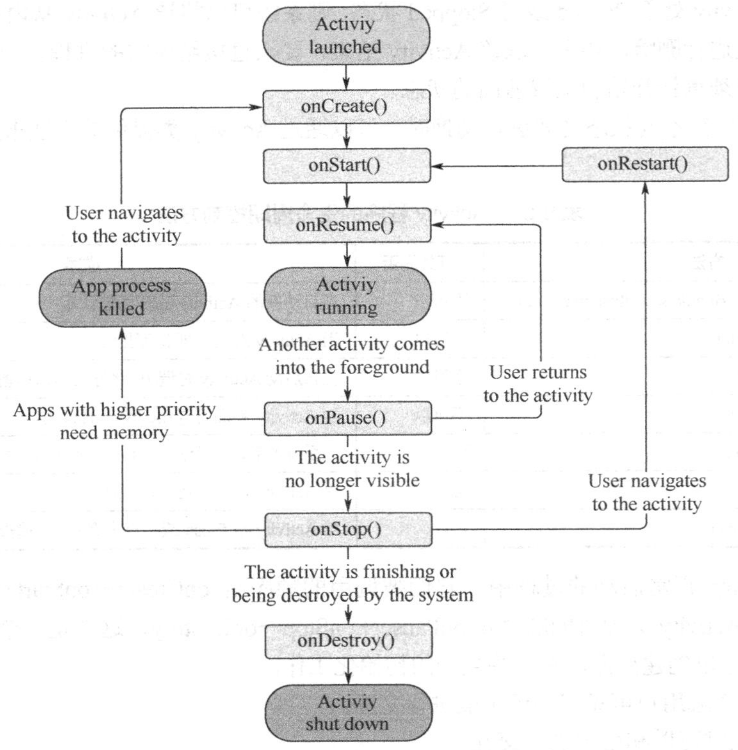

图 6.4 Activity 程序的生命周期图

下面通过一个具体的例子来说明 Activity 的生命周期。两个 Activity 分别命名为 Activity1 和 Activity2，其中 Activity1 是默认启动的，界面中有一个按钮，单击它会跳转到 Activity2，Activity2 中也有一个按钮，单击它会跳转到 Activity1，如图 6.5 所示。

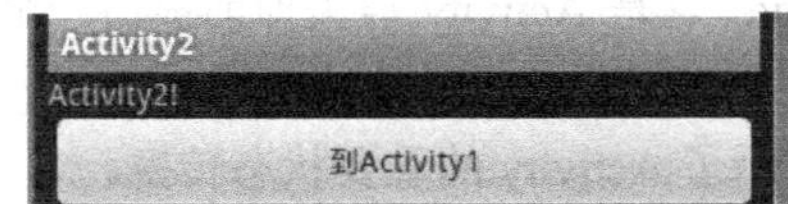

图 6.5 说明 Activity 程序生命周期的图

（1）启动该程序，Activity1 将按顺序调用 onCreate→onStart→onResume 方法。

（2）当单击 Activity1 的按钮跳转到 Activity2 时，执行的是 Activity1 的 onPause 方法，这样可以将 Activity1 程序中未完成的数据保存起来或终止某些操作，Activity2 按顺序调用 onCreate→onStart→onResume 方法，这三个方法过后，Activity2 会完全覆盖 Activity1，Activity1 将不可见，因此 Activity1 会调用 onStop 方法（注意，这里如果不覆盖 Activity1，则不会调用 onStop 方法）。

（3）当单击 Activity2 按钮跳转到 Activity1 时，与（2）执行情况完全类似。

（4）在 Activity2 中单击手机上的“返回”按钮到 Activity1 时，首先执行的是 Activity2 的 onPause 方法，因为是返回操作，所以会读取缓存中的 Activity1，执行 onRestart 方法，Activity1 按顺序调用 onStart→onResume 方法；Activity2 执行 onStop 方法，最后会执行 Activity2 的 onDestroy 方法，因为 Android 的缓存是不可逆的，只能后退而不能向前。一旦程序关掉了，要再次启动就必须重新调用：onCreate→onStart()→onResume()。

6.5 Android 中的消息处理机制

Windows 程序是采用消息驱动的模式，并且有全局的消息循环系统。Google 参考了 Windows 的消息循环机制，在 Android 系统中通过 Looper、Handler 来实现消息处理机制。Android 的消息处理是针对线程的，每个线程都可以有自己的消息队列和消息循环。

6.5.1 消息处理机制基础

要了解 Android 的消息处理机制，必须了解以下几个概念。

（1）消息类：Message，理解为线程间通信的数据单元，主要功能是进行消息的封装，同时指定消息的操作形式。

（2）消息队列类：Message Queue，用来存放通过 Handler 发布的消息，按照先进先出的顺序执行。

（3）消息操作类：Handler，它是 Message 的主要处理者，Message 对象封装了所有的消息，而这些消息的处理需要用 Handler 类完成。

（4）消息通道类：Looper，是 Message Queue 和 Handler 之间的桥梁，循环取出 Message Queue 中的 Message，交给相应的 Handler 进行处理。

（5）线程：UI thread 通常就是 main thread，每一个线程中可含有一个 Looper 对象以及一个 Message Queue 数据结构。在应用程序中，可以定义 Handler 的子类别来接收 Looper 所送出的消息。

整个消息处理的大概流程如下。

（1）包装 Message 对象（指定 Handler、回调函数和携带数据等）。

（2）通过 Handler 的 sendMessage()等方法将 Message 发送出去。

（3）在 Handler 的处理方法中将 Message 添加到 Handler 绑定的 Looper 的 MessageQueue 上。

（4）Looper 的 loop()方法通过循环不断地从 MessageQueue 中提取 Message 进行处理，并移除处理完毕的 Message。

（5）调用 Message 绑定的 Handler 对象的 dispatchMessage()方法完成对消息的处理。

Looper 本身提供的就是一个消息队列的集合，而每个消息都可以通过 Handler 增加和取出，而操作 Handler 的对象就是主线程和子线程。

大致可以这样理解：假设一个隧道就是一个消息队列（Looper），那么其中的每一部汽车就是一个消息（Message），隧道里的管理者（Handle）保证秩序，告知哪辆车进入队列，哪辆车离开队列，确保车能实现先进先出。

1. 消息类：Message

它的主要功能是进行消息的封装，可以定义为一个包含任意类型的描述数据的对象，此对象可以发送给 Handler。对象包含两个额外的 int 字段和一个额外的对象字段，这样可以在很多情况下不用分配。其常用的变量和方法如表 6-9 所示。

表 6-9　Message 类常用的变量和方法

变量或方法	描述
public int what	变量，用于定义此 Message 属于何种操作
public Object obj	变量，用于定义此 Message 传递的信息数据
public int arg1	变量，传递一些整型数据时使用，一般很少使用
public int arg2	变量，传递一些整型数据时使用，一般很少使用
public Messenger replyTo	指定此 Message 发送到哪个 Messenger 对象

续表

变量或方法	描述
public Runnable getCallback()	获取回调对象，此对象会在 Message 处理时执行
public Bundle getData()	获取附加在此事件上的任意数据的 Bundle 对象
public Handler getTarget()	取得操作此消息的 Handler 对象
public long getWhen()	返回此消息的传输时间，以毫秒为单位
public void sendToTarget()	向 Handler 发送此消息
public void setTarget(Handler target)	设置将接收此消息的 Handler 对象
public String toString()	返回一个 Message 对象的描述信息
public static Message obtain(Handler h,int what,int arg1,int arg2,Object obj)	从全局池中分配的一个 Message 对象，可不带参数

其中，参数 h 表示设置的 target 值；what 表示设置的 what 值；arg1 表示设置的 arg1 值；arg2 表示设置的 arg2 值；obj 表示设置的 obj 值。

获取 Message 对象的最好方法是调用 Message.obtain()或者 Handler.obtainMessage()，以节省资源。如果一个 Message 只需要携带简单的 int 型信息，则应优先使用 Message.arg1 和 Message.arg2 属性来传递信息，这比使用 Bundle 更节省内存，尽可能使用 Message.what 来标识信息，以便用不同方式处理 Message。

要在 Android 的 Java 程序中使用 Message 类，必须在程序中使用下面的语句。

```
import android.os.Message;          //导入 os. Message 类
```

2. 消息操作类：Handler

Handler 主要用来与 Looper 进行沟通，增加新消息到 Message Queue 中，或者接收 Looper 所送来的消息。其常用的方法如表 6-10 所示。

表 6-10　Handler 类常用的方法

方法	描述
public Handler()	创建一个新的 Handler 实例
public Handler(Looper looper)	使用指定的队列创建一个新的 Handler 实例
public Handler(Handler.Callback callback)	通过回调对象创建一个新的 Handler 实例
public final Message obtainMessage(int what, Object obj)	获得一个 Message 对象
public final Message obtainMessage(int what, int arg1, int arg2, Object obj)	获得一个带指定参数的 Message 对象
public void handleMessage(Message msg)	处理消息的方法，子类要覆写此方法
public final boolean hasMessages(int what)	判断是否有指定的 Message
public final boolean hasMessages(int what, Object object)	判断是否有指定的 Message
public final void removeMessages(int what)	删除指定的 Message
public final void removeMessages(int what, Object object)	删除指定的 Message
public final boolean sendEmptyMessage(int what)	发送一个空消息
public final boolean sendEmptyMessageAtTime(int what, long uptimeMillis)	在指定的日期时间发送消息
public final boolean sendEmptyMessageDelayed(int what, long delayMillis)	等待指定的时间之后发送消息
public final boolean sendMessage(Message msg)	发送消息

Handler 的作用是把消息加入特定的 Looper 消息队列，并分发和处理该消息队列中的消息。若要使一个线程把消息放入主线程的消息队列，则可以使用 Handler 对象，通过调用 Handler 主线程的 sendMessage 接口，把消息队列放入主线程的消息队列，并使用该 Handler 的 handleMessage()来处理消息。

要在 Android 程序中使用 Handler 类，必须在程序中使用下面的语句。

```
import android.os.Handler;                    //导入 os.Handler 类
```

3．消息通道：Looper

它主要用来管理线程里的 Message Queue，在一个 Activity 类之中，会自动启动 Looper 对象，在一个用户自定义的类中，需要用户在手工调用 Looper 类中的若干方法之后再正常地启动 Looper 对象。

其常用方法如表 6-11 所示。

表 6-11 Looper 类的常用方法

方法	描述
public static final synchronized Looper getMainLooper()	取得主线程
public static final Looper myLooper()	返回当前的线程
public static final void prepare()	初始化 Looper 对象
public static final void prepareMainLooper()	初始化主线程的 Looper 对象
public void quit()	消息队列结束时调用
public static void loop()	启动消息队列
public static MessageQueue myQueue()	返回当前的 MessageQueue
public Thread getThread()	获得 Looper 分配的线程
public String toString ()	成为字符串

要在 Android 程序中使用 Looper 类，必须在程序中使用下面的语句。

```
import android.os.Looper;                    //导入 os.Looper 类
```

6.5.2 一个简单的消息处理实例

实例 6-4：简单的消息处理实例

新建一个项目，项目命名为 Exam6_4，利用 Message 类、Looper 类和 Handler 类实现同线程内不同组件间的消息传递。

（1）新建布局管理文件 activity_main.xml，代码如下：

```
...
<TextView
    android:id="@+id/msg"
    android:layout_width="match_parent"
    android:layout_height="wrap_content" />
<EditText
    android:id="@+id/edit"
    android:layout_width="match_parent"
    android:layout_height="wrap_content"
    android:text="等待接收数据" />
<Button
    android:id="@+id/start"
    android:layout_width="match_parent"
    android:layout_height="wrap_content"
    android:text="传送数据到其他组件" />
</LinearLayout>
```

（2）新建 Activity 文件 MainActivity.java，代码如下：

```
...
```

```
import android.os.Handler;                    //导入 os.Handler 类
import android.os.Looper;                     //导入 os.Looper 类
import android.os.Message;                    //导入 os.Message 类
public class MainActivity extends AppCompatActivity{
 private TextView showmsg;                                  //定义文本显示组件
 private EditText edit;                                     //定义按钮组件
 private Button sendbut;                                    //定义按钮组件
 private static final int SET = 1 ;                         //定义 what 操作码
 public void onCreate(Bundle savedInstanceState) {
     super.onCreate(savedInstanceState);
     super.setContentView(R.layout.activity_main);
     this.showmsg = (TextView) super.findViewById(R.id.msg);   //取得文本组件
     this.sendbut = (Button) super.findViewById(R.id.start);   //取得按钮组件
     this.edit = (EditText) super.findViewById(R.id.edit);//取得编辑框组件
     this.sendbut.setOnClickListener(new OnClickListenerImpl());}//设置事件
 private class OnClickListenerImpl implements OnClickListener {
     public void onClick(View view) {
         switch (view.getId()) {                                //判断操作的组件 ID
         case R.id.start:                                       //表示按钮操作
             Looper loop = Looper.myLooper();                   //取得当前的线程
             MyHandler hand = new MyHandler(loop);              //构造一个 Handler
             hand.removeMessages(0) ;                           //清空所有的消息队列
             String data = "单击按钮后传送的数据";              //设置要发送的数据
             Message mymsg = hand.obtainMessage(SET,1,1,data);//获得 Message 对象
             hand.sendMessage(mymsg);                           //发送消息
             break;}}}
 private class MyHandler extends Handler {                      //定义类继承 Handler
     public MyHandler(Looper looper) {                          //接收 Looper
         super(looper);     }                                   //调用父类构造
     public void handleMessage(Message msg) {                   //处理消息
         switch (msg.what) {                                    //判断操作形式
         case 1:
          MainActivity.this.showmsg.setText(msg.obj.toString());//设置内容
          MainActivity.this.edit.setText(msg.obj.toString());}}}}
```

保存所有文件，运行该项目，进入运行界面，如图 6.6 所示。

图 6.6　简单的消息处理

程序启动时，当前线程(即主线程)已产生了一个 Looper 对象，并且有了一个 MessageQueue 数据结构。

语句 Looper loop = Looper.myLooper()；通过调用 Looper 类的 myLooper()函数，以取得目前线程中的 Looper 对象。

语句 MyHandler hand = new MyHandler(loop)；构造一个 MyHandler 对象来与 Looper 进行沟通。Activity 等对象可以由 MyHandler 对象将消息传给 Looper，然后放入 MessageQueue；MyHandler 对象也扮演着 Listener 的角色，可接收 Looper 对象送来的消息。

语句 Message mymsg = hand.obtainMessage(SET,1,1,data);用于构造一个 Message 对象，并将数据存入对象，形成消息。

语句 hand.sendMessage(mymsg); 通过定义的 MyHandler 对象 hand 将消息 mymsg 传给 Looper，然后放入 MessageQueue。

Looper 对象看到 MessageQueue 中有消息 mymsg，就将它广播出去，hand 对象接到此消息时，会呼叫 handleMessage()函数来处理，于是会按照要求输出指定的数据到文本组件和编辑框组件中。

6.5.3 线程基础知识

当一个 Android 应用程序启动时，系统会启动一个 Linux 进程（Process），并在此进程中开启一个主线程。主线程是应用与界面交互的地方，一般不能阻塞，否则可能会导致进程的关闭。但是运行一些比较费时的程序时，可能会导致主线程阻塞，后续组件不能执行，Android 系统就会自动地把整个进程关闭。因此，需要在进程中增加线程，让一些比较费时的组件运行在其他线程中，从而保证主线程畅通，而不至于让系统关闭应用程序的整个进程。

1. 进程概念

进程是表示资源分配的基本单位，是调度运行的基本单位。例如，用户运行自己的程序，系统会先创建一个进程，并为它分配资源，包括各种内存空间、磁盘空间、I/O 设备等。然后，把该进程放入进程的就绪队列。进程调度程序选中它，为它分配 CPU 以及其他有关资源，该进程才能真正运行。所以，进程是系统中的并发执行的单位。

在 Mac、Windows NT 等采用微内核结构的操作系统中，进程的功能发生了变化，它只是资源分配的单位，而不再是调度运行的单位。在微内核系统中，真正调度运行的基本单位是线程。因此，实现并发功能的单位是线程。

2. 线程概念

线程是进程中执行运算的最小单位，即执行处理器调度的基本单位。如果把进程理解为一个任务，那么线程表示完成该任务的许多可能的子任务。线程可以在处理器上独立调度执行，这样，在多处理器环境下就允许几个线程各自在单独的处理器上进行。操作系统提供线程就是为了方便而有效地实现这种并发性。

3. 进程和线程的关系

（1）一个线程只能属于一个进程，而一个进程可以有多个线程。
（2）资源分配给进程，同一进程的所有线程共享该进程的所有资源。
（3）处理器分给线程，即真正在处理器上运行的是线程。
（4）线程在执行过程中需要协作同步。线程间要利用消息通信的办法实现同步。
（5）二者均可并发执行。

4. 为什么要用多线程

移动开发的原则是不让用户等，要及时响应用户的需要就要求 Android 中的 Main 线程的事件处理不能太耗时，要将所有可能耗时的操作都放到其他线程中处理。

一般来说，Activity 的 onCreate()、onStart()、onResume()方法的执行时间决定了应用首页打开的时间，这里要尽量把不必要的操作放到其他线程中处理。

当用户与应用交互时，一般分为同步和异步两种情况。

（1）同步，需要等待返回结果。例如，用户单击了登录按钮，需要等待服务端返回结果，此时就需要有一个进度条来提示用户。

（2）异步，不需要等待返回结果。例如，收藏功能，单击收藏按钮，是否成功执行完成通知用户即可，此功能就可以通过异步处理来实现。

5. Android 中线程的创建

Android 中线程的创建一般有以下两种常见方法。

（1）在 Android 中实现 Runnable 类并覆写 Run()方法创建线程，其实该线程和 Android 的 Activity 是同一个线程，而不是单独的线程。

```
Runnable updateThread=new Runnable(){
   public void run(){…}}
```

（2）使用 Android 系统框架提供的 HandlerThread 创建新的线程。这是一个真正的线程。

① 创建一个 MyHandler 类并继承于 Handler 类，并在 MyHandler 的构造函数中使用父类的构造函数来接收线程的 Looper，覆写 handleMessage 来接收消息。

```
class MyHandler extends Handler{
    public MyHandler(Looper looper)
        {super(looper); }
public void handleMessage(Messagemsg){
     super.handleMessage(msg);…}}
```

② 创建一个 HandlerThread 对象，并启动该线程。

```
HanderThread myHandlerThread = new HanderThread("ThreadName");
myHandlerThread.Start();
```

③ 实例化 MyHandler 并把 myHandlerThread 的 Looper 对象传递过去。

```
MyHandler myHandler = new MyHandler(myHandlerThread.getLooper());
```

④ 创建一个 myHandler 的消息对象，并把消息传递给指定的线程。

```
Message msg = myHandler.obtainMessage();
msg.sendToTarget();
```

6. Android 主线程与子线程间的消息传递

当子线程要发送 Message 给主线程时，首先要为此子线程创建一个 Handler 类对象，由 Handler 调用相应的方法，将需要发送的 Message 发送到 MessageQueue(消息队列)中，当 Looper 发现 MessageQueue 中有未处理的消息时，就会将此消息广播出去，当主线程的 Handler 接收到此 Message 时，就会调用相应的方法来处理这条信息，完成主界面的更新。

当子线程要发送 Message 给主线程时，有以下几个关键步骤。

（1）创建两个 Handler 类对象，一个是主线程，一个是子线程。

例如：

```
private Handler mainHandler, subHandler;        //定义两个 Handler 对象
```

（2）定义子线程，在子线程中完成如下操作：初始化 Looper、定义子线程的 Handler 对象、覆写 handleMessage 方法、发送消息，启动子线程的消息队列。例如：

```
class ChildThread implements Runnable {        //定义子线程类
    public void run() {
        Looper.prepare();                      //初始化 Looper
```

```
        this.subHandler = new Handler() {   //子线程的 Handler 对象
            public void handleMessage(Message msg) {//覆写 handleMessage
                switch (msg.what) {             //判断 what 操作
                case SETCHILD:                  //主线程发送给子线程的信息
                    System.out.println("Main to Child Message : "+ msg.obj);
            Message toMain= this.mainHandler.obtainMessage();//创建 Message
                    toMain.obj="这是子线程发给主线程消息: "+ super.getLooper().
                        getThread().getName();//设置发送消息的内容
                toMain.what = SETMAIN;          //设置主线程操作的状态码
                this.mainHandler.sendMessage(toMain); //发送消息
                break;  }}};
        Looper.loop();}}                            //启动该线程的消息队列
```

（3）定义主线程，在主线程中完成如下操作：定义主线程的 Handler 对象、覆写 handleMessage 方法、接收消息、启动子线程。例如：

```
    this.mainHandler = new Handler() {              //主线程的 Handler 对象
            public void handleMessage(Message msg) {     //消息处理
                switch (msg.what) {                      //判断 Message 类型
                case SETMAIN:                        //设置主线程的操作类
                    MyThreadDemo.this.msg.setText("主线程接收数据: "
                            + msg.obj.toString());   //设置文本内容
                    break;   }}};
        new Thread(new ChildThread(), "Child Thread").start(); //启动子线程
```

实例 6-5：线程之间消息的传递实例

新建一个项目，项目命名为 Exam6_5，利用 Message 类、Looper 类和 Handler 类将子线程传递消息给主线程。

（1）新建布局管理文件 activity_main.xml，代码如下：

```
    …
    <TextView
        android:layout_width="match_parent"
        android:layout_height="wrap_content"
        android:id="@+id/msg"
        android:text="等待子线程发送消息。"/>
    <Button
        android:layout_width="match_parent"
        android:layout_height="wrap_content"
        android:id="@+id/start"
        android:text="接收子线程消息同时发送消息给子线程"/>
    </LinearLayout>
```

（2）新建 Activity 文件 MainActivity.java，代码如下：

```
    …
    import android.os.Handler;          //导入 os.Handler 类
    import android.os.Looper;           //导入 os.Looper 类
    import android.os.Message;          //导入 os.Message 类
    public class MainActivity extends AppCompatActivity{
```

```
    public static final int ToMain = 1;              //设置一个 what 标记
    public static final int ToSub = 2;               //设置一个 what 标记
    private Handler mainHandler, subHandler;         //定义 Handler 对象
    private TextView mymsg;                          //文本显示组件
    private Button mybut;                        //按钮组件
    class subThread implements Runnable {        //创建子线程类
        public void run() {
            Looper.prepare();                    //初始化 Looper
            MainActivity.this.subHandler = new Handler() {//定义子线程的 Handler 对象
              public void handleMessage(Message msg) {//覆写 handleMessage
                switch (msg.what) {              //判断 what 操作
                case ToSub:                      //主线程发送给子线程的信息
                System.out.println("Main Child Message: "+ msg.obj);//打印消息
              Message toMain = MainActivity.this.mainHandler.obtainMessage();//创建消息
                    toMain.obj = "This is a message from Sub"; //设置显示文字
                    toMain.what = ToMain;            //设置主线程操作的状态码
                    MainActivity.this.mainHandler.sendMessage(toMain); //发送
                     break; }}};
            Looper.loop(); }}                        //启动该线程的消息队列
    public void onCreate(Bundle savedInstanceState) {
        super.onCreate(savedInstanceState);
        super.setContentView(R.layout.activity_main);
        this.mymsg = (TextView) super.findViewById(R.id.msg);
        this.mybut = (Button) super.findViewById(R.id.start);
        this.mainHandler = new Handler() {           //定义主线程的 Handler 对象
            public void handleMessage(Message msg) {     //消息处理
                switch (msg.what) {                      //判断 Message 类型
                case ToMain:                             //设置主线程的操作类
                    MainActivity.this.mymsg.setText("主线程接收数据是: "+msg.obj.
                        toString());                 //设置文本内容
                    break;}}};
        new Thread(new subThread()).start();             //启动子线程
        this.mybut.setOnClickListener(new OnClickListenerImpl()) ;}//单击事件操作
    private class OnClickListenerImpl implements OnClickListener {
        public void onClick(View view) {
            if (MainActivity.this.subHandler != null) {
              //判断是否已经实例化子线程 Handler
              Message childMsg = MainActivity.this.subHandler.obtainMessage();
                //创建消息
              childMsg.obj =  "This is a message from Main.";//设置消息内容
              childMsg.what = ToSub;                     //操作码
              MainActivity.this.subHandler.sendMessage(childMsg);}}}
             //向子线程发送消息
    protected void onDestroy() {
        super.onDestroy();
        MainActivity.this.subHandler.getLooper().quit();    }}  //结束队列
```

保存所有文件，运行该项目，进入运行界面，单击按钮，会接收到来自子线程的消息，如图 6.7

所示。

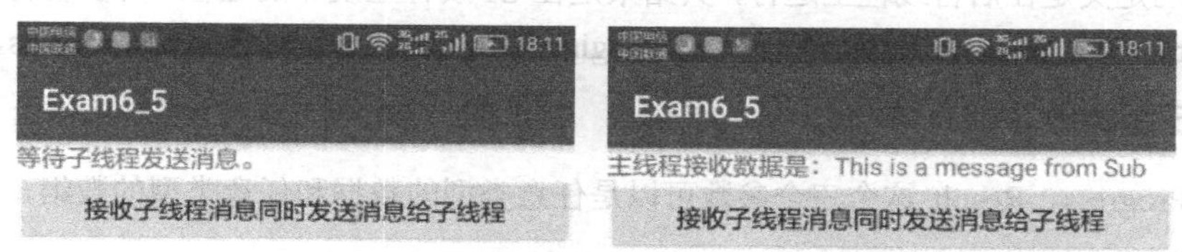

图 6.7　Exam6_5 运行结果

主线程也给子线程发送了消息，打开 Android Monitor 可以发现有一条消息，如图 6.8 所示。

```
05-15 18:10:17.287 10524-10524/org.hnist.cn.exam6_5 I/HwCust: Constructor found for class android.app.HwCustHwWallpaperM
05-15 18:10:17.331 10524-10524/org.hnist.cn.exam6_5 W/art: Before Android 4.1, method android.graphics.PorterDuffColorFi
05-15 18:10:17.578 10524-11021/org.hnist.cn.exam6_5 E/HAL: load: id=gralloc != hmi->id=gralloc
05-15 18:10:17.578 10524-11021/org.hnist.cn.exam6_5 I/OpenGLRenderer: Initialized EGL, version 1.4
05-15 18:10:25.634 10524-11016/org.hnist.cn.exam6_5 I/System.out: Main Child Message : This is a message from Main
6: Android Monitor   0: Messages   4: Run   TODO
```

图 6.8　主线程发送给子线程的消息

这个消息为什么不用 TextView 或者其他组件显示出来呢？注意，子线程不能更新主线程的组件数据，否则会报错。

6.5.4　异步处理工具类

由前面的学习可知道主线程和子线程可以相互传递消息，但是子线程无法直接对主线程里面的组件进行更新，Android 提供的 android.os.AsyncTask 类可以在后台进行操作之后更新主线程的组件，它的层次关系如下：

```
Java.lang.Object
    android.os.AsyncTask<Params, Progress, Result>
```

要在 Android 的 Java 程序中，使用 AsyncTask 类必须在程序中使用下面的语句。

```
import android.os.AsyncTask;                    //导入 os.AsyncTask 类
```

AsyncTask 类中有三个泛型参数，在这个类中定义了一些常用方法，如表 6-12 所示。

表 6-12　AsyncTask 类的常用方法

方法	描述
public AsyncTask ()	创建一个新的异步任务，这个构造函数必须在 UI 线程上调用
public final boolean cancel(boolean mayInterruptIfRunning)	指定是否取消当前线程操作
public final AsyncTask<Params, Progress, Result> execute(Params... params)	执行 AsyncTask 操作
public final boolean isCancelled()	判断子线程是否被取消
protected final void publishProgress(Progress... values)	更新线程进度
public final Result get(long timeout, TimeUnit unit)	等待计算结束并返回结果，最长等待时间为 timeOut
public final AsyncTask.Status getStatus()	获得任务的当前状态
protected abstract Result doInBackground(Params... params)	在后台完成任务执行，可以调用 publishProgress()方法更新线程进度
protected void onProgressUpdate(Progress... values)	在主线程中执行，用于显示任务的进度
protected void onPreExecute()	在主线程中执行，在 doInBackground()之前执行
protected void onPostExecute(Result result)	在主线程中执行，方法参数为任务执行结果
protected void onCancelled()	主线程中执行，在 cancel()方法之后执行

异步任务的定义是在后台线程上运行，其结果是在 UI 线程上发布的过程。异步任务被定义成三个泛型参数(Params、Progress、Result)和四个步骤(begin、doInBackground、processProgress、end)。

1. 三个泛型参数

Params、Progress、Result 这个三个参数可以是任意类型的数据和任意类型的数组，如果不需要，则用 void 代替。

（1）Params：启动时需要的参数类型，例如，HTTP 请求的 URL，对应的方法为 doInBackground (Params...parames)。

（2）Progress：后台执行任务的百分比，例如，进度条需要传递的是 Integer，Progress 对应的方法为 onProgressUpdate() 和 publishProgress(Progress...progress)，用来反映线程执行的进度，其中 publishProgress 方法必须在 doInBackground 方法中调用。

（3）Result：后台执行完毕之后返回的信息，例如，完成数据信息显示传递的是 String。对应的方法为 onPostExecute(Result)，后台进程得出的结果作为参数传递给此方法。

例如：

```
Result doInBackground(){
     A();                //方法 A
   this.publishProgress("state1","I like it");
     B();                //方法 B
   this.publishProgress("state2","for test");
   return result; }
onProgressUpdate(String values) {
     if(values[0].equals("state1"))
         C();            //将 A 读取的数据在 UI 上展现
     else if(values[0].equals("state2"))
       Log.e("value",values[1]);}
```

2. 四个步骤和对应的方法

1）begin 和 onPreExecute()

任务启动后（通过 execute()方法启动任务），这个步骤用来在 UI 线程中做一些初始化的工作。

2）doInBackground 和 doInBackground()

当 onPreExecute()方法执行完成后，立即在后台线程运行，用来处理一些耗时的计算及其他引起 UI 线程阻塞的操作，处理的结果 Result 返回给 onPostExecute(Result)方法，也可以使用 publishProgress() 和 UI 线程进行交互。

3）processProgress 和 onProgressUpdate()

每次在后台线程中调用了 publishProgress()方法后，onProgressUpdate()都会在 UI 线程中执行。在后台线程还未结束时，用来进行 UI 线程和后台线程的交互。

4）end 和 onPostExecute()

当后台线程执行完毕之后，后台线程将得到的结果传递给 onPostExecute()方法，这个步骤在 UI 线程上展现后台线程执行完毕后最终得到的结果。

上面这 4 个方法只有 doInBackground()是在后台线程中执行的，其他都是在 UI 线程中执行的。这 4 个方法都是 protected，必须继承 AsyncTask 类，必须覆写 doInBackground()方法，可能还要覆写 onPostExecute()方法。具体覆写哪些方法应根据实际需要决定，如果要在后台进程尚未执行完成的情况下和 UI 进行交互，就要覆写 onProgressUpdate()方法，如果需要等后台进程执行完毕得到结果后再和 UI 交互，则覆写 onPostExecute()方法即可。

3. 使用 AsyncTask 遵循的线程规则

（1）这个类的实例必须在 UI 线程中创建。

（2）execute()必须在 UI 线程中调用。

（3）不要自己动手调用上面的 4 个方法。

（4）这个任务只能被执行一次，如果尝试多次执行，则会抛出异常。

实例 6-6：利用 AsyncTask 类进行消息传递的实例

新建一个项目，项目命名为 Exam6_6，利用 AsyncTask 类实现线程间消息的传递。

（1）新建布局管理文件 activity_main.xml，代码如下：

```
…
    <TextView
        android:id="@+id/show"
        android:layout_width="match_parent"
        android:layout_height="wrap_content"
        android:text="这里显示计数" />
    <Button android:id="@+id/Start"
        android:layout_width="wrap_content"
        android:layout_height="wrap_content"
        android:text="开始"/>
    <Button android:id="@+id/Stop"
        android:layout_width="wrap_content"
        android:layout_height="wrap_content"
        android:text="停止"/>
    <ProgressBar
        android:id="@+id/progressBar01"
        style="?android:attr/progressBarStyleHorizontal"  //水平进度条
        android:layout_width="match_parent"
        android:layout_height="wrap_content"
        android:max="100"
        android:progress="30"/>
    <ProgressBar
        android:id="@+id/progressBar02"
       style="?android:attr/progressBarStyleLarge"        //大号圆形进度条
        android:layout_width="wrap_content"
        android:layout_height="wrap_content"
        android:max="100"
        android:progress="30"
        android:visibility="gone"/>
</LinearLayout>
```

（2）新建 Activity 文件 MainActivity.java，代码如下：

```
…
public class MainActivity extends AppCompatActivity{
private TextView show;
private Button startBtn,stopBtn;
private ProgressBar progressBar1,progressBar2;
private AsyncTask<Integer,Integer,Integer> task = null;   /*定义 AsyncTask
                                                            对象 task*/
private boolean stop;
```

```
private int count = 0;
  public void onCreate(Bundle savedInstanceState) {
    super.onCreate(savedInstanceState);
    setContentView(R.layout.activity_main);
    show = (TextView)MainActivity.this.findViewById(R.id.show);
    startBtn = (Button)MainActivity.this.findViewById(R.id.Start);
    stopBtn = (Button)MainActivity.this.findViewById(R.id.Stop);
    progressBar1 = (ProgressBar)MainActivity.this.findViewById(R.id.progressBar01);
    progressBar2 = (ProgressBar)MainActivity.this.findViewById(R.id.progressBar02);
    startBtn.setOnClickListener(new StartOnClickListener());
    stopBtn.setOnClickListener(new StopOnClickListener()); }
    class StartOnClickListener implements OnClickListener
   {   public void onClick(View v) {
          if(task == null){
             task = new CounterTask();  //实例化 task
             task.execute(count); //在应用程序中启动一个子线程，执行 doInBackground
             progressBar2.setVisibility(View.VISIBLE); }}} //使第二个进度条可见
   class StopOnClickListener implements OnClickListener
   {   public void onClick(View v) {
         if(task != null){
             stop = true;          //停止线程
             task = null;          //清除线程
             progressBar2.setVisibility(View.GONE); }} } //使第二个进度条消失
   //定义 CounterTask 类继承 AsyncTask
   class CounterTask extends AsyncTask<Integer,Integer,Integer>{
   protected Integer doInBackground(Integer… params){  /*覆写 doInbackground()
                                                        方法*/
        Integer initCounter = params[0];
        Log.v("This is a test","In doInBackground… "); /*打印输出，在 LogCat
                                                        中可以看见*/
        stop = false;
        while(!stop){
            publishProgress(initCounter);//将中间结果传递到 onProgressUpdate 中
        try {
            Thread.sleep(1*200);      //休眠 0.2 秒
        } catch (InterruptedException e) {
            e.printStackTrace();}
        initCounter += 1 ;
        if(initCounter == 100){
            break;   }}
        return initCounter;}            //将返回结果传递到 onPostExecute 中
     protected void onPostExecute(Integer result) { //覆写 onPostExecute()方法
        super.onPostExecute(result);
        Log.v("Hello!","In onPostExecute…");  //打印输出，在 LogCat 中可以看见
        progressBar1.setProgress(result);      //设置第一个进度条的进度值
        progressBar2.setProgress(result);      //设置第二个进度条的进度值
        String text = result.toString();       //将计算结果转换为字符串
        show.setText(text);}                   //更新文本组件的显示内容
     protected void onProgressUpdate(Integer… values) {
         //覆写 onProgressUpdate()方法
        super.onProgressUpdate(values);
        Log.v("Hello!","In onProgressUpdate…"); //打印输出，在 LogCat 中可以看见
```

```
        progressBar1.setProgress(values[0]);   //更新进度条的进度值
        progressBar2.setProgress(values[0]);   //更新进度条的进度值
        String text = values[0].toString();   //将计算结果转换为字符串
        show.setText(text); }                  //更新文本组件的显示内容
    protected void onPreExecute() {            //覆写 onPreExecute()方法
        Log.v("Hello","In onPreExecute…");  //打印输出，在 LogCat 中可以看见
        progressBar1.setProgress(0);           //更新进度条的进度值
        progressBar2.setProgress(0);           //更新进度条的进度值
        super.onPreExecute();} }}              //调用父类的 onPreExecute()方法
```

保存所有文件，运行该项目，进入运行界面，单击按钮，发现 TextView 中的内容被线程中的数据更新了，如图 6.9 所示。

分析程序的代码，可发现 4 个覆写的方法中除了 doInBackground()方法不能更新主线程的组件之外，其他几个方法都能够更新主线程的组件。利用 AsyncTask 类实现线程之间消息的传递比使用前面介绍的方法要简单一些。

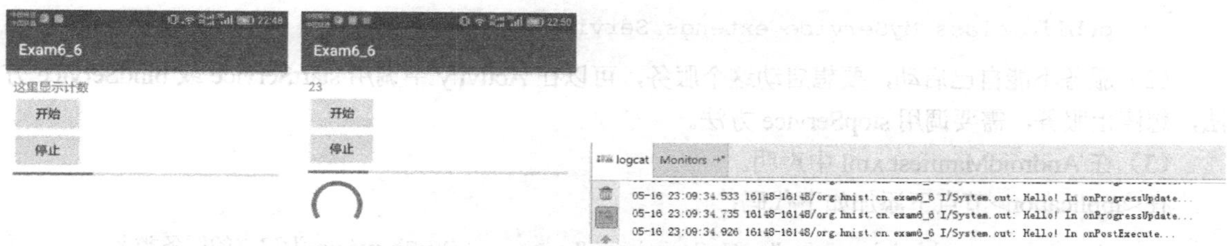

图 6.9 Exam6_6 运行结果

6.6 Service

在 Android 系统开发中，Service 是一个重要的组成部分，它有两个主要目的：后台运行程序和跨进程访问。通过启动一个服务，可以在不显示界面的前提下在后台运行指定的任务，这样可以不影响用户做其他事情。通过 AIDL 服务可以实现不同进程之间的通信，这也是服务的重要用途之一。

6.6.1 Service 基础

Service 的主要功能是为 Activity 程序提供一些必要的支持或者提供一些不需要运行界面的功能。在开发时，用户只需要继承 android.app.Service 类即可完成 Service 程序的开发。

要在 Android 的 Java 程序中使用 Service 组件，必须在程序中使用下面的语句。

```
    import android.app.Service;                //导入 app.Service 类
```

Service 中也有自己的生命周期控制方法，如表 6-13 所示。

表 6-13 Service 的生命周期控制方法

方法及常量	描述
public static final int START_CONTINUATION_MASK	继续执行 Service
public static final int START_STICKY	用于显式地启动和停止 Service
public abstract IBinder onBind(Intent intent)	设置 Activity 和 Service 之间的绑定
public void onCreate()	当一个 Service 创建时调用
public int onStartCommand(Intent intent, int flags, int startId)	启动 Service，由 startService()方法触发
public void onDestroy()	Service 销毁时调用，由 stopService()方法触发

实现 Service 操作需要 Activity 类中一些方法的支持，如表 6-14 所示。

表 6-14　Activity 类中操作 Service 的方法

方法	描述
public ComponentName startService(Intent service)	启动一个 Service
public boolean stopService(Intent name)	停止一个 Service
public boolean bindService(Intent service, ServiceConnection conn, int flags)	与一个 Service 绑定
public void unbindService(ServiceConnection conn)	取消与一个 Service 的绑定

Service 的生命周期控制相对 Activity 的要简单一些，Service 的生命周期中一般只需要使用 startService()和 stopService()两个方法来开始和终止服务。

Service 分为两种：用于应用程序内部的本地服务和用于应用程序之间的远程服务。

创建一个 Service 很简单，按照下面的步骤就可以建立一个 Service。

（1）编写一个服务类，该类必须从 android.app. Service 继承。

```
public class MyService extends Service { … }
```

（2）服务不能自己启动，要想启动这个服务，可以在 Activity 中调用 startService 或 bindService 方法，想停止服务，需要调用 stopService 方法。

（3）在 AndroidManifest.xml 中声明。

在<application>节点下添加如下代码：

```
<service android:name=". MyService " />    //MyService 为建立的服务类名
```

6.6.2　Service 的启动和停止

1．用 startService()方法启动 Service

使用 startService()方法启用服务，调用者与服务之间没有关联，即使调用者退出了，服务仍然运行，在服务未被创建时，系统会先调用服务的 onCreate()方法，再调用 onStart()方法。如果调用 startService()方法前服务已经被创建，则多次调用 startService()方法并不会导致多次创建服务，但会导致多次调用 onStart()方法。采用 startService()方法启动的服务，只能调用 stopService()方法结束服务，服务结束时会调用 onDestroy()方法。

采用 startService()方法启动服务和 stopService()方法停止服务的代码如下：

```
…
        Intent it = new Intent(MyServiceDemo.this, MyService.class);
        startService(it);     //启动 Service
…
        Intent it = new Intent(MyServiceDemo.this, MyService.class);
        stopService(it);      //停止 Service
… }
```

注意：这里的 MyServiceDemo 是主程序的名称，并不是固定不变的值。

2．用 bindService()方法启动 Service

使用 bindService()方法启用服务，调用者与服务绑定在了一起，调用者一旦退出，服务也就终止了，在服务未被创建时，系统会先调用服务的 onCreate()方法，再调用 onBind()方法。此时调用者和服务绑定在一起，调用者退出后，系统就会先调用服务的 onUnbind()方法，再调用 onDestroy()方法。如果调用 bindService()方法前服务已经被绑定，则多次调用 bindService()方法并不会多次调用 onCreate()和 onBind()方法。如果调用者希望与正在绑定的服务解除绑定，则可以调用 unbindService()方法，调用该方法也会导致系统调用服务的 onUnbind()→onDestroy()方法。

采用 bindService()方法启动服务的代码如下：

```
…
private ServiceConnection conn = new ServiceConnection() {  //定义连接名
//连接到 Service
public void onServiceConnected(ComponentName name, IBinder service) {…}
//与 Service 断开连接
public void onServiceDisconnected(ComponentName name) {…}
…
    Intent it = new Intent(MyServiceDemo.this, MyService.class);
    //实现绑定
    MyServiceDemo.this. bindService(it, conn, Context.BIND_AUTO_CREATE);
…   //取消 Service 绑定
    MyServiceDemo.this.unbindService(conn);
…   //启动 Service
    Intent it = new Intent(MyServiceDemo.this, MyService.class);
    MyServiceDemo.this.startService(it));
…   //停止 Service
    Intent it = new Intent(MyServiceDemo.this, MyService.class);
    MyServiceDemo.this.stopService(it));
…}
```

注意：bindService 开启了一个服务后，一定要记得执行停止服务操作，否则系统将会抛出 android.app.ServiceConnectionLeaked 异常。

3．通过 RPC 机制来实现不同进程间 Service 的调用

远程 Service 调用，是 Android 系统为了提供进程间通信而提供的实现方式，这种方式采用一种称为远程进程调用（Remote Procedure Call，RPC）的技术来实现。

RPC 是指在一个进程中，调用另外一个进程中的服务。Android 通过接口定义语言来生成两个进程间的访问代码。Android 接口定义语言（Android Interface Definition Language，AIDL）是 Android 系统的一种接口描述语言，Android 编译器可以将 AIDL 文件编译成一段 Java 代码，生成相对的接口。

6.6.3 绑定 Service

使用 startService()方法启用服务时，调用者与服务之间没有关联，即使调用者退出了，服务仍然运行，直到 Android 系统关闭后服务才会停止。使用 bindService()方法可以将一个 Activity 和 Service 绑定起来，调用者与服务绑定在了一起，调用者一旦退出，服务也就终止了。

Activity 类中专门提供了一个用于绑定 Service 的 bindService()方法，此方法返回的是一个 android.content.ServiceConnection 接口的参数，它主要定义了两个方法，如表 6-15 所示。

表 6-15 ServiceConnection 接口定义的方法

方法	描述
public abstract void onServiceConnected (ComponentName name, IBinder service)	当与一个 Service 建立连接的时候调用
public abstract void onServiceDisconnected(ComponentName name)	当与一个 Service 取消连接的时候调用

ServiceConnection 接口的主要功能是当一个 Activity 程序与 Service 建立连接之后，可以通过 ServiceConnection 接口执行 Service 连接（或取消连接）处理操作，在 Activity 连接到 Service 程序之后，会触发 Service 类中的 onBind()方法，在此方法中要返回一个 IBinder 接口的对象，IBinder 接口定义的常量和方法如表 6-16 所示。

表 6-16　IBinder 接口的常量和方法

常量及方法	描述
public static final int DUMP_TRANSACTION	IBinder 协议的事务码：清除内部状态
public static final int FIRST_CALL_TRANSACTION	用户指令的第一个事务码可用
public static final int FLAG_ONEWAY	transact()方法单向调用的标志位，表示调用者不会等待从被调用者那里返回的结果，而立即返回
public static final int INTERFACE_TRANSACTION	IBinder 协议的事务码：向事务接收端询问其完整的接口规范
public static final int LAST_CALL_TRANSACTION	用户指令的最后一个事务码可用
public static final int PING_TRANSACTION	IBinder 协议的事务码：pingBinder()
public static final int LIKE_TRANSACTION	IBinder 协议的事务码
public static final int TWEET_TRANSACTION	IBinder 协议的事务码，发送一个信号给目标
public abstract void dump(FileDescriptor fd, String[] args)	向指定的数据流输出对象状态
public abstract String getInterfaceDescriptor()	取得被 Binder 对象所支持的接口名称
public abstract boolean isBinderAlive()	检查 Binder 所在的进程是否活着
public abstract void linkToDeath(IBinder.DeathRecipient recipient, int flags)	如果指定的 Binder 消失了，则为通知注册一个新的接收器
public abstract boolean pingBinder()	检查远程对象是否存在
public abstract Interface queryLocalInterface(String descriptor)	取得对一个接口绑定对象的本地实现
public abstract boolean transact(int code, Parcel data, Parcel reply, int flags)	执行一个一般的操作
public abstract boolean unlinkToDeath(IBinder.DeathRecipient recipient, int flags)	删除一个接收通知的接收器

实例 6-7：Service 组件操作举例

新建一个项目，项目命名为 Exam6_7，利用 Service 编程实现服务的启动和绑定服务的启动。

（1）新建布局管理文件 activity_main.xml，代码如下：

```
...
<Button
    android:id="@+id/startSrv"
    android:layout_width="150dp"
    android:layout_height="wrap_content"
    android:layout_gravity="center"
    android:text="启动服务" />
<Button
    android:id="@+id/stopSrv"
    android:layout_width="150dp"
    android:layout_height="wrap_content"
    android:layout_gravity="center"
    android:text="停止服务" />
<Button
    android:id="@+id/bindSrv"
    android:layout_width="150dp"
    android:layout_height="wrap_content"
    android:layout_gravity="center"
    android:text="绑定服务" />
```

```
<Button
    android:id="@+id/unbindSrv"
    android:layout_width="150dp"
    android:layout_height="wrap_content"
    android:layout_gravity="center"
    android:text="解除绑定服务" />
</LinearLayout>
```

（2）编写一个服务类 MyService.java，该类必须继承 android.app.Service，代码如下：

```
import android.app.Service;
import android.content.Intent;
import android.os.Binder;
import android.os.IBinder;
public class MyService extends Service {        //新建类MyService继承于Service
private IBinder myBinder = new Binder(){        //定义IBinder接口
    public String getInterfaceDescriptor() { //取得接口描述信息
        return "MyServiceUtil class.";}} ;      //返回Service类的名称
    public IBinder onBind(Intent intent) {      //绑定时触发
        return myBinder;}
    public void onRebind(Intent intent) {       //重新绑定时触发
        super.onRebind(intent); }
    public boolean onUnbind(Intent intent) { //解除绑定时触发
        return super.onUnbind(intent); }
    public void onCreate() {                    //创建时触发
        super.onCreate();
        Log.i(TAG, "服务已经开启: onCreate()!");}
    public void onDestroy() {                   //销毁时触发
      super.onDestroy();
        Log.i(TAG, "服务已经销毁: onCreate()!");}
    public int onStartCommand(Intent intent, int flags, int startId) {
                                                //启动时触发
      Log.i(TAG, "服务启动: onStart()=>Intent"+intent+",startID="+startId);
      return Service.START_CONTINUATION_MASK;    }}
```

（3）新建 Activity 文件 MainActivity.java，代码如下：

```
…
import android.content.ComponentName;
import android.content.Context;
import android.content.Intent;
import android.content.ServiceConnection;
import android.os.Bundle;
import android.os.IBinder;
public class MainActivity extends AppCompatActivity{
private Button start;
private Button stop;
private Button bind;
private Button unbind;
private ServiceConnection serviceConnection = new ServiceConnection() {
    //连接到Service
    public void onServiceConnected(ComponentName name, IBinder service) {
        try {System.out.println("Service Connect Success");
           } catch (RemoteException e) {
              e.printStackTrace();   }}
    public void onServiceDisconnected(ComponentName name) { //与Service断开连接
```

```
        System.out.println("Service Disconnected."); }};
    public void onCreate(Bundle savedInstanceState) {
        super.onCreate(savedInstanceState);
        super.setContentView(R.layout.activity_main);
        this.start = (Button) super.findViewById(R.id.startSrv);
        this.stop = (Button) super.findViewById(R.id.stopSrv);
        this.bind = (Button) super.findViewById(R.id.bindSrv);
        this.unbind = (Button) super.findViewById(R.id.unbindSrv);
        this.start.setOnClickListener(new StartOnClickListenerImpl()) ;
        this.stop.setOnClickListener(new StopOnClickListenerImpl()) ;
        this.bind.setOnClickListener(new BindOnClickListenerImpl()) ;
        this.unbind.setOnClickListener(new UnbindOnClickListenerImpl()); }
    private class StartOnClickListenerImpl implements OnClickListener {
        public void onClick(View v) {
            //启动 Service
            MainActivity.this.startService(new Intent(MainActivity.this, MyService.
                class)); }}
    private class StopOnClickListenerImpl implements OnClickListener {
        public void onClick(View v) {
            //停止 Service
            MainActivity.this.stopService(new Intent(MainActivity.this, MyService.
                class));}}
    private class BindOnClickListenerImpl implements OnClickListener {
        public void onClick(View v) {
            //绑定 Servic
            MainActivity.this.bindService(new Intent(MainActivity.this, MyService.
                class),
    MainActivity.this.serviceConnection,Context.BIND_AUTO_CREATE);}}
    private class UnbindOnClickListenerImpl implements OnClickListener {
        public void onClick(View v) {
            //取消 Service 绑定
            MainActivity.this.unbindService(MainActivity.this.serviceConnection);}}}
```

（4）在 AndroidManifest.xml 文件中声明。

在<application>节点下添加如下的代码：

```
<service android:name=". MyService " />    //MyService 为建立的服务类名
```

注意：有的模拟器可能没有查看服务的功能，请选择合适的模拟器。

保存所有文件，运行该项目，在 Android Monitor（或手机）上查看 MyService 服务是否被启动，发现服务没有启动。运行该程序，进入如图 6.10 所示界面，单击“启动服务”按钮，再查看，发现 MyService 服务已经启动，如图 6.11 所示。停止 Exam6_7 应用，进入如图 6.12 所示界面。发现 MyService 服务并没有随着程序的停止而终止，单击“停止服务”按钮，进入如图 6.13 所示界面，表示服务已经停止。单击“绑定服务”按钮，然后单击“启动服务”按钮，再查看正在运行的服务情况，可以看到服务已经启动，停止 Exam6_7 应用，发现服务已经随着程序的停止而终止，说明绑定已经发生作用了。

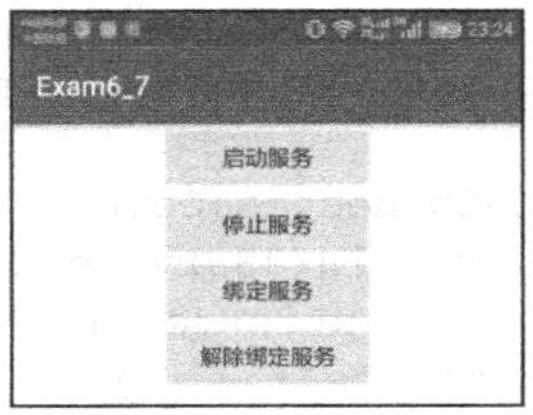

图 6.10　Exam6_7 运行结果图

```
05-17 00:15:16.753 28149-28149/org.hnist.cn.exam6_7 I/ContentValues: 服务已经开启: onCreate()!
05-17 00:15:16.754 28149-28149/org.hnist.cn.exam6_7 I/ContentValues: 服务启动: onStart()=>IntentIntent { cmp=org.hnist.cn.exam6_7/.MyService },startID=1
```

图 6.11　MyService 被启动

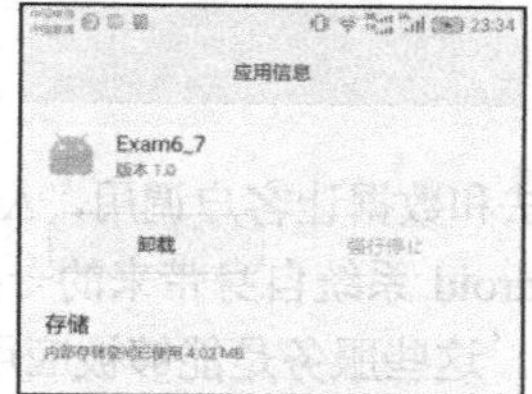

6.12　停止 Exam6_7 应用

```
05-17 00:34:46.002 18934-18934/org.hnist.cn.exam6_7 I/ContentValues: 服务已经开启: onCreate()!
05-17 00:34:46.003 18934-18934/org.hnist.cn.exam6_7 I/ContentValues: 服务启动: onStart()=>IntentIntent { cmp=org.hnist.cn.exam6_7/.MyService },startID=1
05-17 00:34:48.834 18934-18934/org.hnist.cn.exam6_7 I/ContentValues: 服务已经销毁: onCreate()!
```

图 6.13　MyService 被销毁

6.6.4　Service 的生命周期

Service 与 Activity 一样，也有一个从启动到销毁的过程，但 Service 的过程比 Activity 简单得多，如图 6.14 所示。Service 从启动到销毁的过程只会经历如下 3 个阶段：创建服务、开始服务、销毁服务。

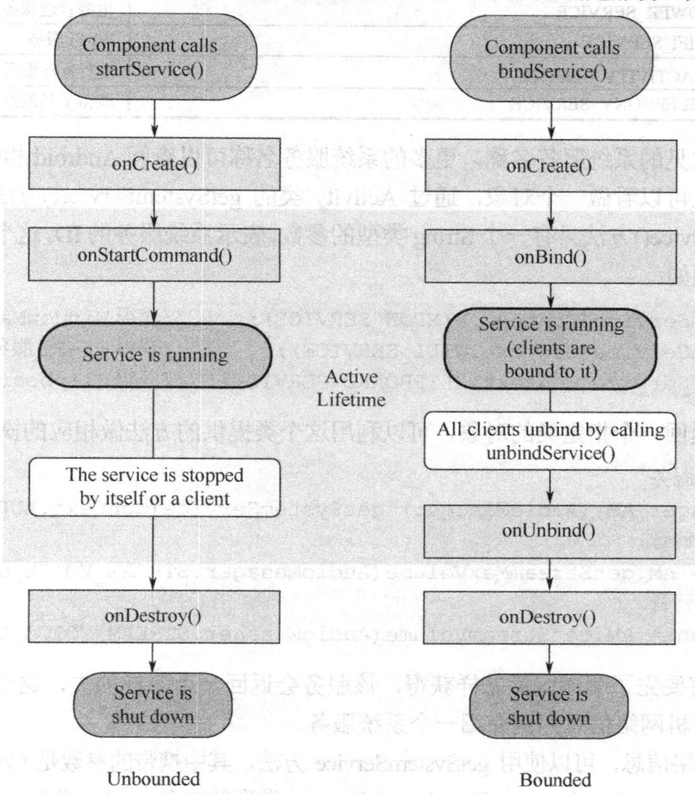

图 6.14　Service 的生命周期示意图

Service 生命周期主要包含以下几个方法：onCreate()、onBind(Intent intent)、onStart(Intent intent,int startId)、onDestory()。

对于使用 startService()方法打开的服务，其生命周期为 onCreate→onStart→onDestroy→服务关闭。

对于使用 bindService()方法打开的服务，其生命周期为 onCreate→onBind→onUnbind→onDestroy→服务关闭。

6.6.5 Service 系统服务

可以利用 Service 组件设计一些方法和数据让客户调用，Android 系统就设计了这样的服务，称为 Service 系统服务。通俗地说，就是 Android 系统自身带来的一些内置的功能，例如，当手机接到来电时，会响铃，会显示对方的号码，等等。这些服务是能够被应用程序调用的。

在 Android 操作系统中，为了方便用户使用系统服务，android.content.Context 类中将所有的系统服务名称以常量的形式进行了绑定，如表 6-17 所示。

表 6-17 Context 类中定义的常见系统服务

常量	描述
public static final String CLIPBOARD_SERVICE	剪贴板服务
public static final String WINDOW_SERVICE	窗口服务
public static final String ALARM_SERVICE	闹铃服务
public static final String AUDIO_SERVICE	音频服务
public static final String NOTIFICATION_SERVICE	Notification 服务
public static final String SEARCH_SERVICE	搜索服务
public static final String POWER_SERVICE	电源管理服务
public static final String WIFI_SERVICE	Wi-Fi 服务
public static final String ACTIVITY_SERVICE	运行程序服务
public static final String TELEPHONY_SERVICE	取得手机网络信息

这里只给出了常见的系统服务名称，更多的系统服务名称可以查阅 Android 相关开发文档获得。

系统服务实际上可以看做一个对象，通过 Activity 类的 getSystemService()方法可以获得指定的系统服务。getSystemService()方法只有一个 String 类型的参数，表示系统服务的 ID，这个 ID 在整个 Android 系统中是唯一的。例如：

```
getSystemService(Context.WINDOW_SERVICE);      //获得 WindowManager 服务
getSystemService(Context.WIFI_SERVICE);        //获得 Wi-Fi 服务
getSystemService(Context.CLIPBOARD_SERVICE);   //获得 ClipboardManager 服务
```

获得服务后会返回一个指定类的对象，可以利用这个类提供的方法做相应的操作。例如：

```
//获得音频服务
AudioManager AM=(AudioManager) getSystemService(Context.AUDIO_SERVICE);
//获得最大铃音
int max = AM.getStreamMaxVolume(AudioManager.STREAM_VOICE_CALL);
//获得当前铃音
int current= AM.getStreamVolume(AudioManager.STREAM_VOICE_CALL); …
```

使用某个服务前要先了解该服务怎样获得，该服务会返回一个怎样的类，这个类有哪些方法和常量等，下面以取得手机网络信息为例介绍一个系统服务。

要想取得手机网络信息，可以使用 getSystemService 方法，其中携带的参数是 Context.TELEPHONY_SERVICE，返回的是 android.telephony.TelephonyManager 类型的对象，这个类有表 6-18 所示的常用常量和表 6-19 所示的常用方法。

表 6-18 TelephonyManager 类的常用常量

常量及方法	描述
public static final int NETWORK_TYPE_CDMA	使用 CDMA 网络
public static final int NETWORK_TYPE_GPRS	使用 GPRS 网络
public static final int PHONE_TYPE_CDMA	使用 CDMA 通信
public static final int PHONE_TYPE_GSM	使用 GSM 通信
public static final String ACTION_PHONE_STATE_CHANGED	电话状态改变
public static final int CALL_STATE_IDLE	空闲（无呼入或已挂机）
public static final int CALL_STATE_OFFHOOK	摘机（有呼入）
public static final int CALL_STATE_RINGING	响铃（接听中）
public static final int DATA_CONNECTED	数据连接状态：已连接
public static final int DATA_DISCONNECTED	数据连接状态：断开

表 6-19 TelephonyManager 类的常用方法

常量及方法	描述
public String getNetworkCountryIso()	获取网络的国家 ISO 代码
public String getDeviceSoftwareVersion()	获得软件版本
public String getDeviceId()	取得设备标识
public String getLine1Number()	取得手机号码
public String getNetworkOperatorName()	取得移动提供商的名称
public int getNetworkType()	取得移动网络的连接类型
public int getPhoneType()	取得电话网络类型
public boolean isNetworkRoaming()	判断电话是否处于漫游状态
public void listen(PhoneStateListener listener, int events)	注册电话状态监听器
public String getSimCountryIso()	获取 SIM 卡中国家 ISO 代码
public String getSimOperator()	获得 SIM 卡中移动国家代码（MCC）和移动网络代码（MNC）
public String getSimOperatorName()	获取服务提供商姓名
public String getSimSerialNumber()	SIM 卡序列号

实例 6-8：取得手机网络信息操作实例

新建一个项目，项目命名为 Exam6_8，利用 Service 获得手机的网络信息。

（1）新建布局管理文件 activity_main.xml，代码如下：

```
...
    <ListView
        android:id="@+id/MyListView"
        android:layout_width="match_parent"
        android:layout_height="wrap_content" />
</LinearLayout>
```

（2）新建 Activity 文件 MainActivity.java，代码如下：

```
...
public class MainActivity extends AppCompatActivity {
    private ListView infolist = null;
    private TelephonyManager TM = null;        //定义 TelephonyManager 对象
    private ListAdapter adapter = null;        //适配器组件
    private List<String> ListInfo = new ArrayList<String>();
    public void onCreate(Bundle savedInstanceState) {
```

```
        super.onCreate(savedInstanceState);
        super.setContentView(R.layout.activity_main);
        this.infolist = (ListView) super.findViewById(R.id.MyListView);
        this.TM =(TelephonyManager)super.getSystemService(Context.TELEPHONY_
         SERVICE);  //取得手机服务
        this.listInfo();        }                              //定义 listInfo 方法列表显示
    private void listInfo() {                                  //执行列表显示操作
        this.ListInfo.add(this.TM.getLine1Number() == null ? "没有手机号码" : "
          手机号码: "
              + this.TM.getLine1Number());                     //取得手机号码
        this.ListInfo.add(this.TM.getNetworkOperatorName() == null ?
              "没有移动网络服务商" : "移动网络服务商: " +   this.TM.getNetwork-
              OperatorName());                                 //取得移动网络商名称
     //判断网络类型
     if (this.TM.getPhoneType()    == TelephonyManager.NETWORK_TYPE_CDMA) {
       this.ListInfo.add("移动网络类型: CDMA");}
       else if (this.TM.getPhoneType() == TelephonyManager.NETWORK_TYPE_GPRS) {
          this.ListInfo.add("移动网络类型: GPRS");
        } else {   this.ListInfo.add("移动网络类型: 未知");}
      //判断电话网络类型
      if (this.TM.getNetworkType() == TelephonyManager.PHONE_TYPE_GSM) {
        this.ListInfo.add("手机网络类型: GSM"); }
        else if (this.TM.getNetworkType() == TelephonyManager.PHONE_TYPE_CDMA) {
             this.ListInfo.add("手机网络类型: CDMA");
        } else {this.ListInfo.add("手机网络类型: UTMS");  }
        //是否漫游
        this.ListInfo.add("是否漫游: " + (this.TM.isNetworkRoaming() ? "漫游" :
          "非漫游"));
        this.ListInfo.add(this.TM.getDeviceId() == null ? "没有设备标识 ID" : "
          设备标识 ID: "+ this.TM.getDeviceId());           //取得设备标识
        this.adapter = new ArrayAdapter<String>(this,    //实例化 ArrayAdapter
             android.R.layout.simple_list_item_1,         //每行显示一条数据
             this.ListInfo);                              //定义显示数据
        this.infolist.setAdapter(this.adapter); }}        //设置适配器
```

（3）修改 AndroidManifest.xml 文件， 增加读取手机状态的权限，添加下面的语句：

```
<uses-permission android:name="android.permission.READ_PHONE_STATE"/>
```

保存所有文件，运行该程序，结果如图 6.15 所示。

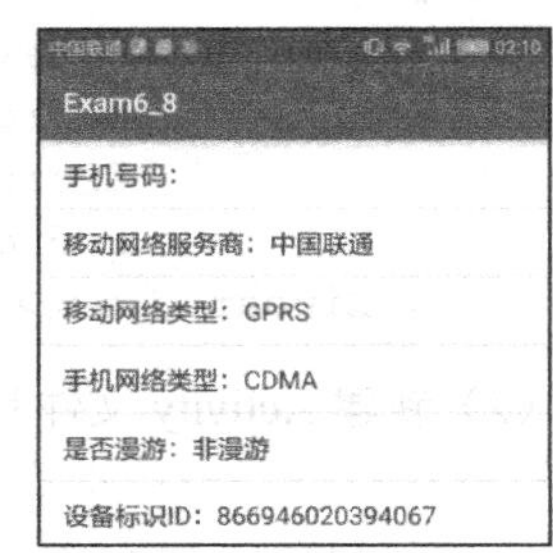

图 6.15　Exam6_8 运行结果

6.7　BroadcastReceiver 的使用

BroadcastReceiver 用来接收来自系统和应用中的广播。

在 Android 系统中，广播在很多地方得到了体现。例如，当网络状态改变时系统会产生一条广播，接收到这条广播就能及时地做出提示和保存数据等操作；当电池电量不足时，系统会产生一条广播，接收到这条广播就能在电量低时告知用户及时保存进度，等等。

Android 中的广播机制设计得非常方便，可以将很多复杂的事情用广播的形式告知开发者，极大地减少了开发工作量。

6.7.1 BroadcastReceiver 基础

广播是一种广泛运用在应用程序之间传输信息的机制。而 BroadcastReceiver 是对发送出来的广播进行过滤接收并响应的组件。BroadcastReceiver 自身并不实现图形用户界面，当它收到某个通知后，BroadcastReceiver 可以启动 Activity 作为响应，或者通过 NotificationMananger 提醒用户，或者启动 Service，等等。广播属于触发式操作，当有了指定操作之后会自动启动广播。

实现一个 BroadcastReceiver 方法的基本步骤如下。

（1）继承 BroadcastReceiver，并重写 onReceive()方法，例如：

```
public class MyBroadcastReceiver extends BroadcastReceiver {
    public MyBroadcastReceiver(){…}                          //构造方法
    public void onReceive(Context context, Intent intent) {…} }//重写 onReceive()
```

在 onReceive 方法内，可以获取随广播而来的 Intent 中的数据，其中包含很多有用的信息，这是广播消息核心部分所在。

创建完 BroadcastReceiver 之后，还需要为它注册一个指定的广播地址。没有注册广播地址的 BroadcastReceiver 就像一个缺少选台按钮的收音机，虽然功能齐全，但无法收到各电台的信号。

（2）为 BroadcastReceiver 注册广播地址。

注册有静态注册和动态注册两种方式。

① 静态注册方式是在 AndroidManifest.xml 文件的 application 中定义 receiver 并设置要接收的 action。

```
<receiver                                               //定义广播处理
    android:name=".MyBroadcastReceiver"                 //广播处理类
    android:enabled="true">                             //启用广播
    <intent-filter>                                     //匹配 Action 操作时广播
        <action android:name="android.intent.action.EDIT" />
    </intent-filter>
</receiver>
```

配置后，只要是 android.intent.action.EDIT 地址的广播，MyBroadcastReceiverUtil 就能够接收到。注意，这种方式的注册是常驻型的，也就是说，当应用关闭后，如果有广播信息传来，MyBroadcastReceiverUtil 也会被系统调用而自动运行。

②动态注册方式是在 Activity 或 Service 中调用函数来注册，有两个关键参数：一个是 receiver，另一个是 IntentFilter，其中是要接收的 action。

```
public class BroadcastReceiverDemo extends Activity {
    private BroadcastReceiver receiver;
      …
      receiver = new CallReceiver();
      //动态注册广播地址
      registerReceiver(receiver,new IntentFilter("android.intent.action.EDIT")); }
      …
      //解除注册广播地址
      unregisterReceiver(receiver);
      …
```

推荐使用静态注册方式，由系统来管理 receiver，动态注册方式不是常驻型的，而且隐藏在代码中，比较难发现，在退出程序前要先调用 unregisterReceiver()方法进行注销。

（3）利用 sendBroadcast 发送广播。

可以根据以上任意一种方法完成注册，当注册完成之后，定义好的广播接收者即可开始正常工作，首先构建 Intent 对象，然后调用 sendBroadcast(Intent)方法将广播发出。

```
Intent it = new Intent("android.intent.action.EDIT ");    //指定 Action
intent.putExtra("msg", "hello receiver.");                 //附加数据
sendBroadcast(it);                                         //进行广播
```

BroadcastReceiver 的生命周期只有十秒左右。

一个 BroadcastReceiver 对象只有在被调用的 onReceive()方法中才有效，当从该函数返回后，该对象就无效了，会结束生命周期。因此，在所调用的 onReceive()方法里，不能有过于耗时的操作，对于耗时的操作，可用 Service 来完成，因为当得到其他异步操作所返回的结果时，BroadcastReceiver 可能已经无效了。

6.7.2 BroadcastReceiver 组件操作实例

实例 6-9：BroadcastReceiver 组件操作举例

建立一个 Android 项目，项目的名称为 Exam6_9，利用 BroadcastReceiver 组件发送广播。

（1）新建布局管理文件 activity_main.xml，代码如下：

```
…
 <Button
      android:id="@+id/mybut"
      android:layout_width="wrap_content"
      android:layout_height="wrap_content"
      android:text="发送广播消息"/>
</LinearLayout>
```

（2）新建 MyBroadcastReceiver.java 文件，继承 BroadcastReceiver 类，并覆写 onReceive()方法，代码如下：

```
…
import android.content.BroadcastReceiver;
import android.content.Context;
import android.content.Intent;
public class MyBroadcastReceiver extends BroadcastReceiver {//继承BroadcastReceiver
     public void onReceive(Context context, Intent intent) { //覆写onReceive()方法
     String msg=intent.getStringExtra("msg");              //获得广播信息内容
     Toast.makeText(context, msg, Toast.LENGTH_LONG).show();}}//显示信息
```

（3）新建 Activity 文件 MainActivity.java，代码如下：

```
…
import android.content.Intent;                    //导入 content.Intent 类
public class MainActivity extends AppCompatActivity {
private Button mybut ;
   public void onCreate(Bundle savedInstanceState) {
       super.onCreate(savedInstanceState);
       super.setContentView(R.layout.activity_main);
       this.mybut = (Button) super.findViewById(R.id.mybut) ;
       this.mybut.setOnClickListener(new SendMsg());}          //设置监听事件
       private class SendMsg implements OnClickListener { //定义事件
```

```
        public void onClick(View v) {
          Intent it = new Intent(Intent.ACTION_EDIT);          //启动 Action
          it.putExtra("msg", "明天上午 9:30 在 16307 开会");
          MainActivity.this.sendBroadcast(it); }}}             //进行广播
```

（4）在 AndroidManifest.xml 文件中注册广播组件，在<application>与</ application >之间写入如下代码：

```
<receiver
    android:name="MyBroadcastReceiver"
    android:enabled="true">
    <intent-filter>
        <action android:name="android.intent.action.EDIT" />
    </intent-filter>
</receiver>
```

保存所有文件，运行该项目，单击按钮，结果如图 6.16 所示。

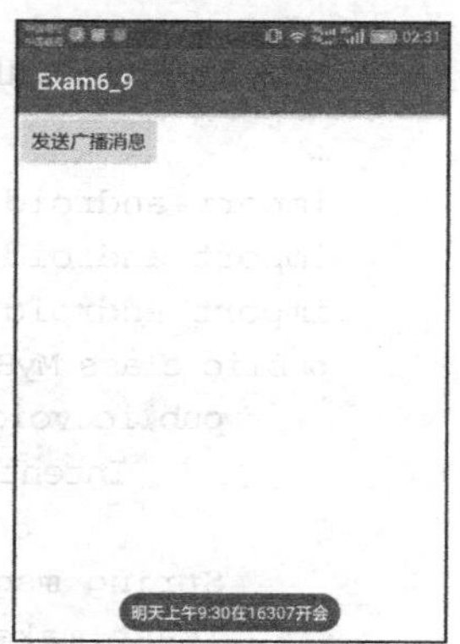

图 6.16　Exam6_9 运行结果

上面的例子只有一个接收者来接收广播，如果有多个接收者都注册了相同的广播地址，又会是什么情况呢？能同时接收到同一条广播吗？相互之间会不会有干扰呢？这就涉及普通广播和有序广播的概念了。

普通广播对于多个接收者来说是完全异步的，通常每个接收者都无需等待即可接收到广播，接收者相互之间不会有影响。对于这种广播，接收者无法终止广播，即无法阻止其他接收者的接收动作。

有序广播比较特殊，它每次只发送到优先级较高的接收者那里，然后由优先级高的接收者再传播到优先级低的接收者，优先级高的接收者有能力终止这个广播。

6.7.3　通过 Broadcast 启动 Service

由前面的学习知道，可以通过 Activity 程序启动 Service，其实 Service 也可以通过 Broadcast 来启动，一个 Service 要通过 Broadcast 来启动很简单，只需要在 Broadcast 中调用 startService()方法即可。

实例 6-10：BroadcastReceiver 组件操作举例

建立一个项目，名称为 Exam6_10，利用 Broadcast 来启动 Service，Service 的主要功能是播放一首歌，如何播放歌曲将在下一章中介绍。在上例的基础上，利用 BroadcastReceiver 组件启动 Service。

（1）在 res 文件夹下建立 raw 文件夹，将 abird.mp3 文件复制到 raw 文件夹中。

（2）建立一个服务 MyService.java，播放 abird.mp3，代码如下：

```
…
import java.io.IOException;
import android.app.Service;
import android.content.Intent;
import android.media.MediaPlayer;
import android.os.IBinder;
public class MyService extends Service     //定义 MyService.java 继承 Service 类
{   private MediaPlayer player;            //定义 MediaPlayer 变量 player
public IBinder onBind(Intent intent)       //定义 onBind 方法
{   return null; }
public void onCreate()                     //覆写 onCreate 方法
```

```
{    super.onCreate();
      player = MediaPlayer.create(MyService.this, R.raw.abird); //获得音频资源
      try {
     MyService.this.player.prepare();          //进入预备状态
   } catch (IllegalStateException e) {
      e.printStackTrace();
   } catch (IOException e) {
      e.printStackTrace();}
       MyService.this.player.start();          //播放文件
     player.setLooping(true);}                 //反复播放
   public void onDestroy()                     //覆写 onDestroy 方法
   {    super.onDestroy();    }
   public void onStart(Intent intent, int startId)     //覆写 onStart 方法
   {    super.onStart(intent, startId);  }}
```

（3）修改 MyBroadcastReceiver.java 文件，斜体是增加的部分，代码如下：

```
…
import android.content.BroadcastReceiver;
import android.content.Context;
import android.content.Intent;
public class MyBroadcastReceiver extends BroadcastReceiver {//继承BroadcastReceiver
    public void onReceive(Context context, Intent intent) { //覆写 onReceive 方法
         Intent serviceIntent = new Intent(context, MyService.class);//定义 Intent
              context.startService(serviceIntent);          //启动指定服务
    String msg=intent.getStringExtra("msg");
    Toast.makeText(context, msg, Toast.LENGTH_LONG).show();/}}/ 显示信息
```

（4）修改 AndroidManifest.xml 文件，在<application>与</application>之间增加如下代码：

```
<service android:enabled="true" android:name=".MyService" />
```

其他文件不做修改，保存所有文件，运行该项目，单击按钮，出现如图 6.16 所示结果的同时还会播放歌曲。

Broadcast 不仅可以用来启动 Service，还可以用来启动 Activity，只要在 Broadcast 中调用 startActivity () 方法即可。

经常开机就要启动某个服务或某个应用程序，要实现这样的功能，可以使用系统“启动完成”广播，接收到这条广播后就可以启动自己的服务或自己的应用程序了，Exam6_11 项目就是利用广播功能实现开机自动启动某个应用的实例，感兴趣的读者可下载学习。

本章小结

本章简要介绍了 Android 提供的 4 种基本组件的基础知识，详细介绍了 Intent 组件的使用，又对 Service、Broadcast 做了介绍。

习题

（1）Android 四大基本组件是什么？简述它们的大致功能。

（2）简要说明 startService()方法与 bindService()方法的区别。

（3）编程实现如下几个页面：登录页面、投票列表页面、投票页面，并能在这几个页面之间顺利跳转，如图 6.17 所示。

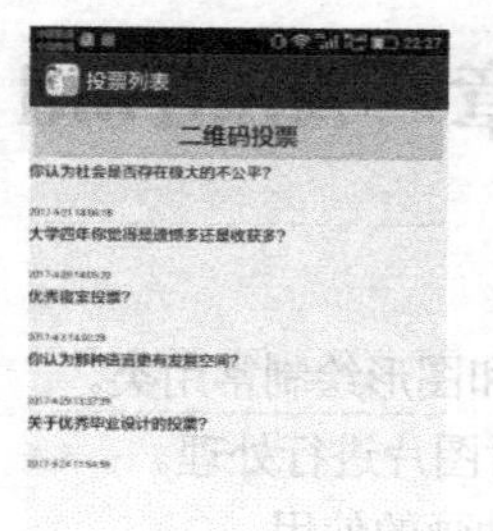

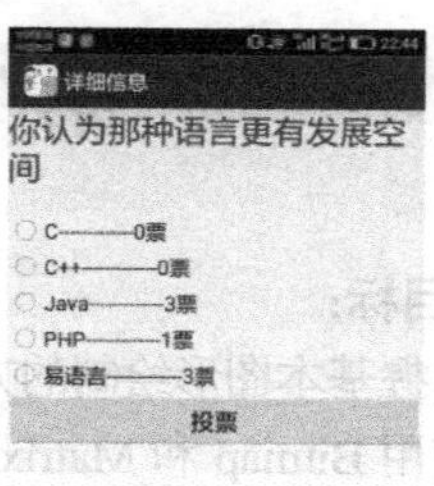

图 6.17　要实现的页面

第 7 章　Android 多媒体技术

学习目标：

- 掌握基本图形绘制的方法和图形绘制常用类。
- 使用 Bitmap 和 Matrix 类对图片进行处理。
- 掌握 Android 中两种常见动画的使用。
- 了解动画操作组件的运用。
- 使用 MediaPlayer 播放音频和视频文件。
- 使用 Camera 捕获 SurfaceView 采集视频数据。
- 掌握 Android 中 MediaRecorder 的常见用法。

多媒体信息包括文本、声音、图形、图像、视频、动画等，多媒体技术就是指处理这些信息的程序和过程。Android 对多媒体信息的处理提供了支持。

7.1　Android 中图形的绘制

7.1.1　图形绘制基础

动态图形绘制的基本思路如下：创建一个 View 类或者 SurfaceView 类，重写 onDraw()方法，使用 Canvas 对象在界面上绘制不同图形，使用 invalidate()方法刷新界面。动态绘制图形的常见类有 Canvas、Paint、Color、Path 等。

（1）Canvas 类：Canvas 类是画布类。其层次关系如下。

```
java.lang.Object
   android.graphics.Canvas
```

要在 Android 的 Java 程序中使用 Canvas 类，必须在程序中使用下面的语句。

```
import android.graphics.Canvas;        //导入 graphics.Canvas 类
```

Canvas 类提供了各种图形的绘制方法，如矩形、圆、椭圆、点、线、文字等，具体方法如表 7-1 所示。

表 7-1　Canvas 类常用方法

方法名称	方法描述
drawText(String text,float x,float y,Paint paint)	画文本
drawPoint(float x,float y,Paint paint)	画点
drawLine(float startX,float startY,float soptX,float stopY,Paint paint)	画线
drawCircle(float cx,float cy,flaot radius,Paint paint)	画圆
drawOval(rectF oval,paint paint)	画椭圆
drawRect(rectF rect,paint paint)	画矩形
drawRoundRect(rectF rect,float rx,float ry,paint paint)	画圆角矩形
clipRect(float left,float top,float right,float bottom)	裁剪矩形
clipRegion(Region region)	裁剪区域

（2）Paint 类：Paint 类是画笔类。其层次关系如下。

```
java.lang.Object
   android.graphics.Paint
```

要在 Android 的 Java 程序中使用 Paint 类，必须在程序中使用下面的语句。

```
import android.graphics.Paint;                    //导入 graphics.Paint 类
```

Paint 类用来描述图形的颜色和风格，如线宽、颜色、字体等信息。Paint 类常用方法如表 7-2 所示。

表 7-2　Paint 类常用方法

方法名称	方法描述
Paint()	构造方法
setColor(int color)	设置颜色
setStrokeWidth(float width)	设置线宽
setTextAlign(Paint.Align align)	设置文字对齐
setTextSize(float textSize)	设置文字尺寸
setShader(Shader shader)	设置渐变
setAlpha(int a)	设置 alpha 值
reset()	复位 Paint 默认设置

（3）Color 类：Color 类中定义了一些颜色常量和创建颜色的方法，其层次关系如下。

```
java.lang.Object
   android.graphics.Color
```

要在 Android 的 Java 程序中使用 Paint 类，必须在程序中使用下面的语句。

```
import android.graphics.Color;                    //导入 graphics.Color 类
```

Color 类定义使用 RGB 颜色模式，常见颜色的描述如表 7-3 所示。

表 7-3　Color 类常见颜色描述

颜色属性名称	描述
BLACK	黑色
BLUE	蓝色
CYAN	青色
DKGRAY	深灰色
GRAY	灰色
GREEN	绿色
LIGRAY	浅灰色
MAGENTA	紫色
RED	红色
TRANSSARENT	透明
WHITE	白色
YELLOW	黄色

（4）Path 类：点与点之间的连线。其层次关系如下。

```
java.lang.Object
   android.graphics.Path
```

要在 Android 的 Java 程序中使用 Path 类，必须在程序中使用下面的语句。

```
import android.graphics.Path;                    //导入 graphics.Path 类
```

Path 类一般用来从某个点到另一个点画线，例如，画梯形需要有点和连线，常用的方法如表 7-4 所示。

表 7-4　Path 类常用方法

方法名称	方法描述
lineTo(float x,float y)	从最后点到指定点画线
moveTo(float x,float y)	移动到某一点
reset()	复位

（5）Shader 类及其子类：用来渲染图像使用的类，子类有 BitmapShader、ComposeShader、LinearGradient、RadialGradient、SweepGradient。各个类的用法可参照 SDK 文档。

7.1.2　图形绘制实例

在一般的图形绘制中，创建一个 View 类，在类中覆写 onDraw()方法，然后在主程序中调用即可实现简单图形的绘制。

实例 7-1：绘制各种图形

新建一个项目，项目命名为 Exam7_1，绘制圆、正方形、长方形、椭圆、三角形、文字，并用不同的方式进行填充输出。

（1）创建一个 MyView.java 继承 View 类，代码如下：

```
import android.view.View;
import android.content.Context;
import android.graphics.Color;
import android.graphics.Paint;
import android.graphics.Rect;
import android.graphics.RectF;
import android.graphics.Path;
import android.graphics.Shader;
import android.graphics.Canvas;
import android.util.AttributeSet;
import android.graphics.LinearGradient;
import android.graphics.SweepGradient;
public class MyView extends View {                       //创建一个 View 类
public MyView(Context context){
  super(context);}
protected void onDraw(Canvas canvas) {                  //覆写 onDraw()方法
super.onDraw(canvas);
canvas.drawColor(Color.WHITE);                          //设置画布颜色
Paint paint = new Paint();
paint.setAntiAlias(true);                               //去除锯齿
paint.setColor(Color.RED);                              //设置线条颜色
paint.setStyle(Paint.Style.STROKE);                     //设置样式
paint.setStrokeWidth(3);                                //设置画笔粗细
canvas.drawCircle(40, 40, 30, paint);                   //画圆
canvas.drawRect(10, 90, 70, 150, paint);                //画矩形
canvas.drawRect(10, 170, 70, 200, paint);
RectF re = new RectF(10,220,70,250);                    //声明矩形区域
canvas.drawOval(re, paint);                             //画椭圆
```

```
//实例化 Path，画一个三角形
Path path = new Path();
path.moveTo(10,330);
path.lineTo(70, 330);
path.lineTo(40, 270);
path.close();
canvas.drawPath(path, paint);
paint.setStyle(Paint.Style.FILL);                    //设置填充
paint.setColor(Color.BLUE);                          //设置填充颜色
canvas.drawCircle(120, 40, 30, paint);               //画圆
canvas.drawRect(90, 90, 150, 150, paint);            //画矩形
canvas.drawRect(90, 170, 150, 200, paint);
RectF re2 = new RectF(90,220,150,250);               //声明矩形区域
canvas.drawOval(re2, paint);                         //画椭圆
//实例化 Path，画一个三角形
Path path2 = new Path();
path2.moveTo(90,330);
path2.lineTo(150, 330);
path2.lineTo(120, 270);
path2.close();
canvas.drawPath(path2, paint);
Shader lShader1 = new LinearGradient(0, 0, 100, 100, new int[]{
Color.RED,Color.GREEN,Color.BLUE,Color.YELLOW}, null, Shader.TileMode.REPEAT );
Shader lShader2 = new LinearGradient(0, 0, 100, 100,Color.RED,Color.GREEN,
                 Shader.TileMode.MIRROR);//设置红、绿、蓝、黄四色填充
Shader sShader1 = new SweepGradient(0, 0,new int[] {
Color.RED,Color.GREEN,Color.BLUE,Color.YELLOW}, null);
Shader sShader2 = new SweepGradient(0, 0,Color.RED,Color.GREEN );
paint.setShader(lShader1);
canvas.drawCircle(200, 40, 30, paint);               //画圆
canvas.drawRect(170, 90, 230, 150, paint);           //画矩形
canvas.drawRect(170, 170, 230, 200, paint);
RectF re3 = new RectF(170,220,230,250);              //声明矩形区域
canvas.drawOval(re3, paint);                         //画椭圆
//实例化 Path，画一个三角形
Path path4 = new Path();
path4.moveTo(170,330);
path4.lineTo(230, 330);
path4.lineTo(200, 270);
path4.close();
canvas.drawPath(path4, paint);
paint.reset();
paint.setColor(Color.BLACK);//
paint.setTextSize(24);
canvas.drawText("圆形", 240, 50, paint);                //绘制输出文字
canvas.drawText("正方形", 240, 120, paint);
canvas.drawText("长方形", 240, 190, paint);
canvas.drawText("椭圆形", 240, 250, paint);
canvas.drawText("三角形", 240, 320, paint);}}
```

（2）创建一个主程序 MainActivity.java 并调用 MyView，代码如下：

```
public class MainActivity extends Activity {
```

```
    public void onCreate(Bundle savedInstanceState) {
        super.onCreate(savedInstanceState);
        setContentView(new MyView(this));    }}     //调用建立的类
```

保存文件，运行该项目，结果如图 7.1 所示。

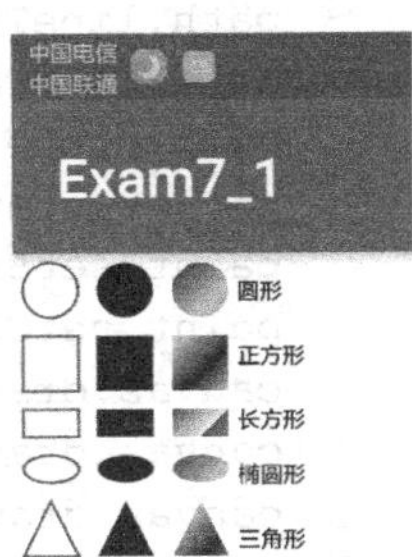

图 7.1　Exam7_1 运行结果

7.2　Android 中图像的处理

Android 可以支持的图像格式有 PNG、JPG、GIF 和 BMP 等。Android 提供了用于操作图像资源的操作类，使用此类可以直接从资源文件之中进行图像资源的读取，并且对这些图像进行一些简单的修改。

7.2.1　图像的获取

访问图像一般有 3 种方式：一是使用在工程中保存的图像；二是使用 XML 定义 Drawable 属性；三是使用构造器来完成。

前面学习了 ImageView 组件，利用它可以方便地获取图像文件，将图像文件 test.jpg 复制到 res\drawable 文件夹下，然后通过如下代码使用该图像。

（1）在程序 MainActivity 中获取该图像。

```
ImageView img = (ImageView)findViewById(R.id.img);
img.setImageResource(R.drawable.test);
```

（2）在布局管理文件 activity_main.xml 中获取该图像。

```
<ImageView
    android:id="@+id/img"
    android:layout_width="fill-parent"
    android:layout_height="wrap_content"
    android:src="@drawable/test"/>
```

（3）也可以利用 Bitmap 和 BitmapFactory 两个类来获取图像文件，这两个类的常用方法如表 7-5、表 7-6 所示。

表 7-5　Bitmap 类的常用方法

方法	描述
public static Bitmap createBitmap (Bitmap src)	复制一个 Bitmap
public static Bitmap createBitmap(Bitmap source, int x, int y, int width, int height, Matrix m, boolean filter)	对一个 Bitmap 进行剪切
public final int getHeight()	取得图像的高
public final int getWidth()	取得图像的宽
public static Bitmap createScaledBitmap(Bitmap src, int dstWidth, int dstHeight, boolean filter)	创建一个指定大小的 Bitmap

表 7-6　BitmapFactory 类的常用方法

方法	描述
public static Bitmap decodeByteArray(byte[] data, int offset, int length)	根据指定的数据文件创建 Bitmap
public static Bitmap decodeFile(String pathName)	根据指定的路径创建 Bitmap
public static Bitmap decodeResource(Resources res, int id)	根据指定的资源创建 Bitmap
public static Bitmap decodeStream(InputStream is)	根据指定的 InputStream 创建 Bitmap

例如：

```
String path = "/sdcard/img.jpg";                      //指定图像的路径
Bitmap bmp = BitmapFactory.decodeFile(path);          //根据指定路径创建图像
ImageView img = new ImageView(this);                  //定义 ImageView 组件
img.setImageBitmap(bmp);                              //获得图像
this.setContentView(img); }}                          //显示图像
```

Bitmap 类还提供了 compress()接口以支持压缩的图像，目前只支持 PNG、JPG 格式的压缩，读者可以自行查看 API 资料进行了解。

7.2.2 对获取的图像进行处理

Android 手机中提供了非常丰富的图像处理功能，例如，图像的缩放、平移、旋转、水印、倒影，等等。这里仅仅介绍对图像进行平移、旋转、缩放、倾斜等变换的操作方法，其他的功能可以参考 Android API。

图像缩放的具体实现有如下几种方法。

（1）将一个位图按照需求重画一遍，画后的位图就是用户需要的，与原位图的显示几乎一样：drawBitmap(Bitmap bitmap, Rect src, Rect dst, Paint paint)。

（2）在原有位图的基础上，缩放原位图，创建一个新的位图：CreateBitmap(Bitmap source, int x, int y, int width, int height, Matrix m, boolean filter)。

（3）借助 Canvas 的 scale(float sx, float sy)，注意此时对整个画布进行了缩放。

（4）借助 Matrix 类实现缩放，这个方法比较灵活。

Android 系统中 Matrix 类中有一个 3×3 的矩阵坐标，Matrix 类的方法如表 7-7 所示，通过这些方法可以实现图像的旋转、平移和缩放等操作。

表 7-7 Matrix 类常用方法

方法	描述
public void reset()	重设矩阵
public void setTranslate(float dx, float dy)	设置矩阵平移(dx, dy)
public void setScale(float sx, float sy)	设置矩阵缩放（sx,sy）
public void setScale(float sx, float sy, float px, float py)	设置矩阵以(px,py)坐标点为中心进行缩放（sx,sy）
public void setRotate(float degrees)	设置矩阵以原点为中心旋转 degrees
public void setRotate(float degrees, float px, float py)	设置矩阵以坐标点（px,py)为中心旋转
public void setSkew(float kx, float ky)	设置矩阵以原点为中心倾斜（kx,ky）
public void setSkew(float kx, float ky, float px, float py)	设置矩阵以坐标点（px,py)为中心倾斜

Matrix 的操作共有 translate(平移)、rotate(旋转)、scale(缩放)和 skew(倾斜)四种，除了 translate 之外，其他 3 种操作都可以指定中心点，每一种变换都提供了 set、post 和 pre 三种操作方式，以 Scale 为例，setScale ()方法、postScale ()方法和 preScale ()方法都可以实现缩放。post、pre 和 set 其实代表了 Matrix 中方法变换的次序，pre 指向前加入队列执行，post 指从后面加入队列执行，set 方法一旦调用就会清空之前 Matrix 中的所有变换，更详细的使用可以查看 Android API 文档。

例如，左边方框的执行顺序等价于右边方框，如下图所示。

Canvas 里 scale、translate、rotate 方法都是 pre 方法，如果要进行更多的变换，则可以先从 Canvas 中获得 Matrix，变换后再设置回 Canvas。

matrix.preScale(0.5f, 1);
matrix.preTranslate(10, 0);
matrix.postScale(0.7f, 1);
matrix.postTranslate(15, 0);
	translate(10, 0) -> scale(0.5f, 1) -> scale(0.7f, 1) -> translate(15, 0)，注意：后调用的 pre 操作先执行，而后调用的 post 操作后

matrix.preScale(0.5f, 1);
matrix.setScale(1, 0.6f);
matrix.postScale(0.7f, 1);
matrix.preTranslate(15, 0);
	translate(15, 0) -> scale(1, 0.6f) ->scale(0.7f, 1) ，matrix.preScale (0.5f, 1)将不起作用

Matrix 进行图像处理的基本步骤如下。

（1）获取 Matrix 对象，可以新建，也可以获取其他对象内封装的 Matrix。

例如：

```
Matrix matrix = new Matrix();
```

（2）调用 Matrix 的方法进行图像的平移、旋转、缩放、倾斜等操作。

例如：

```
matrix.reset();//重置
   matrix.setTranslate()/setSkew()/setRotate()/setScale()//平移/倾斜/旋转/缩放
```

（3）将程序对 Matrix 所做的变换应用到指定图像或组件。

例如：

```
//获取图像（先将图像文件 love.jpg 复制到 res\drawable 文件夹中）
    this.bitmap = BitmapFactory.decodeResource(super.getResources(),love);
    ...
    protected void onDraw(Canvas canvas) {                //覆写 onDraw()方法
      canvas.drawBitmap(this.bitmap, this.matrix, null);} //按照Matrix要求画图
```

7.2.3 图像处理举例

实例 7-2：图像处理实例

新建一个项目，项目命名为 Exam7_2，显示一张指定大小的图片，然后显示对这张图片做缩放、旋转、平移操作的图片。

（1）创建一个 MyView.java 继承 View 类代码如下：

```
   ...
   import android.graphics.Bitmap;
   import android.graphics.BitmapFactory;
   import android.graphics.Canvas;
   import android.graphics.Matrix;
   import android.util.AttributeSet;
   import android.graphics.drawable.BitmapDrawable;
   import android.graphics.Paint;
  public class MyView extends View {                   //继承 View
  private Bitmap bitmap = null ;                       //定义 bitmap
  private Matrix matrix = new Matrix();                //定义 matrix
  public MyView(Context context, AttributeSet attrs) {
     super(context, attrs);
```

```
    //取得项目中的 Bitmap
    this.bitmap=BitmapFactory.decodeResource(super.getResources(),R.drawable.t
estl);
bitmap = Bitmap.createScaledBitmap(bitmap, 1200, 640,true);//创建指定大小图片
     //对图片进行缩放、旋转、平移操作
     this.matrix.preScale(0.5f, 0.5f, 650, 650);           //缩小到原来的一半
     this.matrix.preRotate(10, 750, 750) ;                 //在指定坐标翻转 60°
     this.matrix.preTranslate(150, 800) ;}                 //图像平移
     protected void onDraw(Canvas canvas) {                //覆写 onDraw()方法
        canvas.drawBitmap(this.bitmap, this.matrix, null);//画变换后的图
        canvas.drawBitmap(this.bitmap, 0,0, null); }}     //画原图
```

（2）创建一个布局管理文件 activity_main.xml，包含 MyView，代码如下：

```
    ...
    < org.hnist.demo.MyView                      //引用建立的 MyView 类
        android:layout_width="wrap_content"
        android:layout_height="wrap_content" >
    </LinearLayout>
```

（3）创建一个主程序 MainActivity.java，调用布局管理文件 activity_main.xml。

保存所有文件，运行该项目，结果如图 7.2 所示。

图 7.2 Exam7_2 运行结果

7.3 Android 中的动画

在 Android 中，可以使用 Animation 组件对手机屏幕上的文字或者图像等对象进行旋转、移动、淡入淡出等动画处理，Android 中的动画分为以下两类。

Tweened Animation（渐变动画）：该类的 Animation 可以完成控件的旋转、移动、伸缩、淡入淡出等特效。

Frame Animation（帧动画）：采用帧的方式进行动画效果的编排，将预先定义好的对象按照顺序播放出来。

7.3.1 Tween 动画

Tween 动画可以使视图组件移动、放大、缩小，以及产生透明的变化。例如，在一个 ImageView 组件中，通过 Tween 动画可以使该视图实现缩放、旋转、渐变、平移等效果。Tween 动画相关的类在 android.view.animation 包中，其常用的方法如表 7-8 所示。

（1）Animation：抽象类，其他几个动画类继承自该类。

（2）ScaleAnimation：控制缩放变化动画类。

（3）AlphaAnimation：控制透明变化的动画类。

（4）RotateAnimation：控制旋转变化的动画类。

（5）TranslateAnimation：控制移动变化的动画类。

（6）AnimationSet：定义动画属性集合类。

（7）AnimationUtils：动画的工具类。

表 7-8　Animation 类常用的方法

方法	描述
Public Animation()	创建一个 Animation 对象
public void setDuration（long time）	设置动画持续时间（毫秒）
public void setFillAfter (boolean fillafter)	为 true 时表示该动画在动画结束后被应用
public void setFillBefore (boolean fillbefore)	为 true 时表示该动画在动画开始前被应用
public void setFillEnable (boolean fillenable)	setFillAfter、 setFillBefore 方法是否有效
public void setRepeatCount（int repeatcount）	动画反复执行次数
public void setRepeatMode（int repeatmode）	动画反复执行模式
public void setStartOffset（long startoffset）	设置动画在多少毫秒后开始
public void setInterpolator(Interpolator i)	设置动画变化的速率
public boolean hasStarted()	判断动画是否已运行
public boolean hasEnded()	判断动画是否已结束
public long getDuration ()	返回动画持续时间
public void cancel()	取消动画
public void reset()	动画回到初始状态
public void start()	动画开始执行
public void setAnimationListener(listener)	设置动画监听

其中还有一些常用常量，例如，RESTART——重新运行，INFINITE——反复运行，REVERSE——反转动画效果，等等。

Tween 动画一共有 4 种形式：Scale（缩放动画）、Alpha（渐变动画）、Translate（位置变化动画）、Rotate（旋转变化动画）。

缩放动画由 ScaleAnimation 类来实现，ScaleAnimation 类中常用的方法如下。

Public ScaleAnimation (context,Aattrs)，传入指定的属性以创建 ScaleAnimation 对象。

Public ScaleAnimation(float fromX,float toX,float fromY,float toY)，传入开始点和结束点坐标以创建 ScaleAnimation 对象。其中，fromX 表示动画开始时 X 的坐标值，toX 表示动画结束时 X 的坐标值，fromY 表示动画开始时 Y 的坐标值，toY 表示动画结束时 Y 的坐标值。

渐变动画由 AlphaAnimation 类来实现，AlphaAnimation 类中常用的方法如下。

Public AlphaAnimation(context,Aattrs)，传入指定的属性以创建 AlphaAnimation 对象。

Public AlphaAnimation(float fromAlpha,float,toAlpha)，传入动画开始和结束值以创建 AlphaAnimation 对象。其中，fromAlpha 表示动画开始的透明度，toAlpha 表示动画结束前的透明度（取值为 0.0～1.0）。

平移动画由 TranslateAnimation 类来实现，TranslateAnimation 类中常用的方法如下。

Public TranslateAnimation (context,Aattrs)，传入指定的属性创建 TranslateAnimation 对象。

Public TranslateAnimation(float fromXData,float toXData,float fromYData,float toYData)，传入移动的坐标创建 TranslateAnimation 对象。其中，fromXData 表示动画开始平移前 X 坐标值，toXData 表示动画开始平移后 X 坐标值，fromYData 表示动画开始平移前 Y 坐标值，toYData 表示动画开始平移后 Y 坐标值。

旋转动画由 RotateAnimation 类来实现，RotateAnimation 类中常用的方法如下。

Public RotateAnimation (context,Aattrs)，传入指定的属性创建 RotateAnimation 对象。

Public RotateAnimation(float fromDegrees,float toDegrees)，传入指定旋转角度的范围创建 RotateAnimation 对象。其中，fromDegrees 表示起始旋转角度，toDegrees 表示结束旋转角度，正数表示顺时针旋转，负数表示逆时针旋转。

要创建一个 Tween 动画，一般有以下几个关键步骤。

（1）定义一个 Animation 对象，例如：

```
Animation scaleAnimation = new ScaleAnimation(0.1f,1.0f,0.1f,1.0f);//初始化
```

（2）设置动画运行时间，例如：

```
scaleAnimation.setDuration(500);        //设置动画时间
```

（3）开始运行动画，例如：

```
startAnimation(scaleAnimation);         //开始运行动画
```

7.3.2 创建动画实例

实例 7-3：Tween 动画操作举例

1．简单动画创建

新建一个项目，项目命名为 Exam7_3，通过一个图片缩放的例子来介绍动画的创建，其他的动画形式可以参考此实例实现。

（1）创建一个布局管理文件 activity_main.xml，代码如下：

```
…
<ImageView                                          //添加图片组件
     android:id="@+id/myimg"                        //设置组件 ID，在程序中使用
     android:layout_width="wrap_content"
     android:layout_height="wrap_content"
     android:src="@drawable/love" />                //图片为 drawable\love.jpg 文件
</LinearLayout>
```

（2）建立一个 MainActivity.java 文件，代码如下：

```
…
import android.view.animation.Animation;
import android.view.animation.ScaleAnimation;
import android.widget.ImageView;
public class MainActivity extends AppCompactActivity{
private ImageView img = null;
@Override
public void onCreate(Bundle savedInstanceState) {
     super.onCreate(savedInstanceState);
     super.setContentView(R.layout. activity_main);
     this.img = (ImageView) super.findViewById(R.id.myimg);    //取得组件
     this.img.setOnClickListener(new OnClickListenerImpl());} //设置监听
private class OnClickListenerImpl implements OnClickListener {
public void onClick(View view) {
      Animation scaleAnimation = new ScaleAnimation(0.1f,1.0f,0.1f,1.0f);
                                        //初始化
      scaleAnimation.setDuration(3000) ; //3 秒完成动画
      MainActivity.this.img.startAnimation(scaleAnimation) ; }}}//启动动画
```

保存所有文件，运行该项目，在手机上显示一张图片，点击这张图片，图片会从小逐渐变大，然后停在手机屏幕上。

只要修改 MainActivity.java 中的最后三条语句，就可以实现其他动画效果，读者可以参照下面的语

句自行完成，注意 import 中要做适当的修改。

① 设置 AlphaAnimation：

```
Animation alphaAnimation = new AlphaAnimation(0.1f, 1.0f);    //初始化
alphaAnimation.setDuration(3000);                          //设置动画时间为 3 秒
MainActivity.this.img.startAnimation(alphaAnimation) ; }}}    //启动动画
```

② 设置 RotateAnimation：

```
Animation rotateAnimation = new RotateAnimation(0f, 360f);    //初始化
rotateAnimation.setDuration(1000);                             //设置动画时间
MainActivity.this.img.startAnimation(rotateAnimation) ; }}}   //启动动画
```

③ 设置 TranslateAnimation：

```
Animation tranAnimation=new TranslateAnimation(0.1f,10.0f,0.1f,10.0f);
                                                               //初始化
tranAnimation.setDuration(1000);                               //设置动画时间
MainActivity.this.img.startAnimation(tranAnimation) ; }}}     //启动动画
```

2．复杂动画创建

前面介绍了单个动画效果的实现，能否使上面设置的多个动画效果在一个对象上生效呢？利用 AnimationSet 类就可以轻松实现，AnimationSet 类常用方法如表 7-9 所示。

表 7-9　AnimationSet 类的常用方法

方法名称	描述
public AnimationSet(boolean shareInterpolator)	如果设置为 true，则表示使用 AnimationSet 所提供的 Interpolator（速率）；如果为 false，则使用各个动画效果自己的 Interpolator
public void addAnimation(Animation a)	增加一个 Animation 组件
public List<Animation> getAnimations()	取得所有的 Animation 组件
public long getDuration()	取得动画的持续时间
public long getStartTime()	取得动画的开始时间
public void reset()	重置动画
public void setDuration(long durationMillis)	设置动画的持续时间
public void setStartTime(long startTimeMillis)	设置动画的开始时间

要创建多个 Tween 动画效果，一般有以下几个关键步骤。

（1）建立多个动画对象，例如：

```
Animation tran=new TranslateAnimation(0.1f,10.0f,0.1f,10.0f);
                                              //初始化 Translate 动画
Animation alpha=new AlphaAnimation(0.1f, 1.0f);  //初始化 Alpha 动画
Animation rotate=new RotateAnimation(0f, 360f);  //初始化 Rotate 动画
Animation scale = new  ScaleAnimation(0.1f,1.0f,0.1f,1.0f);
                                              //初始化 Scale 动画
```

（2）建立动画集，例如：

```
AnimationSet set = new AnimationSet(true);    //定义一个动画集
set.addAnimation(tranAnimation);              //增加动画 tran
set.addAnimation(alphaAnimation);             //增加动画 alpha
...
```

（3）设置动画时间，作用到每个动画上，例如：

```
set.setDuration(1000);                                          //设置动画时间
this.startAnimation(set);                                       //启动动画
```

实例 7-4：复杂动画的创建举例

新建一个项目，项目命名为 Exam7_4，通过含有多个 Tween 动画效果的例子来介绍复杂动画的创建。

（1）创建布局管理文件 activity_main.xml，代码与实例 7-3 中的 activity_main.xml 一样。

（2）建立一个 MainActivity.java 文件，代码如下：

```
…
import android.view.View.OnClickListener;
import android.view.animation.AlphaAnimation;
import android.view.animation.Animation;
import android.view.animation.AnimationSet;
import android.view.animation.RotateAnimation;
import android.view.animation.ScaleAnimation;
import android.view.animation.TranslateAnimation;
public class MainActivity extends AppCompatActivity{
private ImageView img = null;
@Override
public void onCreate(Bundle savedInstanceState) {
    super.onCreate(savedInstanceState);
    super.setContentView(R.layout. activity_main);
    this.img = (ImageView) super.findViewById(R.id.myimg);     //取得组件
    this.img.setOnClickListener(new OnClickListenerImpl());} //设置监听
 private class OnClickListenerImpl implements OnClickListener {
    public void onClick(View view) {
    AnimationSet set = new AnimationSet(true);            //定义一个动画集
    Animation tran=new TranslateAnimation(0.1f,10.0f,0.1f,10.0f);
                                                          //初始化 Translate 动画
    Animation alpha=new AlphaAnimation(0.1f, 1.0f);      //初始化 Alpha 动画
    Animation rotate=new RotateAnimation(0f, 360f);      //初始化 Rotate 动画
    Animation scale = new  ScaleAnimation(0.1f,1.0f,0.1f,1.0f);
                                                          //初始化 Scale 动画
        scale.setRepeatCount(2) ;                             //动画重复 2 次
        set.addAnimation(tran) ;                              //增加动画 tran
        set.addAnimation(scale) ;                             //增加动画 scale
        set.addAnimation(alpha) ;                             //增加动画 alpha
        set.addAnimation(rotate) ;                            //增加动画 rotate
        set.setDuration(3000) ;                               //动画持续时间为 3 秒
        MainActivity.this.img.startAnimation(set) ;  }}} //启动动画
```

保存所有文件，运行该项目，在手机上会显示一张图片，点击这张图片，会逐个展示设置的动画效果，然后停在手机屏幕上。

7.3.3 通过 XML 文件来创建动画

前面在 Activity 中通过 Java 代码，实现了 4 种不同的 Tween 动画，其实在 Android 中完全可以通过 XML 文件来实现动画，其可定义的动画效果元素如表 7-10 所示。这样做更加简洁、清晰，也更利

于重用，使程序与配置分离开，建议采用这种形式。

表 7-10　可定义的动画效果元素

可配置的元素	描述
<set>	为根节点，定义全部的动画元素
<alpha>	定义渐变动画效果
<scale>	定义缩放动画效果
<translate>	定义平移动画效果
<rotate>	定义旋转动画效果

<alpha>节点的常用的属性：fromAlpha 表示动画开始的透明度，toAlpha 表示动画结束前的透明度（取值是 0.0～1.0）。

<scale>节点的常用的属性：fromXScale 表示动画开始时的 X 坐标值，toXScale 表示动画结束时的 X 坐标值，fromYScale 表示动画开始时的 Y 坐标值，toYScale 表示动画结束时的 Y 坐标值。

<translate>节点的常用的属性：fromXData 表示动画开始平移前的 X 坐标值，toXData 表示动画开始平移后的 X 坐标值，fromYData 表示动画开始平移前的 Y 坐标值，toYData 表示动画开始平移后的 Y 坐标值。

<rotate>节点的常用的属性：fromDegrees 表示起始旋转角度，toDegrees 表示结束旋转角度，正数表示顺时针旋转，负数表示逆时针旋转。

XML 文件可配置公共属性如表 7-11 所示，Interpolator 对象配置情况如表 7-12 所示。

表 7-11　可以配置的公共属性

可配置的属性	数据类型	描述
android:duration	long	定义动画的持续时间，以毫秒为单位
android:fillAfter	boolean	设置为 true 时表示该动画转化在动画结束后被应用
android:fillBefore	boolean	当设置为 true 时，该动画转化在动画开始前被应用
android:interpolator	String	动画插入器，如 decelerate_interpolator（减速动画）
android:repeatCount	int	动画重复执行的次数
android:repeatMode	String	动画重复的模式（restart、reverse）
android:startOffset	long	动画之间的间隔
android:zAdjustment	int	动画的 Z Order 配置：0（保持 Z Order 不变）、1（保持在最上层）、-1（保持在最下层）
android:interpolator	String	指定动画的执行速率

表 7-12　Interpolator 对象的配置情况

可配置的属性	描述
@android:anim/accelerate_decelerate_interpolator	先加速再减速
@android:anim/accelerate_interpolator	加速
@android:anim/anticipate_interpolator	先回退一小步然后加速前进
@android:anim/anticipate_overshoot_interpolator	在前一基础上超出终点一点再到终点
@android:anim/bounce_interpolator	最后阶段弹球效果
@android:anim/cycle_interpolator	周期运动
@android:anim/decelerate_interpolator	减速
@android:anim/linear_interpolator	匀速
@android:anim/overshoot_interpolator	快速到达终点并超出一点最后到终点

Android 中所有定义好的动画配置 XML 文件都要求保存在 res/anim 文件夹中，然后使用 android.view.animation.AnimationUtils 类中的 loadAnimation 方法来读取这些动画配置文件。

实例 7-5：通过 XML 文件来创建动画

新建一个项目，项目命名为 Exam7_5，通过设置 XML 文件来实现实例 7-3 的效果。

（1）选择“New→Folder→Res Folder”选项，建立一个文件夹 anim 并位于 res 下，选择“New→File”选项，在文件夹 anim 下建立一个 scale.xml 文件，代码如下：

```
<?xml version="1.0" encoding="utf-8"?>
<set xmlns:android="http://schemas.android.com/apk/res/android">
< scale                                     //定义缩放动画
    android:fromXScale="0.1"                //组件从 X 轴缩小到原来的 1/10 开始
    android:toXScale="1.0"                  //组件到 X 轴满屏显示结束
    android:fromYScale="0.1"                //组件从 Y 轴缩小到原来的 1/10 开始
    android:toYScale="1.0"                  //组件到 Y 轴满屏显示结束
    android:duration="3000" />              //动画持续的时间为 3 秒
</set>
```

（2）定义布局管理器文件 activity_main.xml，其配置与实例 7-3 中的 activity_main.xml 一样。

（3）定义 Activity 程序 MainActivity.java，读取 scale.xml 文件，代码如下：

```
…
import android.view.animation.Animation;
import android.view.animation.AnimationUtils;
import android.view.animation.ScaleAnimation;
public class MainActivity extends AppCompatActivity{
private ImageView img = null;
public void onCreate(Bundle savedInstanceState) {
    super.onCreate(savedInstanceState);
    super.setContentView(R.layout. activity_main);
    this.img = (ImageView) super.findViewById(R.id.myimg);   //取得组件
    this.img.setOnClickListener(new OnClickListenerImpl());} //设置监听
private class OnClickListenerImpl implements OnClickListener {
    public void onClick(View view) {
        Animation anim = AnimationUtils.loadAnimation( MainActivity.this,
                    R.anim.scale);    //读取动画配置文件 scale.xml
        MainActivity.this.img.startAnimation(anim) ; }}}        //启动动画
```

保存所有文件，运行结果与实例 7-3 一致。可以设置多个不同的动画配置 XML 文件，然后通过语句 Animation anim = AnimationUtils.loadAnimation(MainActivity.this,R.anim.all);来调用所有的动画配置文件，实现多个动画的效果。

7.3.4 Frame 动画

Frame 动画采用帧的方式进行动画效果的编排，所有的动画会按照事先定义好顺序执行，以这样的顺序播放图像产生的动画效果类似于电影。例如，要实现一个人走路的动画效果，可以通过 3 张图像重复不停播放来实现：第一张是两脚着地；第二张是右脚抬起，左脚着地；第三张是左脚抬起，右脚着地。

如果想使用这种动画，则需要利用 android.graphics.drawable.AnimationDrawable 类来实现。该类中有两个重要的方法：start()和 stop()，分别用来开始和停止动画。

动画一般通过 XML 配置文件来进行配置。在 res/anim 目录中定义 XML 文件，该文件的根元素是<animation-list>，子元素是<item>，子元素可以有多个。

Frame 动画也可以在 XML 文件中进行动画的配置，同样需要将配置文件保存在 res/anim 文件夹中，但是此配置文件的根节点为“<animation-list>”，其中包含多个“<item>”元素，用于定义每一帧动画，其可以配置的属性如表 7-13 所示。

表 7-13　animation-list 常用配置的属性

属性	描述
android:drawable	每一帧动画的资源
android:duration	动画的持续时间
android:oneshot	是否只显示一次，true 表示只显示一次，false 表示重复显示
android:visible	定义 drawable 是否初始可见

要创建一个 Frame 动画，一般有以下几个关键步骤。

（1）定义一个动画配置文件，如 frame.xml。

```
<animation-list                                    //定义动画集合
xmlns:android="http://schemas.android.com/apk/res/android"
android:oneshot="true">                            //默认为显示一次
<item                                              //定义动画帧
    android:drawable="@drawable/tiger1"  //引入的图片文件
    android:duration="200" />                      //动画持续时间为 0.2 秒
  ......                              //可以有多个<item>项，将多个图片文件引入到动画中
</animation-list>
```

（2）定义布局管理文件，例如，定义两个按钮，通过单击按钮启动和停止动画。

（3）定义 Activity 程序，操作帧动画。

```
private class OnClickListenerstart implements OnClickListener {
    @Override
    public void onClick(View view) {
      //设置动画资源
    MainActivity.this.img.setBackgroundResource(R.anim. framedemo);
      //取得 Drawable
    MainActivity.this.draw= (AnimationDrawable)MainActivity.this.
                           img.getBackground();
    MainActivity.this.draw.setOneShot(false);              //动画执行次数
    MainActivity.this.draw.start();}}                      //开始动画
private class OnClickListenerstop implements OnClickListener {
    @Override
    public void onClick(View view) {
        MainActivity.this.draw.stop();}}}                  //停止动画
```

实例 7-6：Frame 动画实例

新建一个项目，项目命名为 Exam7_6，建立帧动画，实现如下功能：单击“开始动画”按钮，老虎开始奔跑，单击“停止动画”按钮，老虎立刻停止不动。

（1）定义一个动画配置文件， frame.xml，放置在 res\drawable 文件夹下，代码如下：

```
<animation-list                                    //定义动画集合
xmlns:android="http://schemas.android.com/apk/res/android"
android:oneshot="true">                            //默认为显示一次
<item                                              //定义动画帧
    android:drawable="@drawable/tiger1"  //引入的图片文件
    android:duration="200" />                      //动画持续时间为 0.2 秒
<item
    android:drawable="@drawable/tiger2"
    android:duration="200" />
```

```
    <item
        android:drawable="@drawable/tiger3"
        android:duration="200" />
    <item
        android:drawable="@drawable/tiger4"
        android:duration="200" />
</animation-list>
```

（2）定义布局管理文件，例如，定义两个按钮，通过单击按钮启动和停止动画，代码如下：

```
…
<ImageView
    android:id="@+id/img"
    android:layout_width="wrap_content"
    android:layout_height="wrap_content"/>
 <Button
    android:id="@+id/start"
    android:layout_width="wrap_content"
    android:layout_height="wrap_content"
    android:text="开始动画"/>
 <Button
    android:id="@+id/stop"
    android:layout_width="wrap_content"
    android:layout_height="wrap_content"
    android:text="停止动画"/>
</LinearLayout>
```

（3）定义 Activity 程序 MainActivity.java 以操作帧动画，代码如下：

```
…
import android.graphics.drawable.AnimationDrawable;
public class MainActivity extends AppCompatActivity{
private ImageView img = null;
private Button start = null;
private Button stop = null;
private AnimationDrawable draw = null;                      //定义动画操作
@Override
public void onCreate(Bundle savedInstanceState) {
    super.onCreate(savedInstanceState);
    super.setContentView(R.layout.main);
    this.img = (ImageView) super.findViewById(R.id.img);
    this.start = (Button) super.findViewById(R.id.start);
    this.stop = (Button) super.findViewById(R.id.stop);
    this.start.setOnClickListener(new OnClickListenerstart()) ;
                                                        //设置开始按钮监听
    this.stop.setOnClickListener(new OnClickListenerstop()) ;}
                                                        //设置停止按钮监听
    private class OnClickListenerstart implements OnClickListener {
    public void onClick(View view) {
                                                        //设置动画资源
    MainActivity.this.img.setBackgroundResource(R.anim.frame);
                                                        //取得背景的 Drawable
    MainActivity.this.draw=(AnimationDrawable)MainActivity.this.
                              img.getBackground();
```

```
    MainActivity.this.draw.setOneShot(false);          //动画执行次数
    MainActivity.this.draw.start(); }}                 //开始动画
    private class OnClickListenerstop implements OnClickListener {
        public void onClick(View view) {
        MainActivity.this.draw.stop();    }}  }//停止
```

保存所有文件，运行该项目，单击“开始动画”按钮，帧动画开始，单击“停止动画”按钮，帧动画停止，如图 7.3 所示。

图 7.3　Exam7_6 运行结果

7.3.5　动画监听器

在对动画的操作过程中，还可以对动画的一些操作状态进行监听，例如，动画是否启动、动画是否正重复执行、动画是否结束，Android 中提供了用于这些动作监听的接口：android.view.animation.Animation.AnimationListener。在此接口中定义了 3 个监听动画的操作方法。

动画开始时触发：public abstract void onAnimationStart (Animation animation)。

动画重复时触发：public abstract void onAnimationRepeat (Animation animation) 。

动画结束时触发：public abstract void onAnimationEnd(Animation animation)。

实例 7-7：动画监听器举例

新建一个项目，项目命名为 Exam7_7，先建立一个缩放动画，动画会重复执行 2 次，设置动画监听器，当缩放动画启动后再增加一个渐变动画。

（1）定义布局管理文件 activity_main.xml，代码如下：

```
…
<ImageView
    android:id="@+id/img"
    android:layout_width="fill_parent"
    android:layout_height="wrap_content"
    android:src="@drawable/love" />
</LinearLayout>
```

（2）建立 Activity 文件 MaiinActivity.java，代码如下：

```
…
import android.view.animation.Animation;
import android.view.animation.Animation.AnimationListener;
import android.view.animation.AnimationSet;
import android.view.animation.RotateAnimation;
import android.view.animation.ScaleAnimation;
public class MainActivity extends AppCompatActivity{
private ImageView img = null;
public void onCreate(Bundle savedInstanceState) {
    super.onCreate(savedInstanceState);
    super.setContentView(R.layout.activity_main);
    this.img = (ImageView) super.findViewById(R.id.img);
    AnimationSet set = new AnimationSet(true);         //定义一个动画集
    Animation scale = new  ScaleAnimation(0.1f,1.0f,0.1f,1.0f);
```

```
                                                        //初始化 Scale 动画
        scale.setRepeatCount(2) ;                       //动画重复 2 次
        set.addAnimation(scale) ;                       //增加动画 scale
        set.setDuration(3000) ;                         //动画持续时间为 3 秒
        MainActivity.this.img.startAnimation(set) ;     //启动动画
        set.setAnimationListener(new AnimationListenerImpl()) ;  //设置动画监听
        this.img.startAnimation(set) ;   }              //启动动画
    private class AnimationListenerImpl implements AnimationListener {
        public void onAnimationEnd(Animation animation) {   }   //动画结束时触发
        public void onAnimationRepeat(Animation animation) {}   //动画重复执行时触发
        public void onAnimationStart(Animation animation) {     //动画开始时触发
            if(animation instanceof AnimationSet) {             //判断类型
                AnimationSet set = (AnimationSet) animation ;
                Animation rotate=new RotateAnimation(0f, 360f); //初始化 Rotate 动画
                rotate.setDuration(3000) ;                      //3 秒完成动画
                set.addAnimation(rotate) ;   }    }    }}       //增加动画
```

保存所有文件，运行该项目，可发现动画开始时会增加旋转的效果，而重复执行动画时，没有旋转效果，这是因为设置了动画开始时触发旋转动画的效果。

7.3.6 动画操作组件

LayoutAnimationController 表示在 Layout 组件上使用动画的操作效果，例如，可以在使用 ListView 组件时增加一些渐变、缩放、旋转、平移动画效果，LayoutAnimationController 可以通过配置文件轻松实现，也可以利用程序代码完成。其常用配置的属性如表 7-14 所示，常用常量及方法如表 7-15 所示。

表 7-14 LayoutAnimationController 常用配置的属性

属性	对应的方法	描述
android:animation	setAnimation(Animation)	要引入的动画配置文件
android:animationOrder	setOrder(int)	动画执行顺序，normal 表示顺序执行，reverse 表示逆序执行，random 表示随机执行
android:delay		多个动画时间间隔
android:interpolator	setInterpolator(Context,int)	配置动画的执行速率

表 7-15 LayoutAnimationController 常用常量及方法

常量及方法	描述
public static final int ORDER_NORMAL	动画采用顺序效果完成
public static final int ORDER_RANDOM	动画采用随机顺序效果完成
public static final int ORDER_REVERSE	动画采用逆序效果完成
public void setDelay(float delay)	设置动画间隔
public void setAnimation(Animation animation)	设置要使用的动画效果
public void setAnimation(Context context, int resourceID)	设置要使用的动画效果的配置文件
public void setOrder(int order)	设置动画的执行顺序
public void start()	开始动画

此项目中实现组件（如 ListView 组件）包含动画效果的步骤如下。

（1）建立动画配置文件 res\anim\anim.xml，例如：

```
<?xml version="1.0" encoding="utf-8"?>
<set xmlns:android="http://schemas.android.com/apk/res/android">
```

```
    <rotate                                           //定义旋转动画
        android:fromDegrees="0"                       //从 0 度开始
        android:toDegrees="360"                       //到 360 度结束
        android:duration="1000" />                    //动画时间 1 秒
    <scale                                            //定义缩放动画
        android:fromXScale="0.1"                      //组件从 X 轴缩小到原来的 0.1 开始
        android:toXScale="1.0"                        //组件到 X 轴满屏显示结束
        android:fromYScale="0.1"                      //组件从 Y 轴缩小到原来的 0.1 开始
        android:toYScale="1.0"                        //组件到 Y 轴满屏显示结束
        android:duration="2000"/>                     //动画时间 2 秒
    </set>
```

（2）建立 LayoutAnimationController 的配置文件 res\anim\layoutanim.xml，应用动画配置文件 anim.xml，代码如下：

```
    <layoutAnimation
    xmlns:android="http://schemas.android.com/apk/res/android"
    android:delay="0.5"                               //动画间隔 0.5 秒
    android:animationOrder="normal"                   //顺序播放动画
    android:animation="@anim/anim" />                 //调用配置文件 anim.xml
```

（3）建立 ListView.xml 文件，显示信息。例如：

```
    …
    <TableRow>
        <TextView
            android:id="@+id/name"
            android:textSize="16sp"
            android:layout_height="wrap_content"
            android:layout_width=" wrap_content "/>
        <TextView
            android:id="@+id/depart"
            android:textSize="16sp"
            android:layout_height="wrap_content"
            android:layout_width=" wrap_content "/>
        <TextView
            android:id="@+id/tel"
            android:textSize="16sp"
            android:layout_height="wrap_content"
            android:layout_width=" wrap_content "/>
     </TableRow>
    </TableLayout>
```

（4）建立布局管理文件 activity_main.xml，代码如下：

```
    …
    <ListView
        android:id="@+id/myListView"
        android:layout_width="fill_parent"
        android:layout_height="wrap_content"
        android:layoutAnimation="@anim/layoutanim" />    //调用动画配置文件
    </LinearLayout>
```

（5）建立 Activity 文件 MainActivity.java，在 ListView 上显示信息，代码如下：

项目 Exam7_8 实现了 ListView 显示的每条信息中的动画效果，即旋转和缩放，感兴趣的读者可以参照此例学习。

7.4 Android 中的媒体播放

智能手机播放音乐或者视频是怎么实现的呢？Android 操作系统中使用 android.media.MediaPlayer 类可以播放音频、视频和流媒体，可以利用该类提供的方法建立自己的音频和视频播放器，其常用方法如表 7-16 所示。

表 7-16 MediaPlayer 类常用的方法

方法名称	描述
public static MediaPlayer create(Context context, Uri uri)	通过 URI 创建一个 MediaPlayer 对象
public static MediaPlayer create(Context context, int resid)	通过资源 ID 创建一个 MediaPlayer 对象
public static MediaPlayer create(Context context, Uri uri, SurfaceHolder holder)	通过 URI 创建一个 MediaPlayer 对象，并显示该视频
public int getCurrentPosition()	返回 Int，得到当前播放位置
public int getDuration()	返回 Int，得到文件的时间
public int getVideoHeight()	返回 Int，得到视频的高度
public int getVideoWidth()	返回 Int，得到视频的宽度
public boolean isLooping()	返回 boolean，是否循环播放
public boolean isPlaying()	返回 boolean，是否正在播放
public void pause()	暂停播放
public void prepare()	准备同步播放，在播放前调用
public void prepareAsync()	准备异步播放，在播放前调用
public void release()	释放 MediaPlayer 对象所占资源
public void reset()	重置 MediaPlayer 对象
public void seekTo(int msec)	指定播放的位置（以毫秒为单位的时间）
public void setAudioStreamType(int streamtype)	指定流媒体的类型
public void setDataSource(String path)	设置多媒体数据来源【根据路径】
public void setDataSource(FileDescriptor fd, long offset, long length)	设置多媒体数据来源【根据文件系统】
public void setDataSource(FileDescriptor fd)	设置多媒体数据来源【根据文件系统】
public void setDataSource(Context context, Uri uri)	设置多媒体数据来源【根据 URI】
public void setDisplay(SurfaceHolder sh)	设置视频显示多媒体信息
public void setLooping(boolean looping)	设置是否循环播放
public void setOnBufferingUpdateListener (MediaPlayer.OnBufferingUpdateListener listener)	监听事件，网络流媒体的缓冲更新时触发
public void setOnCompletionListener (MediaPlayer. OnCompletionListener listener)	监听事件，网络流媒体播放结束触发
public void setOnErrorListener (MediaPlayer.OnErrorListener listener)	监听事件，设置出现错误时触发
public void setOnVideoSizeChangedListener (MediaPlayer.OnVideoSizeChangedListener listener)	监听事件，视频尺寸改变后触发
public void setScreenOnWhilePlaying(boolean screenOn)	设置是否使用 SurfaceHolder 显示
public void setVolume(float leftVolume, float rightVolume)	设置音量
public void start()	开始播放
public void stop()	停止播放

在介绍如何利用 MediaPlayer 类播放音频和视频之前，先来看看 MediaPlayer 的几个状态。

（1）Idle 状态：当使用关键字 new 实例化一个 MediaPlayer 对象或者调用了类中的 reset()方法时会进入此状态，通过 create()方法创建的 MediaPlayer 对象并不处于 Idle 状态，如果成功调用了重载的 create()方法，那么这个对象已经处于 Prepare 状态。

（2）End 状态：当调用 release()方法之后将进入此状态，此时会释放所有占用的硬件和软件资源，并且不会再进入其他的状态，建议一个 MediaPlayer 对象不再被使用时调用 release()方法来释放资源。

（3）Initialized 状态：当 MediaPlayer 对象设置好了要播放的媒体文件（setDataSource()）之后进入此状态。

（4）Prepared 状态：进入预播放状态（prepare()、prepareAsync()），进入此状态时表示目前的媒体文件没有任何问题，可以使用 OnPreparedListener 监听此状态。

（5）Started 状态：正在进行媒体播放（start()），此时可以使用 seekTo()方法指定媒体播放的位置。

（6）Paused 状态：在 Started 状态下使用 Paused 状态可以暂停 MediaPlayer 的播放，暂停之后可以通过 start()方法将其变回 Started 状态，继续播放。

（7）Stop 状态：在 Started 和 Paused 状态下都可以通过 stop()方法停止 MediaPlayer 的播放，在 Stop 状态下要想重新进行播放，则可以使用 prepare()和 prepareAsync()方法进入就绪状态。

（8）PlaybackCompleted 状态：当媒体播放完毕之后会进入此状态，用户可以使用 OnCompletionListener 监听此状态。此时可以使用 start()方法重新播放，也可以使用 stop()方法停止播放，或者使用 seekTo()方法来重新定位播放位置。

（9）Error 状态：当用户播放操作中出现了某些错误（文件格式不正确、播放文件过大等）时进入此状态，用户可以使用 OnErrorListener 来监听此状态，如果 MediaPlayer 进入了此状态，则可以使用 reset()方法重新返回 Idle 状态。

MediaPlayer 操作的生命周期如图 7.4 所示。

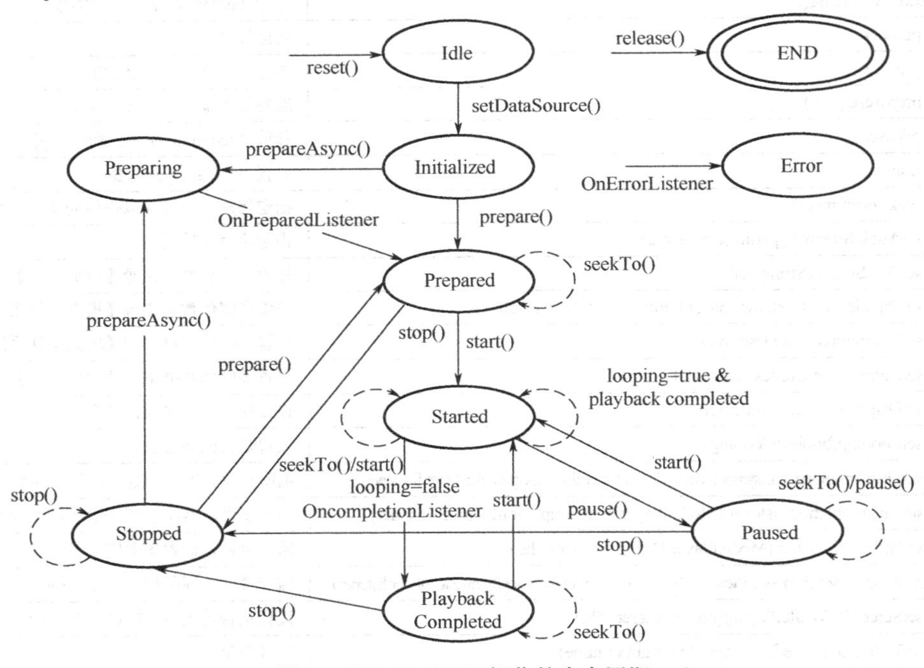

图 7.4 MediaPlayer 操作的生命周期

7.4.1 Android 中的音频播放

MediaPlayer 类支持以下几种不同的音频来源：res 中的音频、文件系统中的音频、网络资源中的音频。其支持的音频格式有：WAV、AAC、MP3、WMA、AMR、OGG、MIDI 等。要在 Android 中实现音频播放的功能，一般有以下几个步骤。

（1）构建 MediaPlayer 对象，例如：

```
MediaPlayer mp = new MediaPlayer();
```

（2）设置数据源，例如：

```
mp.setDataSource("/mnt/sdcard/test.mp3" );                    //从 SD 卡中取得音频文件
MediaPlayer mp = MediaPlayer.create(this, R.raw.test);   //从资源中取得音频文件
Uri uri = Uri.parse( "http://www.xxxx.com/xxx.mp3" ); //定义 URI
player = MediaPlayer.create(this,uri);                        //从网络上取得音频文件
```

（3）准备播放，例如：

```
mp.prepared();              //准备播放
player.prepared();          //网络中的音频准备播放
```

（4）开始播放，例如：

```
mp.start();                 //开始播放
player.start();             //网络中的音频开始播放
```

如果要使用 MediaPlayer 来播放基于互联网内容的流，那么应用程序必须申请互联网访问的权限。

```
<uses-permission android:name="android.permission.INTERNET" />
```

以上 4 步可以实现音乐文件的播放，如果要做音量和进度条控制，则可继续下面的操作。

（5）MediaPlayer 音量管理，例如：

```
//获得音量控制器
AudioManager AM = (AudioManager) getSystemService(Context.AUDIO_SERVICE);
//获得最大音量
int maxVolume = AM.getStreamMaxVolume(AudioManager.STREAM_MUSIC);
//获得当前音量
int currentVolume = AM.getStreamVolume(AudioManager.STREAM_MUSIC);
```

（6）控制音量大小，可以通过触摸触发 dispatchTouchEvent（Activity 中的方法）来控制。

（7）进度控制，可以设置一个 SeekBar 组件，通过拖动在指定位置进行音频播放。

实例 7-8：音频播放实例

新建一个项目，项目命名为 Exam7_9，编写一个 MP3 播放器程序，可以播放程序资源中提供的 MP3 格式的音乐，可以使用开始、暂停、停止按钮来控制音乐的播放。

将 abird.mp3 文件复制到 res\raw 文件夹中，将三张图片——play.jpg、pause.jpg、stop.jpg 复制到 res\drawable 文件夹中。

（1）定义布局管理文件 activity_main.xml，放置一个文本显示框，三个图片按钮——开始、暂停、停止，一个进度控制条。

```
…
<TextView
    android:id="@+id/info"
    android:layout_width="fill_parent"
    android:layout_height="wrap_content"
    android:text="等待播放...... " />
<LinearLayout
    android:orientation="horizontal"
    android:layout_width="wrap_content"
```

```
        android:layout_height="wrap_content">
        <ImageButton
            android:id="@+id/play"
            android:layout_width="wrap_content"
            android:layout_height="wrap_content"
            android:src="@drawable/play" />
        <ImageButton
            android:id="@+id/pause"
            android:layout_width="wrap_content"
            android:layout_height="wrap_content"
            android:src="@drawable/pause" />
        <ImageButton
            android:id="@+id/stop"
            android:layout_width="wrap_content"
            android:layout_height="wrap_content"
            android:src="@drawable/stop" />
    </LinearLayout>
    <SeekBar
        android:id="@+id/seekbar"
        android:layout_width="fill_parent"
        android:layout_height="wrap_content" />
</LinearLayout>
```

（2）定义 Activity 文件 MainActivity.java，控制音频的播放。

```
…
import android.media.MediaPlayer;
import android.media.MediaPlayer.OnCompletionListener;
import android.os.AsyncTask;
import android.view.View.OnClickListener;
import android.widget.SeekBar.OnSeekBarChangeListener;
public class MainActivity extends AppCompatActivity{
private ImageButton play = null;
private ImageButton pause = null;
private ImageButton stop = null;
private TextView info = null;
private MediaPlayer myMediaPlayer = null;                //定义MediaPlayer
private boolean pauseFlag = false;                       //暂停播放标记
private boolean playFlag = true ;                        //是否播放的标记
private SeekBar seekbar = null;                          //定义拖动条
@Override
public void onCreate(Bundle savedInstanceState) {
    super.onCreate(savedInstanceState);
    super.setContentView(R.layout.activity_main);
    this.info = (TextView) super.findViewById(R.id.info);
    this.play = (ImageButton) super.findViewById(R.id.play);
    this.pause = (ImageButton) super.findViewById(R.id.pause);
    this.stop = (ImageButton) super.findViewById(R.id.stop);
    this.seekbar = (SeekBar) super.findViewById(R.id.seekbar);
                                                    //取得组件
    this.play.setOnClickListener(new PlayOnClickListenerImpl()) ;
                                                    //设置播放事件
    this.pause.setOnClickListener(new PauseOnClickListenerImpl());
```

```
                                                    //设置暂停单击事件
    this.stop.setOnClickListener(new StopOnClickListenerImpl());}
                                                    //设置结束事件
  private class PlayOnClickListenerImpl implements OnClickListener {
    @Override
    public void onClick(View view) {
        MainActivity.this.myMediaPlayer = MediaPlayer.create(
            MainActivity.this, R.raw.abird);            //找到指定的资源
  MainActivity.this.myMediaPlayer
              .setOnCompletionListener(new OnCompletionListener() {
                  @Override
                  public void onCompletion(MediaPlayer media) {
                      MainActivity.this.playFlag = false ; //播放完毕
                      media.release();      }});           //释放所有状态
  //设置拖动条长度为媒体长度
  MainActivity.this.seekbar.setMax(MainActivity.this.myMediaPlayer.getDuration());
    UpdateSeekBar update = new UpdateSeekBar() ; //更新拖动条
        update.execute(1000) ;                   //休眠1秒
        MainActivity.this.seekbar.setOnSeekBarChangeListener(new
            OnSeekBarChangeListenerImpl());      //拖动条改变音乐播放位置
        if (MainActivity.this.myMediaPlayer != null) {
            MainActivity.this.myMediaPlayer.stop();  }    //停止播放
        try{ MainActivity.this.myMediaPlayer.prepare(); //进入预备状态
        MainActivity.this.myMediaPlayer.start();         //播放文件
            MainActivity.this.info.setText("正在播放音频文件...");  }
                                                         //设置显示的文字
        catch (Exception e) {                            //异常处理
        MainActivity.this.info.setText("文件播放出现异常，" + e);//设置文字
        }     }   }
  private class UpdateSeekBar extends AsyncTask<Integer, Integer, String> {
    @Override
    protected void onPostExecute(String result) {       }//任务执行完后执行
    protected void onProgressUpdate(Integer... progress) {   //更新之后的数值
    MainActivity.this.seekbar.setProgress(progress[0]) ; }    //更新拖动条
    protected String doInBackground(Integer... params) {     //处理后台任务
        while (MainActivity.this.playFlag) {                 //进度条累加
            try {
                Thread.sleep(params[0]); }                   //延缓执行
            catch (InterruptedException e) {
                e.printStackTrace();     }
            this.publishProgress(MainActivity.this.myMediaPlayer
                    .getCurrentPosition()); }               //修改拖动条
        return null; }    }                                 //返回执行结果
  private class OnSeekBarChangeListenerImpl implements OnSeekBarChangeListener {
    public void onProgressChanged(SeekBar seekBar,
            int progress, boolean fromUser) {        }
    public void onStartTrackingTouch(SeekBar seekBar) {       }
    public void onStopTrackingTouch(SeekBar seekBar) {     //进度条停止拖动
        MainActivity.this.myMediaPlayer.seekTo(seekBar
                .getProgress()); }    }                     //定义播放位置
  private class PauseOnClickListenerImpl implements OnClickListener {
    public void onClick(View view) {
```

```
            if (MainActivity.this.myMediaPlayer != null) {
                if (MainActivity.this.pauseFlag) {    //true 表示由暂停变为播放
                    MainActivity.this.myMediaPlayer.start();      //播放文件
                    MainActivity.this.pauseFlag = false;          //修改标记位
                } else {                                          //false 表示由播放变为暂停
                    MainActivity.this.myMediaPlayer.pause();      //暂停播放
                    MainActivity.this.pauseFlag = true; }}}}      //修改标记位
    private class StopOnClickListenerImpl implements OnClickListener {
        public void onClick(View view) {
            if (MainActivity.this.myMediaPlayer != null) {
                MainActivity.this.myMediaPlayer.stop();               //停止播放
                MainActivity.this.info.setText("停止播放音频文件..."); }}}}
```

保存所有文件，运行该项目，如图 7.5 所示，各个按钮都能实现相应的功能，拖动滚动条能达到预期效果。

使用 MediaPlayer 类可以播放音频，但是该类占用资源较多，这对于游戏应用可能不是很适合，此时一般使用 SoundPool 类(android.media.SoundPool)，它主要用于播放一些较短的声音片段，可以从程序的资源或文件系统中加载。

图 7.5　Exam7_9 运行结果

SoundPool 最大只能支持 1MB 大小的音频文件，但是效率相对来说比 MediaPlayer 高。

SoundPool 播放音频的基本步骤如下。

（1）创建一个 SoundPool。

```
new SoundPool(int maxStreams, int streamType, int srcQuality);
```

其中，maxStream 表示同时播放的流的最大数量；streamType 表示流的类型，一般为 STREAM_MUSIC(具体在 AudioManager 类中列出)；srcQuality 表示采样率转化质量，当前无效果，使用 0 作为默认值。例如：

```
SoundPool soundPool = new SoundPool(3, AudioManager.STREAM_MUSIC, 0);
```

以上语句创建了一个最多支持 3 个流同时播放的、类型标记为音乐的 SoundPool。

（2）从资源或者文件中载入音频流。例如：

```
int  load(Context context, int resId, int priority)    //从 res 资源中载入
int  load(FileDescriptor fd, long offset, long length, int priority)
                                                  //从 FileDescriptor 对象中载入
int  load(AssetFileDescriptor afd, int priority)        //从 Asset 对象中载入
int  load(String path, int priority)                    //以完整文件路径名载入
```

其中，priority 参数表示优先级。

（3）播放声音。例如：

```
play(int soundID, float leftVolume, float rightVolume, int priority, int loop,
      float rate),
```

其中，leftVolume 和 rightVolume 表示左右音量，priority 表示优先级，loop 表示循环次数，rate 表示速率，如速率最低为 0.5，最高为 2，1 代表正常速度。

例如：

```
sp.play(soundId, 1, 1, 0, 0, 1);
```

停止播放可以使用 pause(int streamID)方法，这里的 streamID 和 soundID 均在构造 SoundPool 类的第一个参数中指明了总数量，而 id 从 0 开始。

7.4.2 Android 中的视频播放

MediaPlayer 除了可以对音频进行播放之外，也可以对视频进行播放。它支持以下几种不同的视频来源：res 中的视频、文件系统中的视频、网络资源中的视频。其支持的视频格式有：MP4、H.263 (3GP)、H.264 (AVC)等。

要播放视频只依靠 MediaPlayer 是不够的，还需要依靠 android.view.SurfaceView 组件来完成，SurfaceView 是 View 的子类，其中含有一个专门用于绘制的 Surface 对象，使用它将在视频、3D 图形等需要快速更新的地方有很大的帮助。

SurfaceView 类使用 SurfaceView()方法创建，使用最多的一个操作就是 getHolder()方法，这个方法可以获取 SurfaceHolder 接口。此外，需要重写的方法有以下几个。

（1）在 Surface 的大小发生改变时激发：

```
public void surfaceChanged(SurfaceHolder holder,int format,int width,int
                    height){ }
```

（2）在创建时激发：

```
Public void surfaceCreated(SurfaceHolder holder){ }
```

（3）在销毁时激发：

```
Public void surfaceDestroyed(SurfaceHolder holder) { }
```

这里有一个重要的概念——SurfaceHolder，它可以说是 Surface 的控制器，用来操纵 Surface，处理其 Canvas 中的效果和动画，控制表面、大小、像素等，它的常用方法如表 7-17 所示。

表 7-17　SurfaceHolder 类常用的方法

方法名称	描述
public abstract void addCallback (SurfaceHolder. Callback callback)	给 SurfaceView 当前的持有者一个回调对象
public abstract Canvas lockCanvas()	锁定画布，返回画布对象 Canvas，在其上进行画图等操作
public abstract Canvas lockCanvas(Rect dirty)	锁定画布的某个区域进行画图
public abstract void unlockCanvasAndPost(Canvas canvas)	结束锁定画图，并提交改变
public abstract void setFixedSize(int width,int height)	设置一个 Video 大小区域
public abstract void setType(int type)	设置 SurfaceView 的类型

使用 MediaPlayer 播放视频的步骤如下。

（1）创建 MediaPlayer 的对象，并使其加载指定的视频文件。

```
this.media = new MediaPlayer();                          //创建 MediaPlayer 对象
this.media.setDataSource("/sdcard/test.3gp");//设置播放文件的路径
//或者从资源中读取
this.media=MediaPlayer.create(MainActivity.this, R.raw.future);//读取视频
```

（2）在界面布局文件中定义 SurfaceView 组件，或在程序中创建 SurfaceView 组件。

```
this.surfaceView = (SurfaceView) super.findViewById(R.id.surfaceView);
this.surfaceHolder = this.surfaceView.getHolder(); //取得 SurfaceHolder
//设置 SurfaceView 的类型
this.surfaceHolder.setType(SurfaceHolder.SURFACE_TYPE_PUSH_BUFFERS);
```

（3）调用 MediaPlayer 对象的 setDisplay(SurfaceHolder sh)将所播放的视频图像输出到指定的 SurfaceView 组件中。

```
//设置显示的区域
MainActivity.this.media.setDisplay(MainActivity.this.surfaceHolder);
```

（4）调用 MediaPlayer 对象的 start()、stop()和 pause()方法来控制视频的播放。

实例 7-9：视频播放举例 1

新建一个项目，项目命名为 Exam7_10，使用 MediaPlayer 和 SurfaceView 建立一个简单的视频播放器，播放媒体文件。

（1）建立一个布局管理文件 Activity_main.xml，代码如下：

```
…
    …
    <ImageButton
        android:id="@+id/play"
        …
    <ImageButton
        android:id="@+id/pause"
        …
    <ImageButton
        android:id="@+id/stop"
        …
</LinearLayout>
<SurfaceView
    android:id="@+id/surfaceView"
    android:layout_width="fill_parent"
    android:layout_height="fill_parent" />
</LinearLayout>
```

（2）建立一个 Activity 文件 MainActivity.java，代码如下：

```
…
import android.media.AudioManager;
import android.media.MediaPlayer;
import android.view.SurfaceHolder;
import android.view.SurfaceView;
public class MainActivity extends AppCompatActivity{
private ImageButton play = null;
private ImageButton pause = null;
private ImageButton stop = null;
private MediaPlayer media = null;
private MediaPlayer myMediaPlayer = null;
private SurfaceView surfaceView = null;
private SurfaceHolder surfaceHolder = null;
public void onCreate(Bundle savedInstanceState) {
    super.onCreate(savedInstanceState);
    super.setContentView(R.layout.activity_main);
    this.play = (ImageButton) super.findViewById(R.id.play);
    this.pause = (ImageButton) super.findViewById(R.id.pause);
    this.stop = (ImageButton) super.findViewById(R.id.stop);
    this.surfaceView = (SurfaceView) super.findViewById(R.id.surfaceView);
    this.surfaceHolder = this.surfaceView.getHolder(); //取得 SurfaceHolder
```

```
        //设置 SurfaceView 的类型
        this.surfaceHolder.setType(SurfaceHolder.SURFACE_TYPE_PUSH_BUFFERS);
        this.media = new MediaPlayer();                    //创建 MediaPlayer 对象
        try {
    this.media=MediaPlayer.create(MainActivity.this, R.raw.future);//读取视频
        } catch (Exception e) {
            e.printStackTrace();}
        this.play.setOnClickListener(new PlayOnClickListenerImpl());
                                                        //播放单击事件
        this.pause.setOnClickListener(new PauseOnClickListenerImpl());
                                                        //暂停单击事件
        this.stop.setOnClickListener(new StopOnClickListenerImpl()); }
                                                        //停止单击事件
    private class PlayOnClickListenerImpl implements OnClickListener {
        public void onClick(View arg0) {
         //设置音频类型
        MainActivity.this.media.setAudioStreamType(AudioManager.STREAM_MUSIC);
        //设置显示的区域
        MainActivity.this.media.setDisplay(MainActivity.this.surfaceHolder);
            try {
                MainActivity.this.media.start();           //播放视频
            } catch (Exception e) {
                e.printStackTrace();}}}
    private class PauseOnClickListenerImpl implements OnClickListener {
        public void onClick(View arg0) {
            MainActivity.this.media.pause();}}
    private class StopOnClickListenerImpl implements OnClickListener {
        public void onClick(View arg0) {
    MainActivity.this.media.stop();}}} //停止播放
```

保存所有文件，运行该项目，如图 7.6 所示，各个按钮都能实现相应的功能。

图 7.6　Exam7_10 运行结果

除了可以用 MediaPlayer 来实现视频播放之外，还可通过 android.widget.VideoView 类播放视频文件，它的常用方法如表 7-18 所示，与 VideoView 一起使用的还有 MediaController 类，它的作用是提供一个友好的控制界面，并通过该控件来控制视频的播放。

表 7-18　VideoView 类常用的方法

方法名称	描述
public VideoView (Context context)	创建一个默认属性的 VideoView 实例
Public VideoView (Context context, AttributeSet attrs)	创建一个带有 attrs 属性的 VideoView 实例
Public VideoView (Context context, AttributeSet attrs, int defStyle)	创建一个带有 attrs 属性，并且指定其默认样式的 VideoView 实例
public boolean canPause()	判断是否能够暂停播放视频
public void pause()	播放暂停
public boolean canSeekBackward()	判断是否能够倒退
public boolean canSeekForward()	判断是否能够快进
public int getBufferPercentage()	获得缓冲区的百分比
public int getCurrentPosition()	获得当前的位置
public int getDuration()	获得所播放视频的总时间
public boolean isPlaying()	判断是否正在播放视频
public boolean onKeyDown (int keyCode, KeyEvent event)	如果处理了事件，则返回真。如果允许下一个事件接收器处理该事件，则返回假
public boolean onTouchEvent (MotionEvent ev)	该方法用来处理触屏事件
public void resume()	恢复挂起的播放器
public void seekTo(int msec)	设置播放位置
public void setMediaController (MediaController controller)	设置媒体控制器
public void setOnCompletionListener (MediaPlayer. OnCompletionListener l)	在媒体文件播放完毕时触发
public void setOnErrorListener (MediaPlayer.OnErrorListener l)	出现错误时触发
public void setOnPreparedListener (MediaPlayer.OnPreparedListener l)	在媒体文件加载完毕时触发，可以在播放时调用回调函数
public void setVideoPath(String path)	设置视频文件的路径名
public void setVideoURI(Uri uri)	设置视频文件的统一资源标识符
public void start()	开始播放视频文件
public void stopPlayback()	停止回放视频文件
public void suspend()	挂起视频文件的播放

实例 7-10：视频播放举例 2

新建一个项目，项目命名为 Exam7_11，使用 VideoView 和 MediaController 建立一个简单的视频播放器，播放 SD 卡中的媒体文件。

（1）将视频文件 future.mp4 复制到手机 SD 卡中。

（2）设置布局文件 activity_main.xml，代码如下：

```
...
<VideoView
    android:id="@+id/videoView"
    android:layout_width="fill_parent"
android:layout_height="fill_parent" />
</LinearLayout>
```

（3）建立一个 Activity 文件 MainActivity.java，代码如下：

```
import android.graphics.PixelFormat;import android.os.Bundle;
import android.widget.MediaController;import android.widget.Toast;
import android.widget.VideoView;import java.io.File;
public class MainActivity extends AppCompatActivity{
    private VideoView videoView;            //定义 videoView
```

```
    private MediaController mController;  //定义mController
    private File file;                    //定义file
    public void onCreate(Bundle savedInstanceState){
       super.onCreate(savedInstanceState);
       setContentView(R.layout.activity_main);
       videoView=(VideoView)findViewById(R.id.videoView); //获得videoView
       mController=new MediaController(this);           //获得mController值
       file=new File("/mnt/sdcard/ future.mp4");           //获得file值
       getWindow().setFormat(PixelFormat.TRANSLUCENT); //设置窗口为半透明度
       if(file.exists()){
         videoView.setVideoPath(file.getAbsolutePath());
                                        //设置videoView与file建立联系
         videoView.setMediaController(mController);
                                        //设置videoView与mController联系
         videoView.requestFocus(); }    //使videoView 获取焦点
         else{    Toast.makeText(MainActivity.this, "文件不存在", Toast.
                  LENGTH_LONG).show();  }  }}
```

保存所有文件，运行该项目，运行界面与图 7.6 类似，但开始、暂停和停止按钮是自动生成的。利用 VideoView 类可实现视频播放，只是播放速度可能不太理想。

Android 中还有许多多媒体方面的功能，例如，照相机、录音、手势控制屏幕、多点触控屏幕，等等，感兴趣的读者可以查阅 Android API 进行了解。

本章小结

本章着重介绍了几种常见基本图形的绘制、常见的两种动画处理、使用 MediaPlayer 播放音频和视频文件等技术，简要介绍了动画操作组件、图片的简单处理技术。

习题

（1）绘制一个蓝色的圆需要用到哪些类？给出关键代码。

（2）利用 XML 文件，实现实例 7-4 的动画效果。

（3）播放网络上的音频文件，可用什么语句指定网络上的音频文件？

（4）试着利用 Intent 调用系统的照相机功能，注意应配置 AndroidManifest.xml 文件。

第 8 章　Android 数据存储技术

学习目标：

- 使用 SharedPreferences 存储数据。
- 掌握对文件的保存和读取操作。
- 了解 SQLite 数据库的基本作用。
- 掌握如何在 Activity 文件中操作 SQLite 数据库。
- 掌握 ContentProvider 的使用。

一个应用程序，经常需要与用户进行交互，需要保存用户的设置和用户数据，这些都离不开数据的存储。Android 系统提供了以下 5 种主要的数据存储方式。

（1）使用 SharedPreferences 存储数据。

（2）使用文件存储数据。

（3）使用 SQLite 数据库存储数据。

（4）使用 ContentProvider 存储数据。

（5）使用网络存储数据。

8.1　使用 SharedPreferences 存储数据

对于软件配置参数的保存，Windows 系统通常会采用 ini 文件保存，Java 程序采用 properties 属性文件或 XML 文件保存。类似的，Android 平台为我们提供了 SharedPreferences 接口用于保存参数设置等较为简单的数据，例如，字符串、整型、布尔型等。使用 SharedPreferences 进行保存的数据，采用 key=value 键值对的形式保存一些简单的配置信息，信息以 XML 文件的形式存储在/data/data/<包名>/shared_prefs 目录下。其接口的常用方法如表 8-1 所示。

SharedPreferences 类似 Windows 系统上的 ini 配置文件，但是它有多种权限，可以全局共享访问，以 XML 方式来保存，这样占用的内存资源比较少。

要在 Android 程序中使用 SharedPreferences 组件，必须在程序中使用下面的语句。

```
import android.content.SharedPreferences; //导入content.SharedPreferences类
```

表 8-1　SharedPreferences 接口的常用方法

方法	描述
public abstract SharedPreferences.Editor edit()	使其处于可编辑状态
public abstract boolean contains(String key)	判断某一个 key 是否存在
public abstract Map<String, ?> getAll()	取出全部的数据
public abstract boolean getBoolean(String key, boolean defValue)	取出 boolean 型数据，并指定默认值
public abstract float getFloat(String key, float defValue)	取出 float 型数据，并指定默认值
public abstract int getInt(String key, int defValue)	取出 int 型数据，并指定默认值
public abstract long getLong(String key, long defValue)	取出 long 型数据，并指定默认值
public abstract String getString(String key, String defValue)	取出 String 型数据，并指定默认值

SharedPreferences 对象本身只能获取数据而不支持存储和修改，存储修改是通过 android.

content.SharedPreferences.Editor 接口来实现的。其常用方法如表 8-2 所示。

表 8-2 SharedPreferences.Editor 接口的常用方法

方法	描述
public abstract SharedPreferences.Editor clear()	清除所有的数据
public abstract boolean commit()	提交更新的数据
public abstract SharedPreferences.Editor putBoolean(String key, boolean value)	保存一个 boolean 型数据
public abstract SharedPreferences.Editor putFloat(String key, float value)	保存一个 float 型数据
public abstract SharedPreferences.Editor putInt(String key, int value)	保存一个 int 型数据
public abstract SharedPreferences.Editor putLong(String key, long value)	保存一个 long 型数据
public abstract SharedPreferences.Editor putString(String key, String value)	保存一个 String 型数据
public abstract SharedPreferences.Editor remove(String key)	删除指定 key 的数据

由于 SharedPreferences 和 SharedPreferences.Editor 两个都是接口，所以要想取得 SharedPreferences 接口的实例化对象，还需要 Activity 类中的几个常量和方法的支持，如表 8-3 所示。

表 8-3 取得 SharedPreferences 接口的几个常量和方法

常量及方法	描述
public static final int MODE_PRIVATE	常量，创建的文件只能被一个应用程序调用，或者被具有相同 ID 的应用程序访问
public static final int MODE_WORLD_READABLE	常量，允许其他应用程序读取文件
public static final int MODE_WORLD_WRITEABLE	常量，允许其他应用程序修改文件
public SharedPreferences getSharedPreferences (String name, int mode)	指定保存操作的文件名称，同时指定操作的模式，可以是 MODE_PRIVATE、MODE_WORLD_READABLE、MODE_WORLD_WRITEABLE

其中，String name 用于指定文件名称，不能包含路径分隔符“/”，如果文件不存在，则 Android 会自动创建它。

int mode 用于指定操作模式，有如下几种模式。

（1）Context.MODE_PRIVATE=0 为默认操作模式，表示该文件是私有数据，只能被应用本身访问，在该模式下，写入的内容会覆盖原文件的内容。

（2）Context.MODE_WORLD_READABLE =1 表示当前文件可以被其他应用读取。

（3）Context.MODE_WORLD_WRITEABLE =2 表示当前文件可以被其他应用写入。

8.1.1 使用 SharedPreferences 存储数据

实现 SharedPreferences 存储的步骤如下。

（1）根据 Context 获取 SharedPreferences 对象，例如：

```
SharedPreferences sh= super.getSharedPreferences("my",Activity.MODE_PRIVATE);
```

（2）利用 edit()方法获取 Editor 对象，例如：

```
SharedPreferences.Editor edit = sh.edit();    //获取 Editor 对象
```

（3）通过 Editor 对象存储 key-value 数据，例如：

```
edit.putString("college", "hnist") ;          //保存字符串
edit.putInt("students", 20000);               //保存整型
```

（4）通过 commit()方法提交数据，例如：

```
edit.commit() ;                               //提交更新
```

8.1.2 使用 SharedPreferences 读取数据

实现 SharedPreferences 读取数据的步骤如下。

（1）定义 TextView 对象，用于显示读取的数据，例如：

```
private TextView college = null ;                    //定义文本显示组件
private TextView students = null ;                   //定义文本显示组件
```

（2）根据 Context 获取 SharedPreferences 对象，例如：

```
SharedPreferences sh = super.getSharedPreferences("my",Activity.MODE_PRIVATE);
```

（3）通过 getString()方法获得字符数据，getInt()方法获得数字数据：

```
this. college.setText("学校：" + sh.getString("college ", "没有学校信息。"));
this. students.setText("学生数：" + sh.getInt("students ", 0)); }}
```

实例 8-1：使用 SharedPreferences 存储、读取数据实例

新建一个项目，项目命名为 Exam8_1，包名称为 org.hnist.cn，使用 SharedPreferences 存储、读取指定文件。

（1）修改布局管理文件 activity_main.xml，代码如下：

```
...
<Button
    android:id="@+id/save"
    android:layout_width="wrap_content"
    android:layout_height="wrap_content"
    android:text="保存数据到文件" />
<Button
    android:id="@+id/read"
    android:layout_width="wrap_content"
    android:layout_height="wrap_content"
    android:text="读取文件到页面" />
</LinearLayout>
```

（2）修改 Activity 文件 MainActivity.java，代码如下：

```
import android.content.SharedPreferences;
public class MainActivity extends Activity {
    private Button Save=null;
    private Button Read=null;
    private TextView txt=null;
    @Override
    protected void onCreate(Bundle savedInstanceState) {
        super.onCreate(savedInstanceState);
        setContentView(R.layout.activity_main);
        txt=(TextView)super.findViewById(R.id.txt);
        Save=(Button)super.findViewById(R.id.save);
        Read=(Button)super.findViewById(R.id.read);
        //单击事件，保存 SharedPreferences 数据
        Save.setOnClickListener(new OnClickListener(){
            public void onClick(View v)
            { //获取 SharedPreferences 对象
                SharedPreferences
```

```
share=MainActivity.this.getSharedPreferences("hnist", Activity.MODE_PRIVATE);
            //使用 Editor 保存数据
            SharedPreferences.Editor edit=share.edit();
            edit.putString("college","湖南理工学院");
            edit.putString("students", "25000");
            edit.commit();
            txt.setText("保存成功");} });
      //单击事件，读取 SharedPreferences 数据
      Read.setOnClickListener(new OnClickListener(){
         public void onClick(View v)
         {//获取 SharedPreferences
SharedPreferences     sh     =     MainActivity.this.getSharedPreferences("hnist",
Activity.MODE_PRIVATE);
            //使用 SharedPreferences 读取数据
            String college=sh.getString("college","");
            String students=sh.getString("students","");
            String info="学校: "+college+"\n 学生数: "+students;
            txt.setText(info);}});}}
```

保存文件，运行该项目，运行结果如图 8.1 所示。

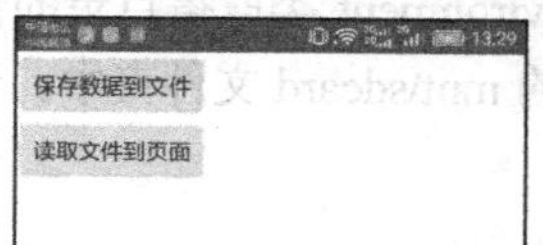

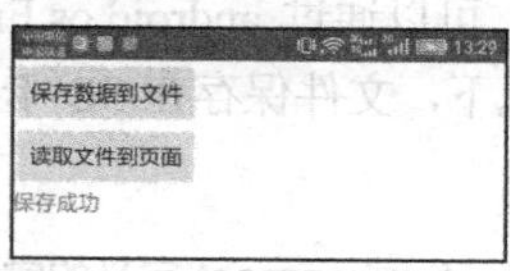

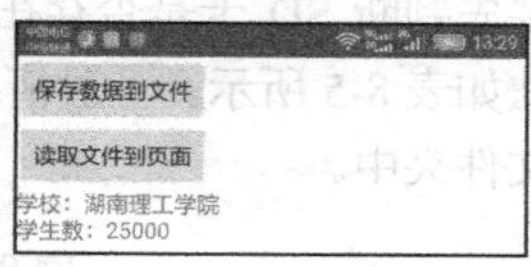

图 8.1 Exam8_1 运行结果

在模拟器上调试时可通过以下步骤查看建立的文件：选择“Tools→Android→Android Device Monitor”选项，选择“File Explorer”选项，找到“/data/data/org.hnist.cn/shared_prefs/”文件，选择工具栏上的“Pull a file from the device”选项，导出指定文件，可用记事本打开查看内容。

真机调试在 File Explorer 中看不到建立的 hnist.xml 文件，要看到的话需要授权 root 权限。

SharedPreferences 对象实现存储和读取的数据比较简单，只能是 boolean、int、float、long 和 String 五种简单的数据类型，如果想存储更多类型的数据，则可以使用文件的存储操作。

8.2 使用文件存储数据

Android 系统基于 Java 语言，Java 语言中提供了一套完整的输入输出流操作体系，与文件有关的有 FileInputStream、FileOutputStream 等，通过这些类可以方便地访问磁盘上的文件，Android 也支持这种方式来访问手机上的文件。表 8-4 是 Activity 类对文件操作的常用方法。

表 8-4 Activity 类对文件操作的常用方法

方法	描述
public FileInputStream openFileInput(String name)	设置要打开的文件输入流
public FileOutputStream openFileOutput(String name, int mode)	设置要打开文件的输出流，指定操作的模式
public Resources getResources()	返回 Resources 对象

其中，String name 用于指定文件名称，不能包含路径分隔符“/”，如果文件不存在，则 Android 会自动创建它；int mode 用于指定操作模式，有以下四种模式。

Context.MODE_PRIVATE=0 为默认操作模式，表示该文件是私有数据，只能被应用本身访问，在

该模式下，写入的内容会覆盖原文件的内容。

Context.MODE_APPEND=32768 表示模式会检查文件是否存在，存在就向文件中追加内容，否则创建新文件再写入内容。

Context.MODE_WORLD_READABLE =1 表示当前文件可以被其他应用读取。

Context.MODE_WORLD_WRITEABLE =2 表示当前文件可以被其他应用写入。

如果希望文件被其他应用读和写，可以输入 Context.MODE_WORLD_READABLE + Context.MODE_WORLD_WRITEABLE，或者直接输入数值 3。这 4 种模式除了 Context.MODE_APPEND 外，其他都会覆盖原文件的内容。

应用程序的数据文件默认保存在/data/data/<包名称>/files 目录下，文件的后缀名随意。

Android 手机中的文件有两个存储位置：内置存储空间和外部 SD 卡，手机内存相对较小，其空间的大小会影响到手机运行速度，不建议将一些大数据保存到手机内存中，通常建议将这些资源存放在外存设备上，最常见的就是 SD 卡。

在模拟器中使用 SD 卡，需要先创建一张 SD 卡（只是镜像文件），在 Android 中并没有提供单独的 SD 卡文件操作类，直接使用 Java 中的文件操作即可，关键是如何确定文件的位置。

在访问 SD 卡的文件之前，需要验证 SD 卡是否已被正确安装，如果没有安装 SD 卡则可能会出现错误。因此要先判断 SD 卡是否存在，可以通过 android.os.Environment 类取得目录的信息来实现，具体常量及方法如表 8-5 所示。一般情况下，文件保存在 SD 卡的 mnt\sdcard 文件夹中。当然，其也可以存放在指定文件夹中。

表 8-5　Environment 定义的常量及方法

常量及方法	描述
public static final String MEDIA_MOUNTED	SD 卡允许进行读/写访问
public static final String MEDIA_CHECKING	SD 卡处于检查状态
public static final String MEDIA_MOUNTED_READ_ONLY	SD 卡处于只读状态
public static final String MEDIA_REMOVED	SD 卡不存在
public static final String MEDIA_UNMOUNTED	没有找到 SD 卡
public static File getDataDirectory()	取得 Data 目录
public static File getDownloadCacheDirectory()	取得下载的缓存目录
public static File getExternalStorageDirectory()	取得扩展的存储目录
public static String getExternalStorageState()	取得 SD 卡的状态
public static File getRootDirectory()	取得 Root 目录
public static boolean isExternalStorageRemovable()	判断扩展的存储目录是否被删除

另外，File 还有下面一些常用的操作，例如：

```
String Name = File.getName();                //获得文件或文件夹的名称
String parentPath = File.getParent();        //获得文件或文件夹的父目录
String path = File.getAbsoultePath();        //绝对路经
String path = File.getPath();                //相对路经
File.createNewFile();                        //建立文件
File.mkDir();                                //建立文件夹
File.isDirectory();                          //判断是文件还是文件夹
File[] files = File.listFiles();             //列出文件夹下的所有文件和文件夹名
File.renameTo(dest);                         //修改文件夹和文件名
File.delete();                               //删除文件夹或文件
```

在使用 SD 卡进行读写的时候 会用到 Environment 类的下面几个静态方法。

（1）getDataDirectory()：获取到 Android 中的 data 数据目录（SD 卡中的 data 文件夹）。

（2）getDownloadCacheDirectory()：获取到下载的缓存目录（SD 卡中的 download 文件夹）。

（3）getExternalStorageDirectory()：获取到外部存储的目录，一般指 SD 卡（/storage/sdcard0）。

（4）getExternalStorageState()：获取外部设置的当前状态，一般指 SD 卡，比较常用的是 MEDIA_MOUNTED（SD 卡存在并且可以进行读写）。

（5）getRootDirectory()：获取到 Android Root 的路径。

8.2.1 读、写 SD 卡文件

读、写 SD 卡上的文件的步骤如下。

（1）调用 Environment 的 getExternalStorageState()方法判断手机上是否插入了 SD 卡，并且应用程序具有读写 SD 卡的权限。Environment.getExternalStorageState()方法用于获取 SD 卡的状态，如果手机装有 SD 卡，并且可以进行读写，那么方法返回的状态等于 Environment.MEDIA_MOUNTED。例如：

```
    if(Environment.getExternalStorageState().equals( Environment.MEDIA_MOUNTED))
```

（2）调用 Environment 的 getExternalStorageDirectory()方法来获取外部存储器，也就是 SD 卡的目录(这里的目录是 mnt\sdcard\hnist，文件是 hnist.txt)。例如：

```
    File file = new File(Environment.getExternalStorageDirectory().toString()
               + File.separator + hnist + File.separator + hnist.txt) ;
               //定义 File 类对象
```

（3）使用 FileInputStream、FileOutputStream、FileReader、FileWriter 等类读、写 SD 卡中的文件。例如：

```
    PrintStream out = null ;                    //打印流对象用于输出
    out = new PrintStream(new FileOutputStream(file, true));//追加文件
    out.println("湖南理工学院信息学院信工 16-2BF");
```

（4）调用 Close()方法，关闭文件输入流。例如：

```
    out.close() ;                               //关闭输入流
```

（5）为了读、写 SD 卡中的数据，必须在应用程序的清单文件（AndroidManifest.xml）中添加读、写 SD 卡的权限。例如：

```
<!-- SD 卡中创建与删除文件权限 -->
<uses-permission android:name="android.permission.MOUNT_UNMOUNT_FILESYSTEMS"/>
<!-- 向 SD 卡写入数据权限 -->
<uses-permission android:name="android.permission.WRITE_EXTERNAL_STORAGE"/>
<!-- 读取 SD 卡文件权限 -->
<uses-permission android:name="android.permission.READ_EXTERNAL_STORAGE"/>
```

实例 8-2：向 SD 卡中写入文件

新建一个项目，项目命名为 Exam8_2，在 SD 卡中建立、删除文件，并向 SD 卡中写入文件。

（1）修改布局管理文件 activity_main.xml，代码较长这里不再给出，界面如图 8.2 所示。

（2）修改 Activity 文件 MainActivity.java，这里只给出了保存文件的代码：

```
...
public class MainActivity extends AppCompatActivity {
   private Button save, read, delete;
   private EditText content;
```

```
    private TextView show;
    @Override
    protected void onCreate(Bundle savedInstanceState) {
        super.onCreate(savedInstanceState);
        setContentView(R.layout.activity_main);
        save = (Button) findViewById(R.id.save);
        read = (Button) findViewById(R.id.read);
        delete = (Button) findViewById(R.id.delete);
        content = (EditText) findViewById(R.id.content);
        show = (TextView) findViewById(R.id.show);
        save.setOnClickListener(new View.OnClickListener() {//事件监听
            @Override
            public void onClick(View v) {
                saveFile();          }    });   //调用 saveFile()保存文件
        read.setOnClickListener(new View.OnClickListener() {   //事件监听
            @Override
            public void onClick(View v) {
                show.setText(readFile());}  }); //调用 readFile()读取文件
        delete.setOnClickListener(new View.OnClickListener() { //事件监听
            @Override
            public void onClick(View v) {
                delete File();      }  });  }   //调用 delete File()删除文件

public void saveFile() {//定义保存文件到 SD 卡的 saveFile()方法
        FileOutputStream fos = null;
        String state = Environment.getExternalStorageState();  //获取 SD 卡状态
        if (!state.equals(Environment.MEDIA_MOUNTED)) {//判断 SD 卡是否就绪
            Toast.makeText(this, "请检查 SD 卡", Toast.LENGTH_SHORT).show();
            return;  }
        File file = Environment.getExternalStorageDirectory();//取得 SD 卡根目录
        try {
            File myFile=new File(file.getCanonicalPath()+"/sd.txt");
            fos=new FileOutputStream(myFile);
            String str = content.getText().toString();
            fos.write(str.getBytes());
            Toast.makeText(this, "保存成功", Toast.LENGTH_SHORT).show();
        } catch (IOException e) {
            e.printStackTrace();
        } finally {
            if (fos != null) {
                try {
                    fos.close();
                    fos.flush();
                } catch (IOException e) {
                    e.printStackTrace();    }       }       }   }
…
```

（3）修改 AndroidManifest.xml 文件，添加读、写 SD 卡的权限。

```
<!-- SD 卡中创建与删除文件权限 -->
<uses-permission android:name="android.permission.MOUNT_UNMOUNT_FILESYSTEMS"/>
<!-- 向 SD 卡写入数据权限 -->
<uses-permission android:name="android.permission.WRITE_EXTERNAL_STORAGE"/>
<!-- 读取 SD 卡文件权限 -->
<uses-permission android:name="android.permission.READ_EXTERNAL_STORAGE"/>
```

保存文件，运行结果如图 8.3 所示。

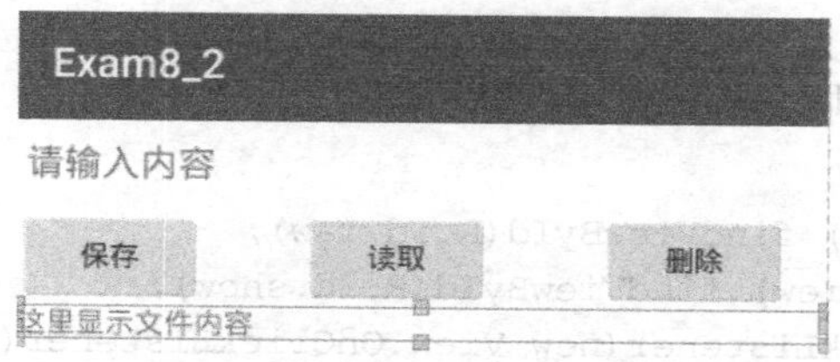

图 8.2　activity_main.xml 界面

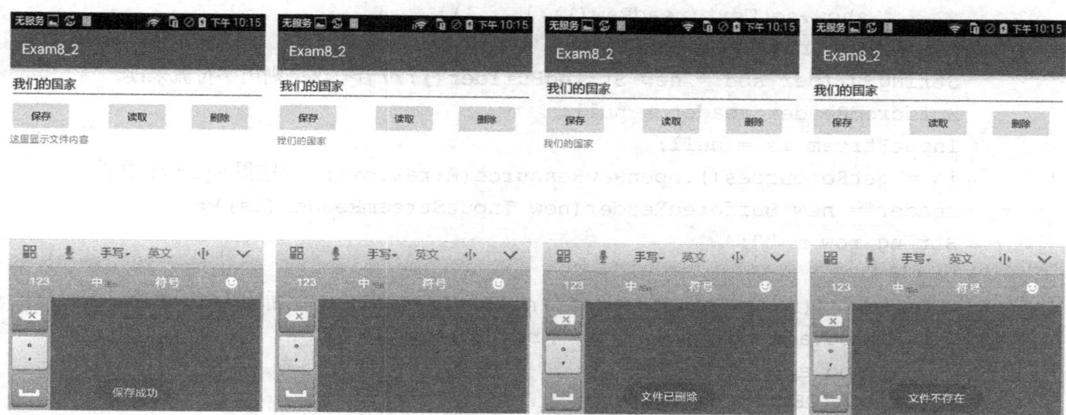

图 8.3　Exam8_2 运行结果

8.2.2　读取资源文件

在 Android 中，还可以对资源文件进行读取（注意，不能写入），这些资源文件的 ID 都会自动通过 R.java 类生成，如果要对这些文件进行读取，使用 android.content.res.Resources 类即可完成，常见的资源文件有两种，使用两种不同的方式打开使用。

raw 中的资源文件使用 getResources().openRawResource(int id)读取，asset 中的资源文件使用 getResources(). getAssets().open(fileName)读取。

实例 8-3：读取 raw 中的资源文件

新建一个项目，项目命名为 Exam8_3，从 resource 的 raw 中读取文件 my.txt 的数据。

（1）在 res 下建立一个文件夹 raw，在 raw 中选择“New→File”选项，建立 my.txt 文件，如图 8.4 所示。

（2）修改布局管理文件 activity_main.xml，代码如下：

```
…
<Button
  android:layout_width="wrap_content"
  android:layout_height="wrap_content"
  android:id="@+id/raw"
  android:text="读取 raw 中的文件"/>
<TextView
  android:layout_width="match_parent"
  android:layout_height="wrap_content"
  android:id="@+id/show"
  android:layout_below="@+id/raw"/>
```

```
</RelativeLayout>
```

（3）修改 Activity 文件 MainActivity.java，代码如下：

```
...
    raw = (Button) findViewById(R.id.raw);
    show = (TextView) findViewById(R.id.show);
    raw.setOnClickListener(new View.OnClickListener() {
        @Override
        public void onClick(View v) {
            show.setText(readRaw());}    });    }
public String readRaw() {
    StringBuilder sbd = new StringBuilder();//若有汉字则用字符流来读
    BufferedReader reader = null;
    InputStream is = null;
    is = getResources().openRawResource(R.raw.my); //取得 my.txt 文件
    reader = new BufferedReader(new InputStreamReader(is));
    String row = "";
    try {
        while ((row = reader.readLine()) != null) {//循环读取赋值给 row
            sbd.append(row);    //将 row 追加到 sbd 中后
            sbd.append("\n");          }
    } catch (IOException e) {
        e.printStackTrace();
    } finally {
        if (reader != null) {
            try {
                reader.close();
            } catch (IOException e) {
                e.printStackTrace();    }    }    }
    return sbd.toString();    }}
```

保存所有文件，运行该项目，结果如图 8.5 所示。

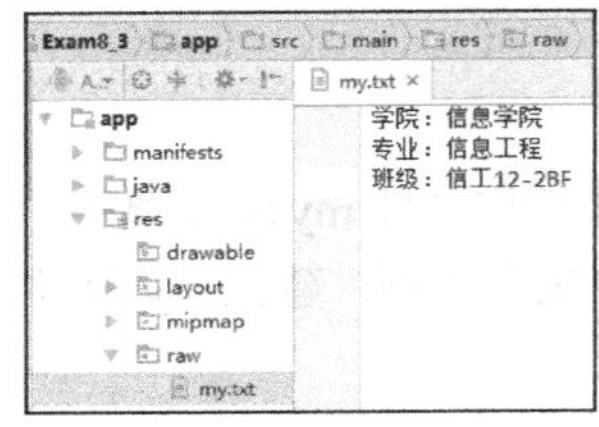

图 8.4　建立资源文件

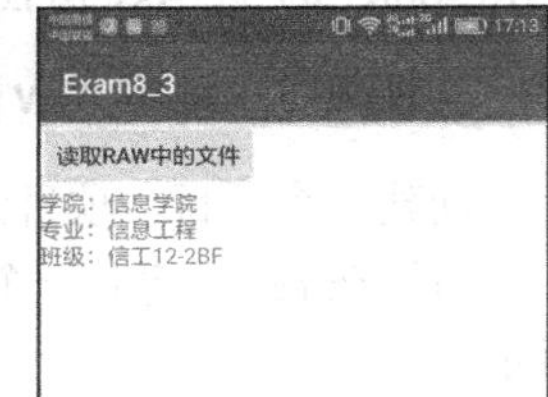

图 8.5　Exam8_3 运行结果

Exam8_4 用于从 assets 中读取资源文件，感兴趣的读者可自行下载学习。

8.3　使用数据库存储数据

对于大量的数据处理，前面介绍的方法就显得力不从心了，Android 平台中集成了一个嵌入式关系型数据库——SQLite，SQLite 支持 SQL，可以方便地实现数据增加、修改、删除、查询等操作，支持的常见的五种数据类型。

SQLite 数据库是 D.Richard Hipp 用 C 语言编写的开源嵌入式数据库，支持的数据库大小为 2TB，具有如下特征。

1．轻量级

SQLite 和 C/S 模式的数据库软件不同，它不存在数据库的客户端和服务器。使用 SQLite 一般只需要带上它的一个动态库，就可以使用它的全部功能，且动态库的文件很小。

2．独立性

SQLite 数据库的核心引擎本身不依赖第三方软件，使用它也不需要“安装”，所以在使用时能够省去不少麻烦。

3．隔离性

SQLite 数据库中的所有信息（如表、视图、触发器）都包含在一个文件内，方便管理和维护。

4．跨平台

SQLite 数据库支持大部分操作系统，除了在电脑上使用的操作系统之外，很多手机操作系统同样可以运行，如 Android、Windows Mobile、Symbian、Palm 等。

5．多语言接口

SQLite 数据库支持很多语言编程接口，如 C/C++、Java、Python、dotNet、Ruby、Perl 等。

6．安全性

SQLite 数据库通过数据库层级上的独占性和共享锁来实现独立事务处理，多个进程可以在同一时间从同一数据库读取数据，但只有一个可以写入数据。在某个进程或线程向数据库执行写操作之前，必须获得独占锁定。

Android 集成了 SQLite 数据库，所以每个 Android 应用程序都可以使用 SQLite 数据库，在 samples 下可以找到关于如何使用数据库的例子。

在 Android 系统中，如果要进行 SQLite 数据库的操作，则主要使用 Android 系统提供的下面几个类或接口。

（1）android.database.sqlite.SQLiteDatabase：完成数据的增、删、修改、查询操作。

（2）android.database.sqlite.SQLiteOpenHelper：完成数据库的创建及更新操作。

（3）android.database.Cursor：保存所有的查询结果。

（4）android.database.ContentValues：对传递的数据进行封装。

下面简要介绍这几个类。

① android.database.sqlite.SQLiteDatabase 类

在 Android 系统中，通过 android.database.sqlite.SQLiteDatabase 类可以执行 SQL 语句，以完成对数据表的增加、修改、删除、查询等操作，在此类之中定义了基本的数据库执行 SQL 语句的操作方法以及一些操作的模式常量，如表 8-6 所示。

表 8-6　SQLiteDatabase 类定义的常用操作方法

常量或方法	描述
public static final int OPEN_READONLY	常量，以只读方式打开数据库
public static final int OPEN_READWRITE	常量，以读/写方式打开数据库
public static final int CREATE_IF_NECESSARY	常量，如果指定的数据库文件不存在，则创建新的文件

续表

常量或方法	描述
public static final int NO_LOCALIZED_COLLATORS	常量，打开数据库时，不对数据进行基于本地化语言的排序
public void beginTransaction()	开始事务
public void endTransaction()	结束事务，提交或者回滚数据
public void close()	关闭数据库
public void execSQL(String sql)	执行 SQL 语句
public void execSQL(String sql, Object[] bindArgs)	执行 SQL 语句，同时绑定参数
public static SQLiteDatabase openDatabase(String path, SQLiteDatabase.CursorFactory factory, int flags)	以指定的模式打开指定路径下的数据库文件
public static SQLiteDatabase openOrCreateDatabase(File file, SQLiteDatabase.CursorFactory factory)	打开或者创建一个指定路径下的数据库
public static SQLiteDatabase openOrCreateDatabase(String path, SQLiteDatabase.CursorFactory factory)	打开或者创建一个指定路径下的数据库
public long insert(String table, String nullColumnHack, ContentValues values)	插入数据，table 为表名称，nullColumnHack 表示传入的 valuesnull，即列被设为 null，values 表示所有要插入的数据
public long insertOrThrow(String table, String nullColumnHack, ContentValues values)	插入数据，但会抛出 SQLException 异常
public int update(String table, ContentValues values, String whereClause, String[] whereArgs)	修改数据，table 为表名称，values 为更新数据，whereClause 指明 WHERE 子句，whereArgs 为 WHERE 子句参数，用于替换 “?”
public int delete(String table, String whereClause, String[] whereArgs)	删除数据，table 为表名称，whereClause 指明 WHERE 子句，whereArgs 为参数，用于替换 “?”
public boolean isOpen()	判断数据库是否已打开
public void setVersion(int version)	设置数据库的版本
public Cursor query (boolean distinct, String table, String[] columns, String selection, String[] selectionArgs, String groupBy, String having, String orderBy, String limit)	执行数据表查询操作，其中的参数有 distinct（是否去掉重复行）、table（表名称）、columns（列名称）、selection（WHERE 子句）、selectionArgs（WHERE 条件）、groupBy（分组）、having（分组过滤）、orderBy（排序）、limit（LIMIT 子句）
public Cursor query(String table, String[] columns, String selection, String[] selectionArgs, String groupBy, String having, String orderBy)	执行数据表查询操作
public Cursor rawQuery(String sql, String[] selectionArgs)	执行指定的 SQL 查询语句

② android.database.sqlite.SQLiteOpenHelper 类

SQLiteDatabase 类本身只是一个数据库的操作类，如果要进行数据库的操作，则需要在 android.database.sqlite.SQLiteOpenHelper 数据库操作辅助类的帮助下才可以进行，SQLiteOpenHelper 类是一个抽象类，它的常用方法如表 8-7 所示。使用的时候需要定义其子类，并且在子类中要覆写相应的抽象方法。

表 8-7　SQLiteOpenHelper 类定义的方法

方法	描述
public SQLiteOpenHelper(Context context, String name, SQLiteDatabase.CursorFactory factory, int version)	通过此构造方法指明要操作的数据库名称以及数据库的版本编号
public synchronized void close()	关闭数据库
public synchronized SQLiteDatabase getReadableDatabase()	以只读的方式创建或者打开数据库
public synchronized SQLiteDatabase getWritableDatabase()	以修改的方式创建或者打开数据库
public abstract void onCreate(SQLiteDatabase db)	创建数据表
public void onOpen(SQLiteDatabase db)	打开数据表
public abstract void onUpgrade(SQLiteDatabase db, int oldVersion, int newVersion)	更新数据表

③ android.database.Cursor 接口

数据库的操作除了上述之外，还有一项数据的查询操作，当 Android 程序需要进行数据查询操作的时候，需要保存全部的查询结果，而保存查询结果可以使用 android.database.Cursor 接口完成，Cursor 接口的常用方法如表 8-8 所示。

表 8-8 Cursor 接口的常用方法

方法	描述
public abstract void close()	关闭查询
public abstract int getCount()	返回查询的数据量
public abstract int getColumnCount()	返回查询结果之中列的总数
public abstract String[] getColumnNames()	得到查询结果之中全部列的名称
public abstract String getColumnName(int columnIndex)	得到指定索引位置列的名称
public abstract boolean isAfterLast()	判断结果集指针是否在最后一行数据之后
public abstract boolean isBeforeFirst()	判断结果集指针是否在第一行记录之前
public abstract boolean isClosed()	判断结果集是否已关闭
public abstract boolean isFirst()	判断结果集指针是否指在第一行
public abstract boolean isLast()	判断结果集指针是否指在最后一行
public abstract boolean moveToFirst()	将结果集指针移到第一行
public abstract boolean moveToLast()	将结果集指针移动到最后一行
public abstract boolean moveToNext()	将结果集指针向下移动一行
public abstract boolean moveToPrevious()	将结果集指针向前移动一行
public abstract boolean requery()	更新数据后刷新结果集中的内容
public abstract int getXxx(int columnIndex)	根据指定列的索引取得指定的数据

④ android.database.ContentValues 类

在 SQLiteDatabase 类之中专门提供了增加、删除、修改、查询等方法，在使用这些方法进行数据库操作的时候，所有的数据必须使用 android.database.ContentValues 类进行封装，ContentValues 类常用的方法如表 8-9 所示。

表 8-9 ContentValues 类常用的方法

方法	描述
public ContentValues()	创建 ContentValues 类实例
public void clear()	清空全部的数据
public void put(String key, 包装类 value)	设置指定字段（key）数据
public Integer getAs 包装类(String key)	根据 key 取得数据，如 getInteger()
public int size()	返回保存数据的个数

8.3.1 创建数据库及表

1. 创建数据库

在 Android 应用程序中使用 SQLite 时，必须自己创建数据库，然后创建表、索引，填充数据。Android 提供了 SQLiteOpenHelper 类帮助用户创建数据库，只要继承 SQLiteOpenHelper 类，就可以创建数据库，在 SQLiteOpenHelper 类的子类中，需要实现下面几个方法。

（1）构造函数，调用父类 SQLiteOpenHelper 类的构造函数，例如：

```
DatabaseHelper(Context context, String name, CursorFactory cursorFactory,
```

```
int version) { super(context, name, cursorFactory, version); }
```

这个方法需要四个参数：上下文环境（如一个 Activity）、数据库名称、一个可选的游标工厂（通常是 Null）、一个正在使用的数据库模型版本的整数。

（2）onCreate()方法，生成相应的数据表，它需要一个 SQLiteDatabase 对象作为参数，根据需要对这个对象填充表和初始化数据，例如：

```
public void onCreate(SQLiteDatabase db) {…} //创建数据库后，对数据库的操作
```

（3）onUpgrade()方法，当数据库版本要升级时会调用此方法，例如：

```
public void onUpgrade(SQLiteDatabase db, int oldVersion, int newVersion) { …}
```

它需要 3 个参数：一个 SQLiteDatabase 对象、一个旧的版本号和一个新的版本号，一般可以在此方法中将数据表删除。

（4）onopen()方法，当数据库打开时会调用这个方法，例如：

```
public void onOpen(SQLiteDatabase db) {super.onOpen(db); }}
                    //每次成功打开数据库后先被执行
```

2. 创建数据表

调用 getReadableDatabase()或 getWriteableDatabase()方法，可得到 SQLiteDatabase 实例，具体调用哪个方法，取决于是否需要改变数据库的内容，例如：

```
db=(new DatabaseHelper(getContext())).getWritableDatabase();
return (db == null) ? false : true;
```

上面这段代码会返回一个 SQLiteDatabase 类的实例，使用这个对象就可以查询或者修改数据库。当完成了对数据库的操作（如 Activity 已经关闭）时，需要调用 SQLiteDatabase 的 Close()方法来释放数据库连接。

创建表和索引

为了创建表和索引，需要调用 SQLiteDatabase 的 execSQL()方法来执行相应的语句，例如，创建一个名为 mytable 的表，表有一个列名为_id，并且是主键，此列的值会自动增长。另外，还有两列：title(字符)和 value(浮点数)。其代码如下：

```
db.execSQL("CREATE TABLE mytable(_id INTEGER PRIMARY KEY AUTOINCREMENT, title
        TEXT, value REAL);");
```

通常情况下，SQLite 会自动为主键列创建索引。

实例 8-4：创建数据库及表实例

新建一个项目，项目命名为 Exam8_5，创建一个数据库及表，为了后续使用的方便，这里专门建立一个类来创建数据库及表，然后在主程序中调用这个类中的方法。

（1）建立一个类继承 SQLiteOpenHelper 类，右击该项目的包，选择“New→Java Class”选项，输入类的名称，如 DataBaseHelp.java。在其中定义数据库的名称、表名称以及表中各字段的名称、类型、长度等，具体代码如下：

```
package org.hnist.demo;
import android.content.Context;
import android.database.sqlite.SQLiteDatabase;
import android.database.sqlite.SQLiteOpenHelper;
//定义 DataBaseHelp 继承 SQLiteOpenHelper 类
```

```
public class DataBaseHelp extends SQLiteOpenHelper {
private static final String DATABASENAME = "hnist.db" ;   //定义数据库名称
private static final int DATABASEVERSION = 1 ;             //定义数据库版本
private static final String TABLENAME = "personinfo" ;    //定义数据表名称
public DataBaseHelp(Context context) {                     //定义构造
  super(context, DATABASENAME, null, DATABASEVERSION); }   //调用父类构造
public void onCreate(SQLiteDatabase db) {                  //创建数据表
    String sql = "CREATE TABLE " + TABLENAME + " (" +
        "id          INTEGER      PRIMARY KEY ," +         //设置为自动增长列
        "name        VARCHAR(50)  NOT NULL ," +
        "sex        VARCHAR(10)  NOT NULL ," +
        "DateofBirth     DATE          NOT NULL ," +
        "email      VARCHAR(50)  NOT NULL)";               //定义 SQL 语句
    db.execSQL(sql) ;   }                                  //执行 SQL 语句
public void onUpgrade(SQLiteDatabase db, int oldVersion, int newVersion) {
    String sql = "DROP TABLE IF EXISTS " + TABLENAME ;     //SQL 语句
    db.execSQL(sql);                                       //执行 SQL 语句
    this.onCreate(db); }}                                  //创建表
```

（2）修改 Activity 文件 MainActivity.java，代码如下：

```
package org.hnist.demo;
import android.app.Activity;
import android.database.sqlite.SQLiteOpenHelper;
import android.os.Bundle;
public class MainActivity extends Activity {
   public void onCreate(Bundle savedInstanceState) {
      super.onCreate(savedInstanceState);
      super.setContentView(R.layout.activity_main);
      SQLiteOpenHelper helper = new DataBaseHelp(this); //定义数据库辅助类
      helper.getWritableDatabase() ; }}              //以修改方式打开数据库
```

保存所有文件，运行该项目，会创建数据库 hnist.db，并建立含有指定字段的数据表 personinfo，怎么才能查看到这个数据库及数据表呢？

使用模拟器测试，在 Android Studio 中选择“Tools→Android→Android Device Monitor ”选项即可进入 DDMS 界面后选择“File Explorer→data→data→ org.hnist.cn →databases”选项即可看到数据库文件，将 hnist.db 导出，可以使用 adb shell 命令来查看，也可以利用 SQLite Manager 工具来查看。下载安装 SQLite Manager，运行它，选择“File→Open”选项，找到导出的 hnist.db 文件，选择其中的“Table”选项，可以看到表中的字段信息，如图 8.6 所示。

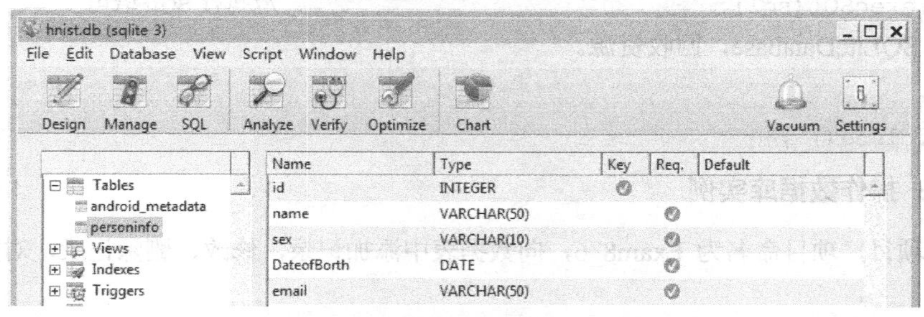

图 8.6　新建数据表的结构

Android Studio 真机上看不到这个数据库文件，因为 APP 的数据库文件是被保护的，只能 APP 自

己访问，APP 以外只有获得了 Root 权限才可以访问。

8.3.2 操作数据库

数据库以及表建立好了，表中还没有记录，怎样向其中添加记录并对记录做一些简单的操作呢？这些可以借助 SQL 语句来实现。

插入语句：

```
insert into 表名(字段列表) values(值列表)。
```

例如：

```
insert into personinfo(name, sex) values("张小君","男")
```

更新语句：

```
update 表名 set 字段名=值 where 条件子句。
```

例如：

```
update personinfo set name='张小君' where id=10
```

删除语句：

```
delete from 表名 where 条件子句。
```

例如：

```
delete from personinfo  where id=10
```

查询语句：

```
select * from 表名 where 条件子句 group by 分组字句 having ... order by 排序子句
```

例如：

```
select * from personinfo                    //获取表中的所有记录
   select * from Account limit 5 offset 2   //获取 5 条记录，跳过前面 2 条记录
```

使用 SQLiteDatabase 进行数据库操作的步骤如下。

（1）获取 SQLiteDatabase 对象，它代表了与数据库的连接，例如：

```
   MainActivity.this.helper.getWritableDatabase(); //取得可写的数据库
```

（2）定义要执行的 SQL 语句。

例如：

```
String sql = "INSERT INTO " + TABLENAME + " (name,sex,DateofBirth,email) VALUES
('"+ name + "','" + sex + "','" + DateofBorth + "','" + email + "')";
```

（3）执行定义好的 SQL 语句。

例如：

```
this.db.execSQL(sql);                                   //执行 SQL 语句
```

（4）关闭 SQLiteDatabase，回收资源。

例如：

```
this.db.close() ;
```

实例 8-5：操作数据库实例

新建一个项目，项目命名为 Exam8_6，向数据表中添加记录，修改、删除记录，对数据库进行简单操作。

（1）建立布局管理文件 activity_main.xml，定义三个按钮，代码如下：

```
…
```

```
<Button
    android:id="@+id/insert"
    android:layout_width="fill_parent"
    android:layout_height="wrap_content"
    android:text="增加数据" />
<Button
    android:id="@+id/update"
    android:layout_width="fill_parent"
    android:layout_height="wrap_content"
    android:text="修改数据" />
<Button
    android:id="@+id/delete"
    android:layout_width="fill_parent"
    android:layout_height="wrap_content"
    android:text="删除数据" />
</LinearLayout>
```

（2）建立一个类 DataBaseHelp.java 继承 SQLiteOpenHelper 类，定义数据库的名称、表名称以及表的结构，具体代码与 Exam8_5 的 DataBaseHelp.java 一致。

（3）建立一个表的操作类 OperateTable.java，进行表中数据的添加、更新和删除操作，具体代码如下：

```
…
import android.database.sqlite.SQLiteDatabase;
public class OperateTable {
private static final String TABLENAME = "personinfo" ;      //表名称
private SQLiteDatabase db = null ;                           //SQLiteDatabase
public  OperateTable(SQLiteDatabase db) {                    //构造方法
    this.db = db ;  }
public void insert(String name, String sex,String DateofBirth,String email) {
    String sql = "INSERT INTO " + TABLENAME + " (name,sex,DateofBirth,email)
            VALUES ('"+ name + "','" + sex + "','" + DateofBirth + "',
            '" + email + "')";                               //SQL 语句
    this.db.execSQL(sql);                                    //执行 SQL 语句
    this.db.close() ;    }                                   //关闭数据库操作
public void update(int id, String name, String sex,String DateofBirth,String
    email) {String sql = "UPDATE " + TABLENAME + " SET name='" + name+ "',sex=
    '" + sex + "',email='" + email   + "',DateofBirth='" + DateofBirth + "'
    WHERE id=" + id;                                         //SQL 语句
    this.db.execSQL(sql);                                    //执行 SQL 语句
    this.db.close() ;        }                               //关闭数据库操作
public void delete(int id) {
String sql = "DELETE FROM " + TABLENAME + " WHERE id=" + id;//SQL 语句
    this.db.execSQL(sql) ;                                   //执行 SQL 语句
    this.db.close() ;    }}                                  //关闭数据库操作
```

（4）定义 Activity 文件 MainActivity.Java，代码如下：

```
…
import android.database.sqlite.SQLiteOpenHelper;
public class MainActivity extends AppCompatActivity {
private DataBaseHelp helper = null ;                     //数据库操作
```

```
    private OperateTable mytable = null ;                   //mytable 表操作类
    private Button insBut = null ;
    private Button updBut = null ;
    private Button delBut = null ;
    private static int count = 0 ;                          //计数统计
       @Override
       public void onCreate(Bundle savedInstanceState) {
           super.onCreate(savedInstanceState);
           super.setContentView(R.layout.activity_main);
           this.insBut = (Button) super.findViewById(R.id.insert);
           this.delBut = (Button) super.findViewById(R.id.delete);
           this.updBut = (Button) super.findViewById(R.id.update);
           this.helper = new DataBaseHelp(this) ;           //定义数据库辅助类
             //注册监听
           this.insBut.setOnClickListener(new InsOnClickListenerImpl()) ;
           this.delBut.setOnClickListener(new DelOnClickListenerImpl());
           this.updBut.setOnClickListener(new UpnClickListenerImpl()) ; }
    private class InsOnClickListenerImpl implements OnClickListener {//实现监听
        public void onClick(View view) {
            MainActivity.this.mytable = new OperateTable (MainActivity.this.
            helper.getWritableDatabase());                 //取得可写的数据库
            MainActivity.this.mytable.insert("zhang" + count ++ ,
            "男","1978-10-12","zxj@163.com") ; }}          //添加记录
    private class DelOnClickListenerImpl implements OnClickListener {
        public void onClick(View view) {
            MainActivity.this.mytable = new OperateTable (MainActivity.this.
                helper.getWritableDatabase());             //取得可写的数据库
          MainActivity.this.mytable.delete(2) ; }}         //删除第二条记录数据
    private class UpnClickListenerImpl implements OnClickListener {
        public void onClick(View view) {
            MainActivity.this.mytable = new OperateTable (MainActivity.this.
                helper.getWritableDatabase());             //取得可写的数据库
    //更新第一条记录数据
    MainActivity.this.mytable.update(1, "胡晓莲", "女","1981-06-27",
"hxl@163.com") ; }}}
```

保存所有文件，运行项目，进入界面，如图 8.7 所示，先单击三次“添加记录”按钮，然后分别单击另外两个按钮，最后导出 hnist.db，用 SQLite Manager 打开表，如图 8.8 所示。

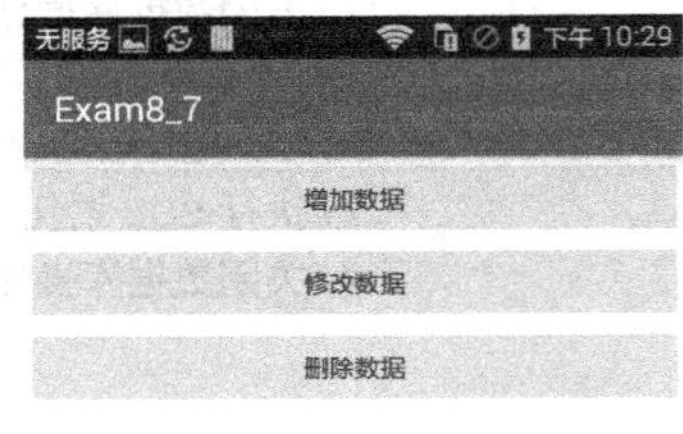

图 8.7 新建数据表的结构

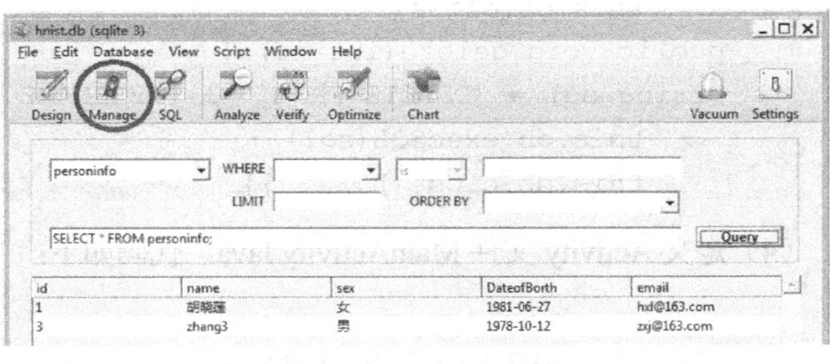

图 8.8 对表进行添加、修改和删除操作后的数据

8.3.3 数据查询操作

数据库的操作除了上述简单的操作之外，还有一个很重要的数据查询操作，当 Android 程序需要进行数据查询操作的时候需要保存全部的查询结果，而保存查询结果可以使用 android.database.Cursor 接口完成。

Cursor 接口没有提供方法而可直接控制指针移动，如果想从前到后依次取得全部数据记录，就要使用 isAfterLast()、moveToNext()、moveToFirst()三个方法，按照如下步骤进行。

（1）使用 moveToFirst()方法将结果集的指针放在第一行数据中。

（2）使用 isAfterLast()方法判断是否还有数据，如果还有数据，则进行取出。

（3）利用 moveToNext()方法将指针向下移动，并继续使用 isAfterLast()方法判断。

记录集指针控制如图 8.9 所示。

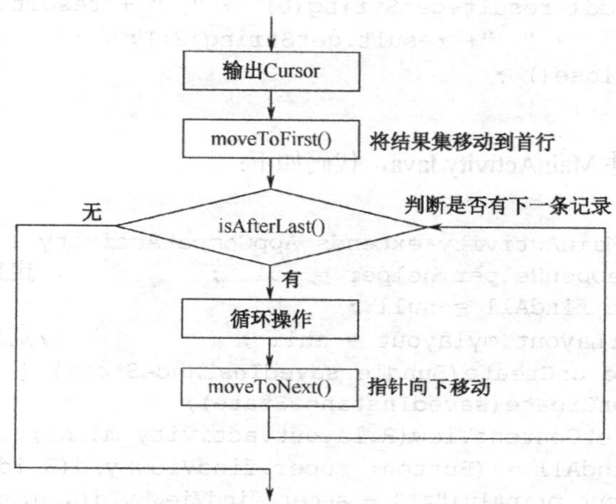

图 8.9 记录集指针控制示意图

代码如下：

```
for (result.moveToFirst(); !result.isAfterLast(); result.moveToNext()){循环体;}
```

其中，result 表示查询到记录的集合。

实例 8-6：数据库查询操作实例

新建一个项目，项目命名为 Exam8_7，将 hnist.db 中的 personinfo 表中所有记录的姓名、性别、邮箱字段内容全部显示出来。

（1）建立布局文件 activity_main.xml，定义一个按钮，代码如下：

```
…
<Button
    android:id="@+id/findall"
    android:layout width="fill parent"
    android:layout_height="wrap_content"
    android:text="查询全部记录" />
</LinearLayout>
```

（2）建立一个类 DataBaseHelp.java 继承 SQLiteOpenHelper 类，定义数据库的名称、表名称以及表的结构，具体代码与 Exam8_5 的类 DataBaseHelp.java 一致。

（3）建立一个表的查询操作类 SearchTable.java，定义查询操作，代码如下：

```
import java.util.ArrayList;
import java.util.List;
import android.database.Cursor;
import android.database.sqlite.SQLiteDatabase;
public class SearchTable {
private static final String TABLENAME = "personinfo" ;    //数据表名称
private SQLiteDatabase db = null ;                        //SQLiteDatabase
public SearchTable(SQLiteDatabase db) {                   //构造方法
     this.db = db ; }                                     //接收 SQLiteDatabase
public List<String> find() {                              //查询数据表
     List<String> all = new ArrayList<String>() ;         //定义 List 集合
     String sql = "SELECT name,sex,email FROM " + TABLENAME; //定义 SQL
       Cursor result = this.db.rawQuery(sql,null);        //不设置查询参数
     for (result.moveToFirst(); !result.isAfterLast(); result.moveToNext()) {
         all.add( result.getString(0)  + "  " + result.getString(1)
                 + "  "+ result.getString(2));    }      //设置集合数据
     this.db.close() ;                                    //关闭数据库连接
     return all ; }}
```

（4）定义 Activity 文件 MainActivity.Java，代码如下：

```
…
public class MainActivity extends AppCompatActivity {
private SQLiteOpenHelper helper = null ;           //定义数据库辅助类组件
private Button findAll = null ;
private LinearLayout mylayout = null ;             //定义布局管理器组件
   public void onCreate(Bundle savedInstanceState) {
      super.onCreate(savedInstanceState);
      super.setContentView(R.layout.activity_main);
      this.findAll = (Button) super.findViewById(R.id.findall) ;
      System.out.println("**" + super.findViewById(R.id.mylayout).getClass()) ;
      this.mylayout = (LinearLayout) super.findViewById(R.id.mylayout) ;
      this.helper = new DataBaseHelp(this) ;     //定义数据库辅助类
      this.findAll.setOnClickListener(new OnClickListenerfindall()) ; }
                                                 //设置监听
private class OnClickListenerfindall implements OnClickListener {
public void onClick(View view) {
     MainActivity.this.helper = new DataBaseHelp(MainActivity.this) ;
     ListView listView = new ListView(MainActivity.this) ; //定义 ListView
     listView.setAdapter(new ArrayAdapter<String>(MainActivity.this,
android.R.layout.simple_list_item_1, //每行显示一条数据
      new  SearchTable(MainActivity.this.helper.getReadableDatabase()).find()));

      MainActivity.this.mylayout.addView(listView) ; }}} //追加组件
```

保存所有文件，运行该项目，单击按钮，得到如图 8.10 所示的结果。

11:34
查询全部记录
胡晓莲 女 hxl@163.com
zhang1 男 zxj@163.com
zhang2 男 zxj@163.com

图 8.10　Exam8_7 运行结果

如果要查询含有性别为“女”的记录，应该如何操作呢？只需要把查询操作类 SearchTable.java 中的代码

```
String sql = "SELECT name,sex,email FROM " + TABLENAME;
```

修改为

```
String sql = "SELECT name,sex,email FROM " + TABLENAME
+ " WHERE sex = '女'";
```

其他语句不用做任何修改。读者可以在上面的基础上自行完成其他条件的查询操作。

项目 Exam8_8 是一个关于数据库的增、删、改、查综合运用的例子，读者可自行下载学习。

8.4 使用 ContentProvider 存储数据

在 Android 中，每一个应用程序的数据都是采用私有的形式进行操作的，一般不能被外部应用程序访问。为了使其他应用程序操作本程序的数据，可以通过 ContentProvider 提供数据操作的接口。可以将底层数据封装成 ContentProvider，有效地屏蔽底层操作的细节，并且使程序保持良好的扩展性和开放性。

例如，打电话程序和发短信程序都需要使用联系人的数据，这些数据就是以 ContentProvider 形式存放的。

8.4.1 ContentProvider 基础

ContentProvider（内容提供者）是一个类，这个类主要是对 Android 系统中共享的数据进行包装，并提供统一的访问接口供其他程序调用。这些被共享的数据可以是系统提供的，也可以是某个应用程序中的数据，ContentProvider 使用表的形式来组织数据。

一个程序可以通过实现一个 Content Provider 的抽象接口将自己的数据暴露出去。外界可以通过这些接口和程序里的数据交互读取程序的数据，也可以删除程序的数据，当然，其中也会涉及一些权限的问题。

1. ContentProvider 类

ContentProvider 类主要用于制定数据的操作标准，常用的操作方法如表 8-10 所示。

表 8-10 ContentProvider 类常用的方法

方法	描述
public abstract boolean onCreate()	当启动此组件的时候调用
public abstract int delete(Uri uri, String selection, String[] selectionArgs)	根据指定的 URI 删除数据，并返回删除数据的行数
public final Context getContext()	返回 Context 对象
public abstract String getType(Uri uri)	根据指定 URI，返回操作的 MIME 类型
public abstract Uri insert(Uri uri, ContentValues values)	根据指定的 URI 进行增加数据的操作，并且返回增加后的 URI，在此 URI 中会附带新数据的_id
public abstract Cursor query(Uri uri, String[] projection, String selection, String[] selectionArgs, String sortOrder)	根据指定的 URI 执行查询操作，所有的查询结果通过 Cursor 对象返回
public abstract int update(Uri uri, ContentValues values, String selection, String[] selectionArgs)	根据指定的 URI 进行数据的更新操作，并返回更新数据的行数

2. URI 类

使用 ContentProvider 类进行数据操作时，用 URI 的形式可进行数据的交换，URI 是标识资源的逻辑位置，并不提供资源的具体位置，URI 代表了要操作的数据，URI 主要包含了两部分信息：①需要操作的 ContentProvider；②对 ContentProvider 中的什么数据进行操作。

一个 URI 一般由以下几部分组成。

协议部分：ContentProvider（内容提供者）访问协议，已经由 Android 规定为 content://。

主机名(或 Authority)：用于唯一标识 ContentProvider，外部调用者可以根据这个标识来找到它，一般为程序的“包.类”名称，要采用小写字母的形式表示。

Path 部分：访问的路径，一般为要操作的数据表的名称。

例如：

content://org.hnist.demo.personprovider/person 的含义就是访问 person 表中的所有记录。

content://org.hnist.demo.personprovider/person/5 的含义就是访问 person 表中 id 为 5 的记录。

content://org.hnist.demo.personprovider/person/5/name 的含义就是访问 person 表中 id 为 5 的记录的 name 字段。

当然，要操作的数据不一定来自数据库，也可以是文件、XML 或网络等其他存储方式，例如，要操作 XML 文件中 person 节点下的 name 节点，可以构建这样的路径：/person/name。

前面介绍的 URI 都是以字符串的形式出现的，在 Android 中要利用 URI 类对这些字符串进行封装才能被访问，URI 类常用的操作方法如表 8-11 所示。

表 8-11　URI 类常用的操作方法

方法	描述
public static String decode(String s)	对字符串进行编码
public static String encode(String s)	对编码后的字符串进行解码
public static Uri fromFile(File file)	从指定的文件之中读取 URI
public static Uri withAppendedPath(Uri baseUri, String pathSegment)	在已有地址之后添加数据
public static Uri parse(String uriString)	将给出的字符串地址变为 URI 对象

要把一个字符串转换成 URI，可以使用 URI 类中的 parse()方法，例如：

```
Uri uri = Uri.parse("content://org.hnist.demo.personprovider/person/5")
```

3．使用 ContentResolver 操作 ContentProvider 中的数据

当外部应用需要对 ContentProvider 中的数据进行添加、删除、修改和查询操作时，可以使用 ContentResolver 类来完成，其常用方法如表 8-12 所示，要获取 ContentResolver 对象，可以使用 Activity 提供的 public ContentResolver getContentResolver()方法。

表 8-12　ContentResolver 类常用方法

方法	描述
public final int delete(Uri url, String where, String[] selectionArgs)	调用指定 ContentProvider 对象中的 delete()方法
public final Uri insert(Uri url, ContentValues values)	调用指定 ContentProvider 对象中的 insert()方法
public final Cursor query(Uri uri, String[ ] projection, String selection, String[ ] selectionArgs, String sortOrder)	调用指定 ContentProvider 对象中的 query()方法
public final int update(Uri uri, ContentValues values, String where, String[ ] selectionArgs)	调用指定 ContentProvider 对象中的 update()方法

使用 ContentResolver 对 ContentProvider 中的数据进行添加、删除、修改和查询操作，例如：

```
ContentResolver resolver = getContentResolver();
Uri uri = Uri.parse("content://org.hnist.demo.personprovider/person");
ContentValues values = new ContentValues();//添加一条记录
values.put("name", "xiaowang");
values.put("sex", "男");
resolver.insert(uri, values);
//获取person表中所有记录
Cursor cursor = resolver.query(uri, null, null, null, "personid desc");
while(cursor.moveToNext()){
```

```
    Log.i("ContentTest","personid="+cursor.getInt(0)+ ",name="+ cursor.getString(1)); }
//把 id 为 1 的记录的 name 字段值更新为 zhangsan
ContentValues updateValues = new ContentValues();
updateValues.put("name", "zhangsan");
Uri updateIdUri = ContentUris.withAppendedId(uri, 2);
resolver.update(updateIdUri, updateValues, null, null);
Uri deleteIdUri = ContentUris.withAppendedId(uri, 2); //删除 id 为 2 的记录
resolver.delete(deleteIdUri, null, null);
```

URI 代表了要操作的数据，经常需要解析 URI，并从 URI 中获取数据。Android 系统提供了两个用于操作 URI 的工具类，分别为 UriMatcher 和 ContentUris 。

4．UriMatcher 类

在使用 ContentProvider 类操作的时候，某个方法可能要传递多种 URI，必须对这些传递的 URI 进行判断后才可以决定最终的操作形式，为了方便用户的判断，系统专门提供了 android.content.UriMatcher 类进行 URI 的匹配，其常用方法如表 8-13 所示。

表 8-13 UriMatcher 类常用方法

方法	描述
public static final int NO_MATCH	表示一个–1 的整型数据，在实例化对象时使用
public UriMatcher(int code)	实例化 UriMatcher 类对象
public void addURI(String authority, String path, int code)	增加一个指定的 URI 地址
public int match(Uri uri)	与传入的 URI 进行比较，如果匹配成功，则返回相应的 code；如果匹配失败，则返回–1

UriMatcher 类用于匹配 URI，它的用法如下。

首先，把需要匹配的 URI 路径全部列出来，例如：

```
//常量 UriMatcher.NO_MATCH 表示不匹配任何路径的返回码
UriMatcher  sMatcher = new UriMatcher(UriMatcher.NO_MATCH);
/*如果 match()方法匹配"content://org.hnist.demo.personprovider/person"
                        路径，则返回匹配码为 1*/
sMatcher.addURI("Content://org.hnist.demo.personprovider", "person", 1);
                        //添加需要匹配 URI，如果匹配则会返回匹配码 1
/*如果 match()方法匹配"content://org.hnist.demo.personprovider/person/5"
                        路径，则返回匹配码为 2*/
sMatcher.addURI("content://org.hnist.demo.personprovider","person/#", 2);
                        //#为通配符
switch (sMatcher.match(Uri.parse("content://org.hnist.demo.personprovider/
person/10"))) {    case 1
    break;
case 2
    break;
default://不匹配
    break;}
```

注册完需要匹配的 URI 后，就可以使用 sMatcher.match(uri)方法对输入的 URI 进行匹配了，如果匹配就返回相应的匹配码，匹配码是调用 addURI()方法传入的第三个参数，假设匹配 content://org.hnist.demo.personprovider/person 路径，则返回的匹配码为 1。

5. ContentUris 类

由于所有的数据都要通过 URI 进行传递，以增加操作为例，当用户执行完增加数据操作后，往往需要将增加后的数据 ID 通过 URI 返回，当接收到这个 URI 的时候就需要从其中取出增加的 ID，Android 中提供了 android.content.ContentUris 辅助工具类，帮助用户来完成这些操作。其常用方法如表 8-14 所示。

表 8-14　ContentUris 类常用方法

方法	描述
public static long parseId(Uri contentUri)	从指定 URI 之中取出 ID
public static Uri withAppendedId(Uri contentUri, long id)	在指定的 URI 之后增加 ID 参数

例如：

```
Uri uri = Uri.parse("content://org.hnist.demo.personprovider/person")
Uri resultUri = ContentUris.withAppendedId(uri, 5);
```

生成后的 URI 为

```
content://org.hnist.demo.personprovider/person /5
Uri uri = Uri.parse("content://org.hnist.demo.personprovider/person /5")
long personid = ContentUris.parseId(uri);       //获取的结果为：5。
```

8.4.2　创建自己的 ContentProvider

如果希望数据共享，要么建立一个自己的 ContentProvider，要么将自己的数据添加到已经存在的 ContentProvider 中。

创建一个自己的 ContentProvider，至少要实现以下 3 个方面的作用。

（1）继承 ContentProvider 并重写以下方法：

```
public class PersonContentProvider extends ContentProvider{
public boolean onCreate()
public Uri insert(Uri uri, ContentValues values)
public int delete(Uri uri, String selection, String[] selectionArgs)
public int update(Uri uri, ContentValues values, String selection, String[]
        selectionArgs)
public Cursor query(Uri uri, String[] projection, String selection, String[]
        selectionArgs, String sortOrder)
public String getType(Uri uri)  }
```

（2）实现 SQLiteOpenHelper 类，用于创建和删除 member 表；

```
private static class DatabaseHelper extends SQLiteOpenHelper{
    DatabaseHelper(Context context) {
        super(context, DATABASE_NAME, null, DATABASE_VERSION); }
    public void onCreate(SQLiteDatabase db) {
    //创建用于存储数据的表
    db.execSQL("Create table " + TABLE_NAME + " ( _id INTEGER PRIMARY KEY
            AUTOINCREMENT, USER_NAME TEXT); "); }
    public void onUpgrade(SQLiteDatabase db, int oldVersion, int newVersion) {
    //创建用于删除数据的表
    db.execSQL("DROP TABLE IF EXISTS " + TABLE_NAME);
        onCreate(db); }  }
```

（3）在 AndroidManifest.xml 文件中使用<provider>对该 ContentProvider 进行配置，为了能让其他应用找到该 ContentProvider ，ContentProvider 采用了 authorities（主机名/域名）对它进行唯一标识：

```
<provider
    android:name=". MyContentProvider " //ContentProvider 的名称
    android:authorities=" org.hnist.cn.mycontentprovider"/> //主机名
```

ContentProvider 类的开发比较麻烦，因为涉及的类很多，这里给出一个实例 Exam8_9，限于篇幅，具体的代码这里就不再给出。在实际应用中，一般不会要求自己开发 ContentProvider，系统为用户开发了许多 ContentProvider，只要通过 URI 操作这些 ContentProvider 即可。

8.4.3 操作联系人的 ContentProvider

前面介绍了，如果希望数据共享，要么建立一个自己的 ContentProvider，要么将自己的数据添加到已经存在的 ContentProvider 中，如果存在包含着和想公开数据相同类型数据的 ContentProvider，并且拥有权限，就可以将这些数据写入到 ContentProvider 中，从而实现数据共享。

Android 系统提供了多种数据类型的 ContentProvider(声音、视频、图片、联系人等)，它们大都位于 android.provider 包中，例如：

Browser：读取或修改书签，浏览历史或网络搜索。

CallLog：查看或更新通话历史。

Contacts：获取、修改或保存联系人信息。

LiveFolders：由 Content Provider 提供内容的特定文件夹。

MediaStore：访问声音、视频和图片。

Setting：查看和获取蓝牙设置、铃声和其他设置偏好。

SearchRecentSuggestions：该类为应用程序创建简单的查询建议提供者。

SyncStateContract：用于使用数据数组账号关联数据的 ContentProvider 约束。

UserDictionary：在可预测文本输入时，提供用户定义的单词给输入法使用，等等。

一般情况下没有必要开发自己的 ContentProvider 类，编程人员只需要操作好这些系统提供的 ContentProvider 就足够了，前提是已获得适当的权限。

android.provider 包中的 Contacts 就是 ContentProvider 对外提供联系人数据及操作的接口。

1. 获取所有联系人信息

Android 系统中的联系人是通过 ContentProvider 来对外提供数据的，其数据库为/data/data/com.android.providers.contacts/database/contacts2.db，如图 8.11 所示。

⊟ com.android.providers.contacts		2013-07-27 01:41	drwxr-x--x
⊟ databases		2013-07-27 01:41	drwxrwx--x
contacts2.db	83968	2013-07-27 01:41	-rw-rw----

图 8.11 查询所有记录

导出这个数据库，用 SQLite Manager 打开它，发现里面有很多表，如图 8.12 所示，其中 raw_contacts 表保存了联系人 ID，data 表和 raw_contacts 是多对一的关系，保存了联系人的各项数据，mimetypes 表为数据类型。操作时，先查询 raw_contacts 得到每个联系人的 ID，然后使用 ID 从 data 表中查询对应数据，根据 mimetype 分类数据。这里要注意的是，Android 版本不同，表的结构也不完全相同，这里的结构是 Android 4.4 版本中的结构。

其中主要是查询“contacts”表，其结构如图 8.13 所示。

要想进行 ContentProvider 程序访问，需要一个 CONTENT_URI 常量，这些常量在 android.provider.ContactsContract.Contacts 类中有定义，访问联系人的 CONTENT_URI 常量为 ContactsContract.Contacts.CONTENT_URI。

表中的字段较多，这里用到了下面几个。

人员 ID：ContactsContract.Contacts._ID。

人员姓名：ContactsContract.Contacts.DISPLAY_NAME。

读取手机号码信息的 URI：ContactsContract.CommonDataKinds.Phone.CONTENT_URI。

取得某个用户的电话信息和保存的内容，需要得到以下内容。

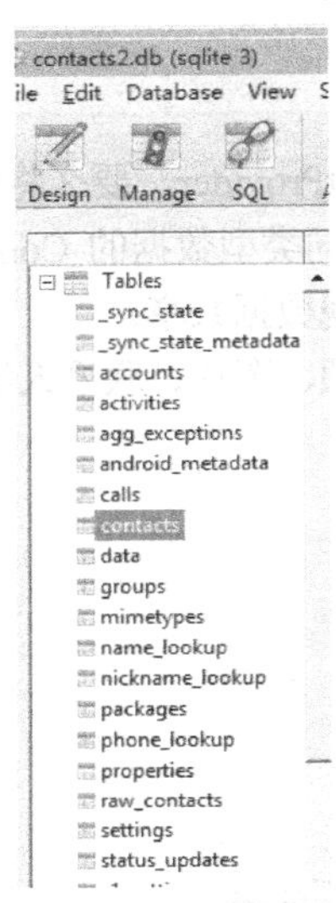

图 8.12 contacts2 的所有表

Alter Table contacts

Table name: contacts

Column Name	Datatype	PriKey	Unique	NotNull	AutoInc
_id	INTEGER	✓			✓
display_name	TEXT				
photo_id	INTEGER				
custom_ringtone	TEXT				
send_to_voicemail	INTEGER			✓	
times_contacted	INTEGER			✓	
last_time_contacted	INTEGER				
starred	INTEGER			✓	
in_visible_group	INTEGER			✓	
has_phone_number	INTEGER			✓	
lookup	TEXT				
status_update_id	INTEGER				
single_is_restricted	INTEGER			✓	
name_raw_contact_id	INTEGER				

Foreign Key

图 8.13 contacts 表结构

手机号码：ContactsContract.CommonDataKinds.Phone.NUMBER。

邮箱地址：ContactsContract.CommonDataKinds.Email.DATA。

所属联系人 ID：ContactsContract.CommonDataKinds.Phone.CONTACT_ID。

要获取手机中联系人信息，应按照下面的步骤进行。

（1）利用 getContentResolver()方法得到一个 ContentResolver 实例。

例如：

```
ContentResolver cr = this.getContentResolver(); //得到 ContentResolver 对象
```

（2）通过查询方式获取联系人的所有信息。

例如：

```
Cursor cursor = cr.query(ContactsContract.Contacts.CONTENT_URI, null, null, null,
null);
```

（3）循环读取每条记录的信息，逐个读出姓名、电话、Email 等信息。

例如：

```
for (cursor.moveToFirst(); !cursor.isAfterLast(); cursor.moveToNext()){
    Int nameIndex = cursor.getColumnIndex(ContactsContract.Contacts.
          DISPLAY_NAME);
    String name = cursor.getString(nameIndex);      //取得联系人名字
    sbLog.append("name=" + name + ";");             //取得联系人 ID
    String contactId=cursor.getString(cursor.getColumnIndex(Contacts
```

```
        Contract.Contacts._ID));          //根据联系人 ID 查询对应的电话号码
    Cursor phoneNumbers = cr.query(ContactsContract.CommonDataKinds.
        Phone.CONTENT_URI,null, ContactsContract.CommonDataKinds.
        Phone.CONTACT_ID + " = "+ contactId, null, null);
        //根据联系人 ID 查询对应的 E-mail
    Cursor emails = cr.query(ContactsContract.CommonDataKinds. Email.CONTENT_URI,
        null, ContactsContract.CommonDataKinds.Email.CONTACT_ID + " = "
        + contactId, null, null);}
```

（4）显示获取的信息。

例如：

```
text.setText(sb.toString());
```

（5）修改 AndroidManifest.xml 文件权限，在文件中添加如下代码：

```
<uses-permission android:name="android.permission.READ_CONTACTS"/>
```

实例 8-7：显示操作联系人信息实例

新建一个项目，项目命名为 Exam8_9，编写程序获得手机中的联系人信息，并将姓名、电话、邮箱信息显示出来。

（1）建立布局管理文件 activity_main.xml，代码如下：

```
...
   <TextView
      android:id="@+id/text"
      android:layout_width="fill_parent"
      android:layout_height="wrap_content" />
   <Button
      android:id="@+id/but"
      android:layout_width="wrap_content"
      android:layout_height="wrap_content"
      android:text="获取联系人信息" />
</LinearLayout>
```

（2）建立 Activity 文件 MainActivity.java，代码如下：

```
...
public class MainActivity extends AppCompatActivity {
   private Button but=null;
   private TextView text=null;
   public void onCreate(Bundle savedInstanceState) {
      super.onCreate(savedInstanceState);
      setContentView(R.layout.activity_main);
      text=(TextView) this.findViewById(R.id.text);
      but=(Button) this.findViewById(R.id.but);
      this.but.setOnClickListener(new dispOnListener());}
      private class dispOnListener implements OnClickListener{
        public void onClick(View view){
           StringBuilder sb=getContacts();
           text.setText(sb.toString()); } }
   private StringBuilder getContacts() {
      StringBuilder sbLog = new StringBuilder();
      ContentResolver cr = this.getContentResolver(); //得到ContentResolver 对象
        //取得通信录信息,主要是查询“Contacts”表
```

```
        Cursor cursor = cr.query(ContactsContract.Contacts.CONTENT_URI, null,
            null, null, null);
        //循环读取通信录中的信息
        for (cursor.moveToFirst(); !cursor.isAfterLast(); cursor.moveToNext())
         { //取得联系人名字(显示出来的名字)，实际内容在 ContactsContract.Contacts 中
         int nameIndex = cursor.getColumnIndex(ContactsContract.Contacts.
           DISPLAY_NAME);
            String name = cursor.getString(nameIndex);
            sbLog.append("name=" + name + ";");
            String contactId=cursor.getString(cursor.getColumnIndex
(ContactsContract.Contacts._ID));   //取得联系人 ID
            //根据联系人 ID 查询对应的电话号码
            Cursor phoneNumbers = cr.query(ContactsContract.CommonDataKinds.
                Phone.CONTENT_URI,null, ContactsContract.CommonDataKinds.
                Phone.CONTACT_ID + " = "+ contactId, null, null);
            //取得电话号码(可能存在多个号码)
            while (phoneNumbers.moveToNext())
            {String strPhoneNumber = phoneNumbers.getString(phoneNumbers.
                getColumnIndex(ContactsContract.CommonDataKinds.Phone.NUMBER));
                sbLog.append("Phone=" + strPhoneNumber + ";"); }
            phoneNumbers.close();
            //根据联系人 ID 查询对应的 E-mail
            Cursor emails = cr.query(ContactsContract.CommonDataKinds.
                Email.CONTENT_URI, null, ContactsContract.CommonDataKinds.
                Email.CONTACT_ID + " = " + contactId, null, null);
            //取得 E-mail(可能存在多个 E-mail)
            while (emails.moveToNext())
            {String strEmail = emails.getString(emails.getColumnIndex
                (ContactsContract.CommonDataKinds.Email.DATA));
                sbLog.append("Email=" + strEmail + ";");   }
    emails.close();}
        cursor.close();
         return sbLog;     } }
```

（3）在 AndroidManifest.xml 文件中添加如下内容：

```
    <uses-permission
android:name="android.permission.READ_CONTACTS"/>
```

（4）保存所有文件，运行该程序，单击按钮，运行结果如图 8.14 所示。

图 8.14　Exam8_9 运行结果

2．添加联系人信息

添加联系人信息到数据库中的步骤如下。

（1）要得到一个 ContentResolver 实例，需要利用 getContentResolver()方法。例如：

```
ContentResolver resolver = this.getCon-
                         tentResolver();
```

（2）定义一个 ContentValues 类，用于封装要添加的数据。例如：

```
    ContentValues values = new ContentValues();
```

（3）向 raw_contacts 表中插入一条数据，得到主键值。

例如：`long contactid = ContentUris.parseId(resolver.insert(uri, values));`

（4）逐个添加姓名、电话、邮箱等信息。例如：

```
uri = Uri.parse("content://com.android.contacts/data");
values.put("raw_contact_id", contactid );
values.put("mimetype", "vnd.android.cursor.item/name");
values.put("data2", "黄腾飞");
resolver.insert(uri, values);       //将信息写入
…
```

联系人相关的 URI：

```
content://com.android.contacts/contacts 操作的数据是联系人信息 URI
content://com.android.contacts/data/phones 联系人电话 URI
content://com.android.contacts/data/E-mails 联系人 E-mail URI
```

（5）修改 AndroidManifest.xml 权限，在文件中添加如下内容：

```
<uses-permission android:name="android.permission.READ_CONTACTS"/>
<uses-permission android:name="android.permission.WRITE_CONTACTS" />
```

实例 8-8：添加操作联系人信息实例

新建一个项目，项目命名为 Exam8_10，在实例 8-7 的基础上，增加一条记录到操作联系人数据库中。

（1）建立布局管理文件 activity_main.xml，代码如下：

```
…
    <TextView
        android:id="@+id/text"
        android:layout_width="fill_parent"
        android:layout_height="wrap_content" />
    <Button
        android:id="@+id/but"
        android:layout_width="wrap_content"
        android:layout_height="wrap_content"
        android:text="获取联系人信息" />
    <Button
        android:id="@+id/insertbut"
        android:layout_width="wrap_content"
        android:layout_height="wrap_content"
        android:text="添加联系人信息" />
</LinearLayout>
```

（2）建立 Activity 文件 MainActivity.java，在实例 8-7 的 MainActivity.java 的基础上，增加如下代码：

```
this.insertbut=(Button) this.findViewById(R.id.insertbut);
this.insertbut.setOnClickListener(new insertOnClickListener());
private class  insertOnClickListener  implements OnClickListener{
    public void onClick(View view){
      try { testAddContacts(); }
catch (Exception e) { e.printStackTrace(); }
        text.setText("添加成功"); } }
public void testAddContacts() throws Exception{
       Uri uri = Uri.parse("content://com.android.contacts/raw_contacts");
      ContentResolver resolver = this.getContentResolver();
```

```
        ContentValues values = new ContentValues();
        //向 raw_contacts 表中插入一条数据，得到主键值
        long contactid = ContentUris.parseId(resolver.insert(uri, values));
        //添加姓名
        uri = Uri.parse("content://com.android.contacts/data");
        values.put("raw_contact_id", contactid );
        values.put("mimetype", "vnd.android.cursor.item/name");
        values.put("data2", "黄腾飞");
        resolver.insert(uri, values);
        //添加电话
        values.clear();
        values.put("raw_contact_id", contactid);
        values.put("mimetype", "vnd.android.cursor.item/phone_v2");
        values.put("data2", "2");
        values.put("data1", "13807304561");
        resolver.insert(uri, values);
        //添加 E-mail
        values.clear();
        values.put("raw_contact_id", contactid);
        values.put("mimetype", "vnd.android.cursor.item/email_v2");
        values.put("data2", "2");
        values.put("data1", "htf@qq.com");
        resolver.insert(uri, values); }
```

（3）在 AndroidManifest.xml 文件中添加如下内容：

```
    <uses-permission android:name="android.permission.READ_CONTACTS"/>
    <uses-permission android:name="android.permission.WRITE_CONTACTS" />
```

保存所有文件，运行该项目，单击“添加联系人信息”按钮，会显示“添加成功”，再查看所有联系人信息，会发现已经多了一条记录，结果如图 8.15 所示。

这里为了程序的简单起见，将信息直接显示在了 TextView 中，实际上可以使信息在 ListView 中显示出来，读者可以适当修改程序。

Android 系统对所有的多媒体数据库接口进行了封装，不用自己创建数据库，直接利用 ContentResolver 去调用封装好的接口就可以进行多媒体数据库的操作。限于篇幅，这里不做介绍了，感兴趣的读者可以查找 Android API 自己学习。

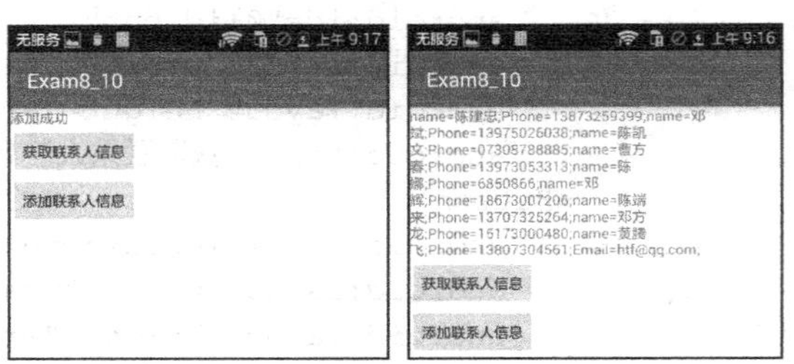

图 8.15　Exam8_10 运行结果

8.5　JSON 数据

有的时候需要在不同平台间进行数据交换，例如，手机客户端 APP 可能会与网站服务器中的数据进行交互，它们是不能直接进行数据传递的，此时就要进行数据的转换了。

JSON（JavaScript Object Notation）是一种轻量级的数据交换格式，具有良好的可读性和便于编写的特征，可以在不同平台间进行数据交换，它具有如下特点：

（1）JSON 数据是一系列键值对的集合。

（2）JSON 已经在网络数据的传输当中应用得非常广泛了。

（3）JSON 相对于 XML 来讲解析会方便一些，相对于 XML 而言，JSON 简单的语法格式和清晰的层次结构明显比 XML 容易阅读，并且在数据交换方面，JSON 数据体积小，与 JavaScrpit 的交互更加方便，执行速度要远远快于 XML。

8.5.1 JSON 基础

在 JS 语言中，一切都是对象。因此，任何支持的类型都可以通过 JSON 来表示，例如，字符串、数字、对象、数组等。用 JSON 表示的基本原则：对象表示为键值对，数据由逗号分隔，花括号保存对象，方括号保存数组。

JSON 最常用的格式是对象的键值对，例如：

```
{"Name": "张珊", "Sex": "女"}
```

JSON 数组方式要使用方括号 []，例如：

```
{"student:[{"Name":"张珊", "Sex": "女"},{"Name":"李斯", "Sex": "女"}]}
```

JSON 键值对是用来保存 JS 对象的一种方式，和 JS 对象的写法也大同小异，键值对组合中的键名写在前面并用双引号 "" 包裹，使用冒号:分隔，后面加上值，例如，{"Name": "张珊"}相当于 JavaScript 语句{Name = "张珊"}。

JSON 是 JS 对象的字符串表示法，它使用文本表示一个 JS 对象的信息，本质是一个字符串。例如，var obj = {Name: "张珊", Sex: "女"}是一个对象，var json = '{"Name": "张珊", "Sex": "女"}'是一个 JSON 字符串。

JSON 可以将 JavaScript 对象中表示的一组数据转换为字符串，然后在网络或者程序之间传递这个字符串，并在需要的时候将它还原为各编程语言所支持的数据格式。

将对象转换为 JSON 字符串，可使用 JSON.stringify() 方法，例如：

```
var json = JSON.stringify({Name: "张珊", Sex: "女"});
```

结果是 json = '{"Name": "张珊", "Sex": "女"}'

将 JSON 字符串转换为对象，可使用 JSON.parse() 方法：

```
var obj = JSON.parse('{"Name": "张珊", "Sex": "女"}');
结果是 obj ={Name: "张珊", Sex: "女"}
```

JSON 数据量小时比较适用，当数据量大时，就会陷入烦琐复杂的数据节点查找，这里介绍一些在线工具，如 BeJson、SoJson，它们可以让新接触 JSON 格式的程序员更快地了解 JSON 的结构，更快地精确定位 JSON 格式错误，读者可以尝试使用。

8.5.2 JSON 的使用

Android 平台自带了 4 个与 JSON 相关的类和一个 Exceptions：JSONArray、JSONObject、JSONStringer、JSONTokener 以及 JSONException。

（1）JSONObject 类是系统中有关 JSON 定义的基本单元，包含一对(Key/Value)数值。它对外部调用的响应体现为一个标准的字符串，例如，{"Name": "张珊", "Sex": "女"}，最外被大括号包裹，其中的 Key 和 Value 被冒号“:”分隔。内部行为的操作格式有些差异，例如，初始化一个 JSONObject 实例，引用内部的 put()方法添加数值，即 new JSONObject().put(“JSON”，“Hello, World!”)，在 Key 和 Value 之间以“，”,”分隔。Value 的类型包括 Boolean、JSONArray、JSONObject、Number、String 或者默认值 JSONObject.NULL object。

有两个不同的取值方法：get()和 opt()。get()在确定数值存在的条件下使用，当无法检索到相关 Key 时，将会抛出一个 Exception 信息。opt()方法相对比较灵活，当无法获取所指定数值时，将会返回一个默认数值，并不会抛出异常。

（2）JSONArray 类代表一组有序的数值，可将其转换为 String 输出(toString)所表现的形式，用方括号包裹，数值以“,”分隔（如[value1,value2,value3]）。这个类的内部同样具有查询行为，get()和 opt()两种方法都可以通过 index 返回指定的数值，put()方法用来添加或者替换数值。

同样，这个类的 value 类型可以包括 Boolean、JSONArray、JSONObject、Number、String 或者默认值 JSONObject.NULL object。

（3）JSONStringer 类可以帮助用户快速和便捷地创建 JSONtext。它最大的优点是可以减少由于格式的错误而导致的程序异常，引用这个类可以自动严格按照 JSON 语法规则创建 JSONtext。每个 JSONStringer 实体只能对应创建一个 JSONtext。

例如：

```
String myString = new JSONStringer().object()
.key("name")
.value("张珊")
.endObject()
.toString();
```

其相当于一组标准格式的 JSON text：{"name" : "张珊"}。

其中的.object()和.endObject()方法必须同时使用，这是为了按照 Object 标准给数值添加边界。类似的，针对数组也有一组标准的方法来生成边界——.array()和.endArray()方法。

（4）JSONTokener 是系统为 JSONObject 和 JSONArray 构造器解析 JSON source string 的类，它可以从 source string 中提取数值信息。

（5）JSONException 是 JSON.org 类抛出的异常信息。

可以将文件、输入流中的数据转化为 JSON 字符串，然后从对象中获取 JSON 保存的数据内容，项目 Exam8_11 就是这样一个实例，运行结果如图 8.16 所示。只是它的实现比较烦琐，远没有 Gson 简单，限于篇幅本书不再赘述，感兴趣的读者可以对照 Exam8_11 进行学习。

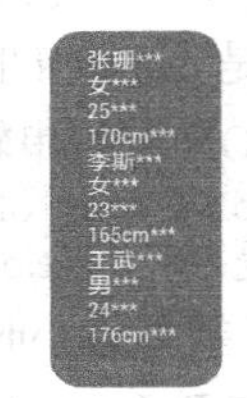

图 8.16　Exam8_11 运行结果

8.5.3　Gson 的基本操作

Gson 是 Google 提供的用来在 Java 对象和 JSON 数据之间进行映射的 Java 类库，它可以轻松地将一个 JSON 字符串转换成一个 Java 对象，或者将一个 Java 对象转换成一个 JSON 字符串。

Gson 有两个重要的方法，一个是 tojson，将 bean 中的内容转换为 json 内容；另一个就是 fromjson，从 JSON 对象封装出一个个的 bean 对象。

1．导入 Gson JAR 包

使用 Gson 解析 JSON 数据时需要下载 Gson JAR 包，并导入 libs 文件夹，具体步骤如下。

（1）选择“File→Project Structure”选项，如图 8.17 所示，在弹出的对话框中选择“app→Dependencies→"+" →Library dependency”选项，如图 8.18 所示。

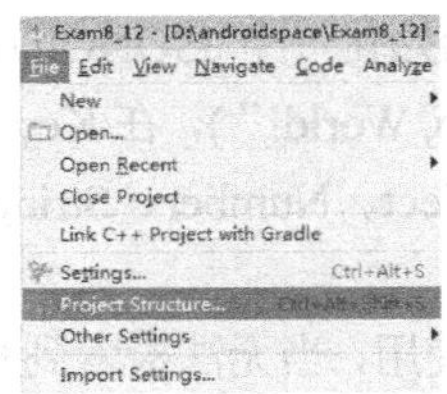

图 8.17　选择 Project Structure

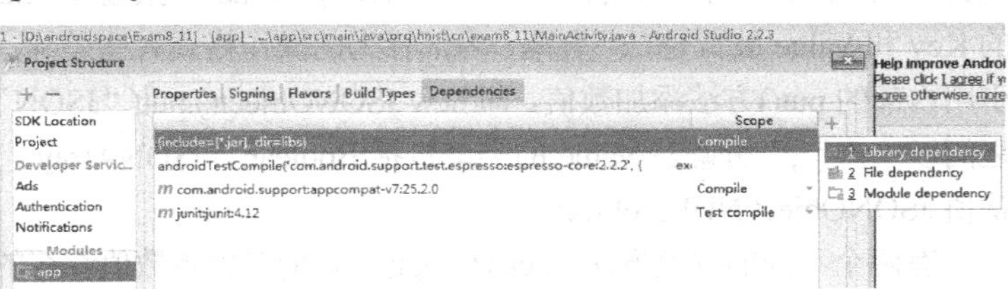

图 8.18　选择 Library dependency

（2）在弹出的对话框中查找 Gson，选择如图 8.19 所示的两个依赖包。单击“OK”按钮，进入如图 8.20 所示界面，单击“OK”按钮， Gson 包即可添加成功。

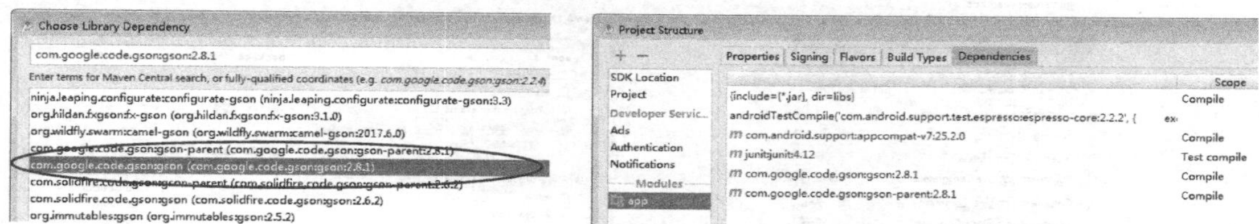

图 8.19　选择 Gson 依赖包　　图 8.20　Gson 依赖包添加成功

（3）可以在 Project 结构类型中查看是否添加成功，如图 8.21 所示。

2. 使用 GsonFormat 的生成实体类

（1）选择“File→Settings”选项，如图 8.22 所示。在弹出的对话框中选择“Plugins”选项，单击“Browse repositories...”按钮，如图 8.23 所示。在搜索框中输入“GsonFormat”，找到“GsonFormat”，单击“Install”按钮， 如图 8.24 所示，安装完毕后，单击“Restart Android Studio”按钮，重新启动 Android Studio。

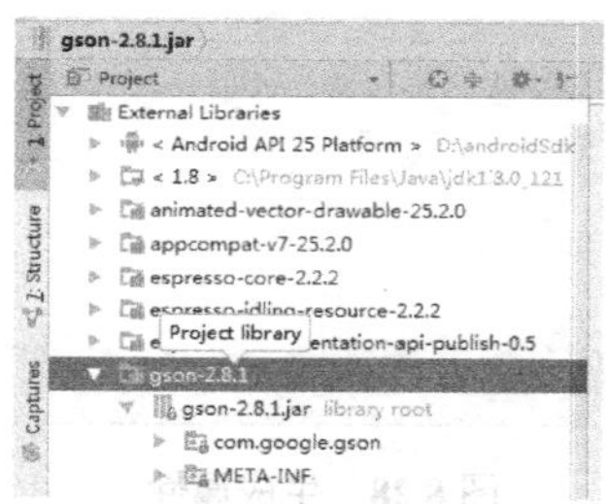

图 8.21　查看 Gson 是否添加成功

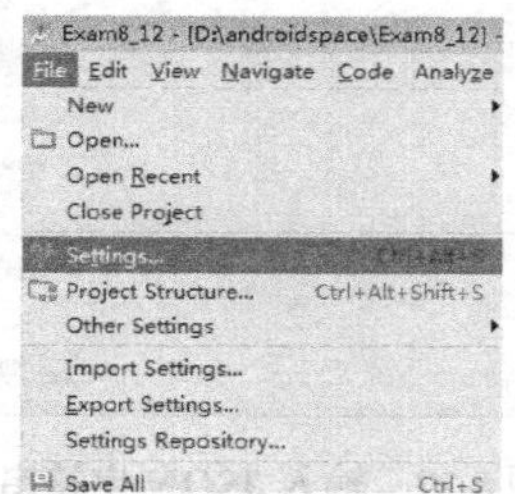

图 8.22　选择 Settings

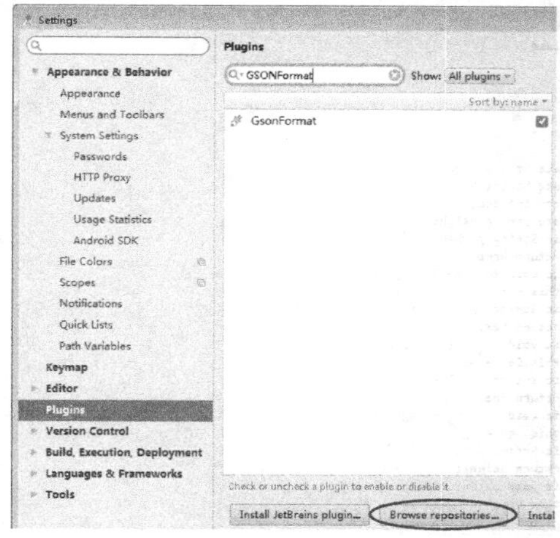

图 8.23　单击“Browse repositories…”按钮

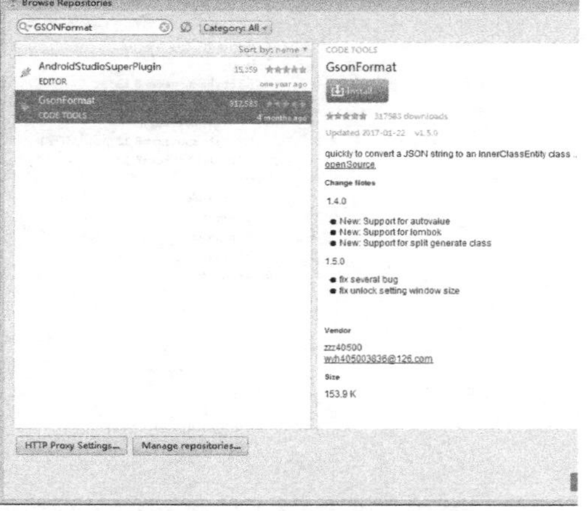

图 8.24　安装 GsonFormat

（2）重新打开程序，右击包名，选择“New→Java Class”选项，新建一个 bean.java 文件，光标停在大括号内，然后选择“Code→Generate”选项，如图 8.25 所示。在弹出的对话框中选择“GsonFormat”选项，如图 8.26 所示。

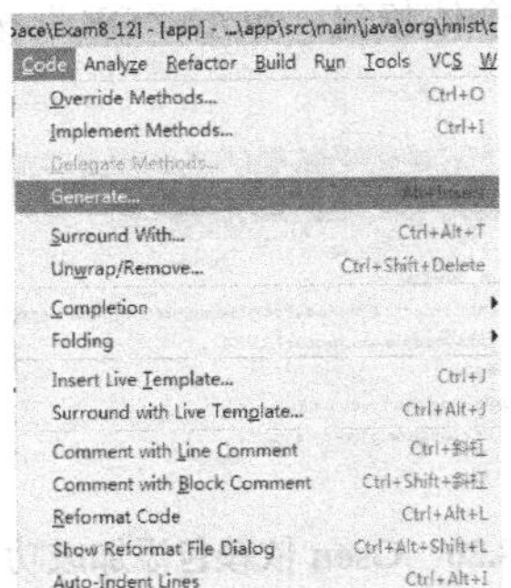

图 8.25　选择“Code→Generate”选项

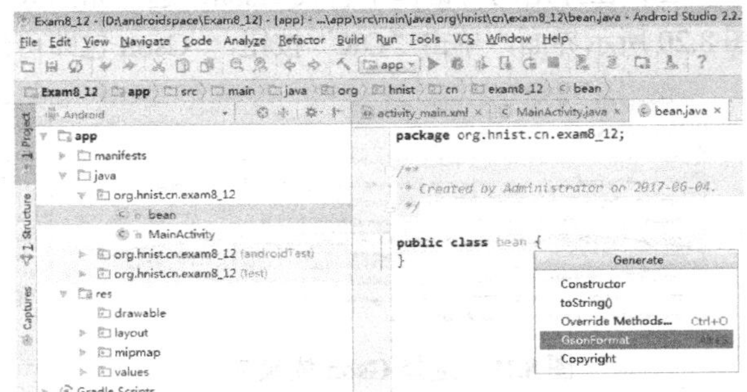

图 8.26　选择 GsonFormat

（3）在弹出的对话框中输入 JSON 字符串内容：{"name": "张珊","Sex": "女","age": 25,"height": "170cm"}，如图 8.27 所示。单击“OK”按钮，结果如图 8.28 所示，再次单击“OK”按钮。

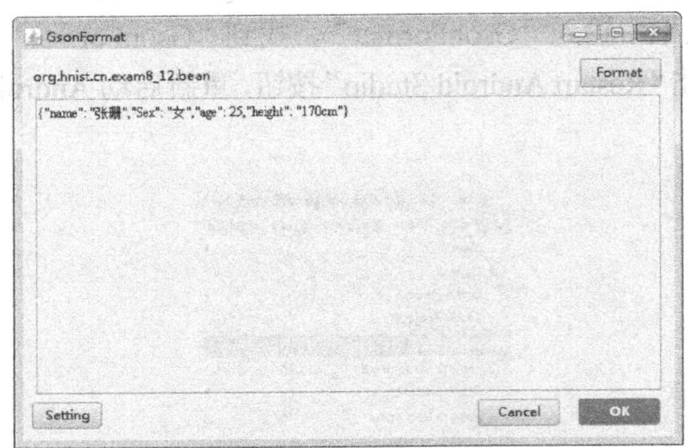

图 8.27　输入 JSON 字符串

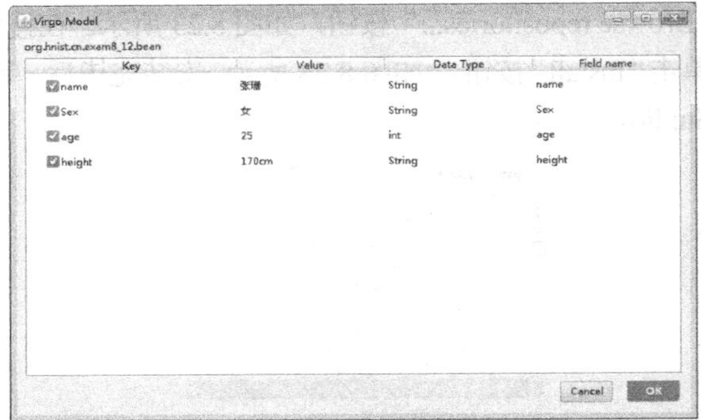

图 8.28　生成属性

（4）回到 bean.java，发现其中的代码已经自动生成了，如图 8.29 所示。

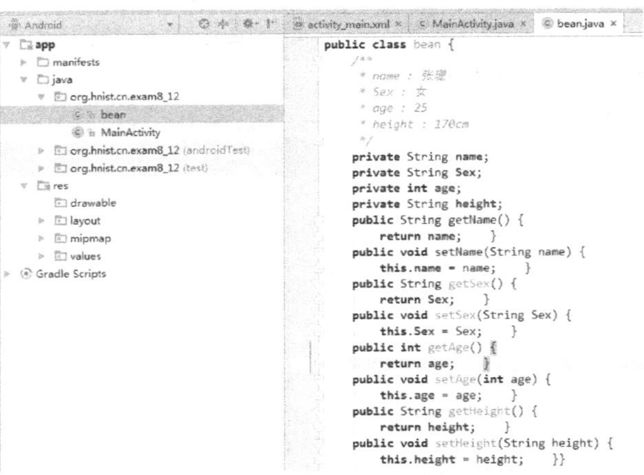

图 8.29　自动生成实体类

3. 使用 Gson 实现 Java 对象和 JSON 字符串转换

Gson 有两个重要的方法：一个是 tojson，将 bean 中的对象转换为 JSON；一个是 fromjson，将 JSON 内容封装成 bean 对象。

```
    //把 Java 对象转换成 JSON
```

```
bean user=new bean("fashion", "男",25,"180cm"); //定义对象并赋值
Gson gson1=new Gson();                          //定义 Gson 对象 gson1
String obj=gson1.toJson(user);                  //将 user 对象转换成 JSON 字符串
//把 JSON 转换成 Java 对象
jsonData="{\"name\":\"张珊\",\"Sex\":\"女\",\"age\":25,\"height\":170cm}";
 Gson gson2=new Gson();                         //定义 Gson 对象 gson2
bean user=gson2.fromJson(jsonData, bean.class); //将 JSON 字符串转换成 user 对象
```

实例 8-10：使用 Gson 实现 Java 对象和 JSON 字符串转换实例

新建一个项目，项目命名为 Exam8_12，使用 Gson 实现 Java 对象和 JSON 字符串的转换。

（1）按照前面的介绍，添加 Gson JAR 包到项目中。

（2）手动或者使用 GsonFormat 的生成实体类，代码如下：

```
package org.hnist.cn.exam8_12;
/**
 * Created by Administrator on 2017-06-04
 */
public class bean {
    /**
     * name : 张珊
     * Sex : 女
     * age : 25
     * height : 170cm
     */
    private String name;
    private String Sex;
    private int age;
    private String height;
    public String getName() {
        return name;    }
    public void setName(String name) {
        this.name = name;    }
    public String getSex() {
        return Sex;    }
    public void setSex(String Sex) {
        this.Sex = Sex;    }
    public int getAge() {
        return age;    }
    public void setAge(int age) {
        this.age = age;    }
    public String getHeight() {
        return height;    }
    public void setHeight(String height) {
        this.height = height;    }
    public bean() {
        super();    }
    public bean(String name, String Sex,int age,String height) {
        super();
        this.name = name;
        this.age = age;
        this.Sex=Sex;
        this.height=height;    }}
```

（3）建立布局管理文件 activity_main.xml，代码如下：

```
…
<Button android:id="@+id/btn01"
    android:layout_width="fill_parent"
    android:layout_height="wrap_content"
    android:gravity="center_horizontal"
    android:text="解析单条"/>
<Button android:id="@+id/btn02"
    android:layout_width="fill_parent"
    android:layout_height="wrap_content"
    android:gravity="center_horizontal"
    android:text="解析多条"/>
<Button android:id="@+id/btn03"
    android:layout_width="fill_parent"
    android:layout_height="wrap_content"
    android:gravity="center_horizontal"
    android:text="Java To Json"/>
</LinearLayout>
```

（4）在包名上右击，选择“New→Java Class”选项，建立辅助操作类 JsonUtils.java，代码如下：

```
package org.hnist.cn.exam8_12;
/**
 * Created by Administrator on 2017-06-04.
 */
import java.io.StringReader;
import java.lang.reflect.Type;
import java.util.Iterator;
import java.util.LinkedList;
import com.google.gson.Gson;
import com.google.gson.reflect.TypeToken;
import com.google.gson.stream.JsonReader;
public class JsonUtils {
    public void parseJson01(String jsonData){
        Gson gson=new Gson();
        bean user=gson.fromJson(jsonData, bean.class);
        System.out.println("name--->" + user.getName());
        System.out.println("sex--->" + user.getSex());
        System.out.println("age---->" + user.getAge());
        System.out.println("height--->" + user.getHeight()); }
    public void parseJson02(String jsonData){
        Type listType = new TypeToken<LinkedList<bean>>(){}.getType();
        Gson gson=new Gson();
        LinkedList<bean> users=gson.fromJson(jsonData, listType);
        for(bean user:users){
            System.out.println("name--->" + user.getName());
            System.out.println("sex--->" + user.getSex());
            System.out.println("age---->" + user.getAge());
            System.out.println("height--->" + user.getHeight()); }
        System.out.println("==================");
        for (Iterator iterator = users.iterator(); iterator.hasNext();) {
            bean user = (bean) iterator.next();
```

```
            System.out.println("name--->" + user.getName());
            System.out.println("sex--->" + user.getSex());
            System.out.println("age---->" + user.getAge());
            System.out.println("height--->" + user.getHeight()); }    }    }
```

（5）建立 Activity 文件 MainActivity.java，在前一个实例 MainActivity.java 的基础上，增加如下代码：

```
    …
    import com.google.gson.Gson;
    public class MainActivity extends AppCompatActivity {
      private Button btn01=null;
      private Button btn02=null;
      private Button btn03=null;
      private String jsonData01 ="[{\"name\":\"张珊\",\"Sex\":\"女\",\"age\":25,
\"height\":\"170cm\"}," +
                                "{\"name\":\"李斯\",\"Sex\":\"女\",\"age\":23,
\"height\":\"165cm\"}," +
                                "{\"name\":\"王武\",\"Sex\":\"男\",\"age\":24,
\"height\":\"176cm\"}]";
      private String jsonData02 = "{\"name\":\"张珊\",\"Sex\":\"女\",\"age\":25,
\"height\":170cm}";
          @Override
          public void onCreate(Bundle savedInstanceState) {
             super.onCreate(savedInstanceState);
             setContentView(R.layout.activity_main);
             btn01=(Button)findViewById(R.id.btn01);
             btn02=(Button)findViewById(R.id.btn02);
             btn03=(Button)findViewById(R.id.btn03);
             btn01.setOnClickListener(listener);
             btn02.setOnClickListener(listener);
             btn03.setOnClickListener(listener);  }
          View.OnClickListener listener=new View.OnClickListener(){
             public void onClick(View v) {
                JsonUtils jsonUtils=new JsonUtils();
                switch (v.getId()) {
                   case R.id.btn01:
                      jsonUtils.parseJson02(jsonData02);
                      break;
                   case R.id.btn02:
                      jsonUtils.parseJson03(jsonData01);
                      break;
                   case R.id.btn03:
                      //把 Java 对象转换成 JSON
                      bean user01=new bean("fashion", "男",25,"180cm");
                      Gson gson=new Gson();
                      String obj=gson.toJson(user01);
                      System.out.println(obj);//输出对象信息
                      break;    }       }        };    }
```

保存所有文件，运行该项目，结果如图 8.30 所示，分别单击按钮，查看 Android Monitor 界面，提示信息分别如图 8.31、图 8.32、图 8.33 所示。

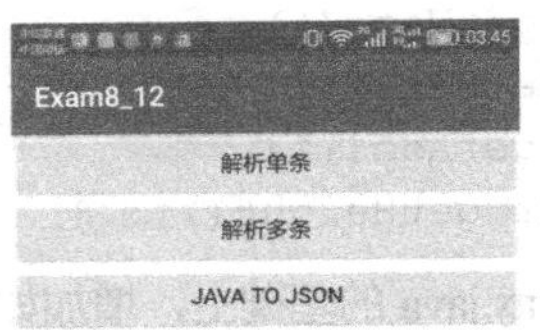

图 8.30　Exam8_12 运行结果

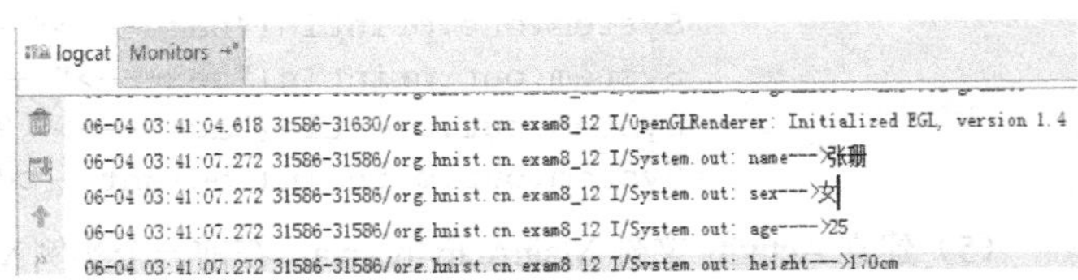

图 8.31　解析 JSON 字符串

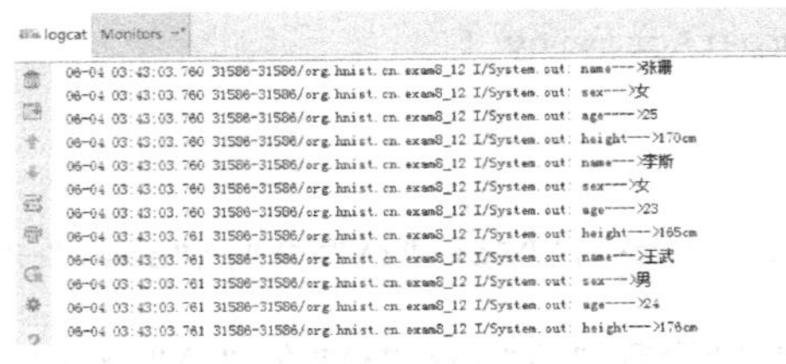

图 8.32　解析 JSON 数组

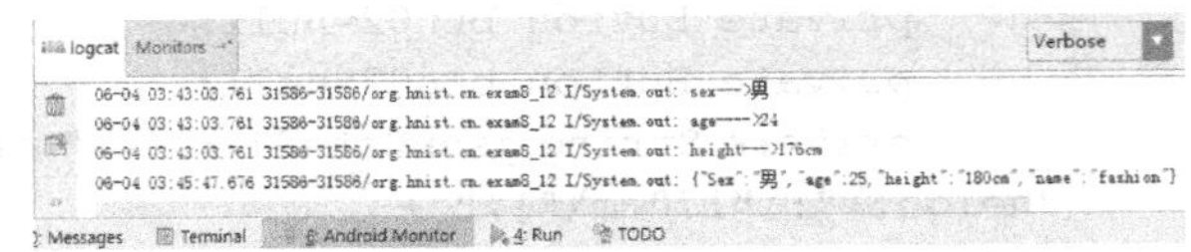

图 8.33　对象转换为 JSON 字符串

对比 Exam8_11、Exam8_12 可以发现使用 Gson 实现 JSON 数据转换要轻松得多，建议采用这种形式。前面介绍的几种存储都是将数据存储在本地设备上，其实，还有一种存储/读取数据的方式，通过网络来实现数据的存储和读取，将在第 9 章中介绍。

本章小结

本章介绍了几种数据的存储形式，着重介绍了如何操作 SQLite 数据库和数据表，以及查询操作；介绍了运用系统提供的 ContentProvider 如何进行数据的共享，着重介绍了利用 Gson 如何进行 JSON 数据解析与转换。

习题

（1）要在 AndroidManifest.xml 文件中怎样设置才能将一个文件保存到 SD 卡上？

（2）如何才能将信息从 SD 卡读出？给出实现的步骤和关键代码。

（3）如何才能将信息写入 SD 卡保存？给出实现的步骤和关键代码。

（4）查找资料，了解通话记录的 ContentProvider。

（5）编程实现，将实例 8-10 的信息显示在一个 ListView 组件中，写出关键代码。

（6）利用 Gson JAR 包将 JSON 字符串内容{"name": "张珊","Sex": "女"}转换成对象。

第 9 章　Android 网络通信技术

学习目标：

- 掌握 Android 中 WebView 组件的使用。
- 掌握 Android 中 HTTP 协议通信技术的使用。
- 掌握 Android 中 Socket 协议通信技术的使用。
- 了解 Android 中蓝牙通信技术。
- 了解 Android 中 WiFi 通信技术。

手机发展到今天，功能已不仅仅是打电话了，手机看新闻、手机炒股、手机缴费、手机银行、手机地图，等等，新的应用层出不穷，事实上，在 Android 中，掌握了网络通信技术就可以开发出这些网络应用程序。

9.1　Android 网络通信技术基础

手机能够上网是因为手机底层使用了 TCP/IP 协议，可以使手机终端通过无线网络建立 TCP 连接。TCP 协议可以对上层网络提供接口，使上层网络数据的传输建立在“无差别”的网络之上。

所谓无线网络就是采用无线传输媒介（如无线电波、红外线等）的网络，它包括远距离无线连接的全球语音和数据网络、近距离无线连接的蓝牙技术及射频技术。

Android 基于 Linux 内核，目前，Android 平台提供的与网络相关的包如表 9-1 所示。

表 9-1　Android SDK 中一些与网络有关的包

包	描述
java.net	提供与网络通信相关的类，包括流和数据包 socket、Internet 协议和常见 HTTP 处理。该包是一个多功能网络资源。有经验的 Java 开发人员可以立即使用这个熟悉的包创建应用程序
java.io	虽然没有提供现实网络通信功能，但是仍然非常重要。该包中的类由其他 Java 包中提供的 socket 和链接使用。它们还可用于与本地文件的交互
java.nio	包含表示特定数据类型的缓冲区的类，适用于两个基于 Java 语言的端点之间的通信
org.apache.*	表示许多为 HTTP 通信提供精确控制和功能的包。可以将 Apache 视为流行的开源 Web 服务器
android.net	除核心 java.net.*类以外，包含额外的网络访问 socket。该包包括 URI 类，后者频繁用于 Android 应用程序开发，而不仅仅是传统的联网
android.net.http	包含处理 SSL 证书的类

Android 通信可以利用系统提供的 WebView 组件来实现一些网络功能，也可以利用 HTTP 方式、Socket 方式、Web Service 方式、WiFi 方式通信，如果是近距离的通信，则还可以借助蓝牙通信(注意，APP 网络编程中所有网络连接必须在子线程中完成)。

9.1.1　Android 中的 HTTP 协议基础

超文本传输协议(Hyper Text Transfer Protocol，HTTP)是 Web 联网的基础，也是手机联网常用的协议之一，HTTP 协议是建立在 TCP 协议之上的一种应用，用于传送 WWW 方式的数据。HTTP 协议采用了请求/响应模式，是一个属于应用层的面向对象的协议。

HTTP 大致工作流程如下：客户端向服务器发出 HTTP 请求，服务器接收到客户端的请求后，处理客户端的请求，处理完成后再通过 HTTP 应答返回给客户端。这里的客户端是指 Android 手机端，服务器一般指 HTTP 服务器，HTTP 请求方法有 POST、GET 等方法。

HTTP 协议主要用于 Web 浏览器和 Web 服务器之间的数据交换，当在地址栏中输入 http://host:port/path 时，其中，http 表示要通过 HTTP 协议来定位网络资源，相当于通知浏览器使用 HTTP 协议来和 host 所确定的服务器进行通信；host 表示合法的 Internet 主机域名或者 IP 地址；port 指定一个端口号，为空则使用默认端口 80；path 指定请求资源的 URI。

HTTP 通信编程可以使用 Java 的 java.net.URL 类。

Android 中的 HTTP 协议版本是 HTTP1.1，采用了请求/响应模式：在客户端向服务器发送一个请求并接收一个响应后，只要不关闭网络连接，就可以继续向服务器发送 HTTP 请求，客户端发送的请求都需要服务器回送响应，在请求结束后，会主动释放连接。

9.1.2 Android 中的 Socket 基础

如果双方建立的是 HTTP 连接，则服务器需要等到客户端发送一次请求后才能将数据传回给客户端，客户端定时向服务器端发送连接请求，不仅可以保持在线，还可以“询问”服务器是否有新的数据，如果有就将数据传给客户端。如果进行多人联网的游戏，那么 HTTP 不能很好地满足要求，Socket（也称为套接字）通信可以很好地解决这个问题。

Socket 是一种低级、原始的通信方式，要编写服务器端代码和客户端代码，自己设置端口，自己设置通信协议、验证数据安全和合法性，而且通常是多线程的，开发起来比较麻烦。但是它也有其优点：灵活，不受编程语言、设备、平台和操作系统的限制，通信速度快而高效。

应用层通过传输层进行数据通信时，TCP 会遇到同时为多个应用程序进程提供并发服务的问题。多个 TCP 连接或多个应用程序进程可能需要通过同一个 TCP 协议端口传输数据。为了区分不同的应用程序进程和连接，许多计算机操作系统为应用程序与 TCP/IP 协议交互提供了套接字接口。应用层可以和传输层通过 Socket 接口，区分来自不同应用程序进程或网络连接的通信，实现数据传输的并发服务。

Socket 是通信的基石，是支持 TCP/IP 协议的网络通信的基本操作单元。它包含进行网络通信必需的 5 种信息：连接使用的协议、本地主机的 IP 地址、本地进程的协议端口、远地主机的 IP 地址、远地进程的协议端口。

Socket 有两种主要的操作方式：面向连接的操作方式和无连接的操作方式。

面向连接的操作使用 TCP 协议，这个模式下的 Socket 必须在发送数据之前与目的地的 Socket 取得连接。一旦连接建立了，Socket 就可以使用一个流接口进行打开、读、写、关闭等操作。所有发送的消息都会在另一端以同样的顺序被接收。它的优点是数据安全性较高，缺点是操作的效率较低。

无连接的操作使用 UDP 协议，这个模式下的 Socket 不需要连接一个目的 Socket，它只是简单发送数据报，多个数据报的到达顺序可能和出发时的顺序不一样。它的优点是操作的效率较高，缺点是数据安全性较低。

要建立 Socket 连接至少需要一对套接字：其中一个运行于客户端，称为 ClientSocket；另一个运行于服务器端，称为 ServerSocket。

Socket 之间的连接过程分为 3 个步骤：服务器监听、客户端请求、连接确认。

服务器监听：服务器端套接字并不定位具体的客户端套接字，而是处于等待连接的状态，实时监控网络状态，等待客户端的连接请求。

客户端请求：客户端的套接字提出连接请求，要连接的目标是服务器端的套接字。为此，客户端的套接字必须先描述它要连接的服务器的套接字，指出服务器端套接字的地址和端口号，然后向服务器端套接字提出连接请求。

连接确认：当服务器端套接字监听到或者接收到客户端套接字的连接请求时，就响应客户端套接字的请求，建立一个新的线程，把服务器端套接字的描述发给客户端，一旦客户端确认了此描述，双方就正式建立了连接。而服务器端套接字继续处于监听状态，继续接收其他客户端套接字的连接请求。

创建 Socket 连接时，可以指定使用的传输层协议，Socket 可以支持不同的传输层协议（TCP 或 UDP），当使用 TCP 协议进行连接时，该 Socket 连接就是一个 TCP 连接。

Java 中，Socket 相关类都在 java.net 包中，其中主要的类是 Socket 和 ServerSocket。

9.1.3 Android 中的蓝牙基础

蓝牙是目前使用最广泛的无线通信协议之一，用于实现近距离无线通信。其最初由爱立信公司创建，蓝牙的标准是 IEEE 802.15.1(如今已不再维持该标准)，蓝牙协议工作在无需许可的 ISM(Industrial Scientific Medical)频段的 2.4～2.485GHz。其中蓝牙 4.0 最高速度可达 3Mb/s。

蓝牙的优势是无需驱动程序，小型化无线，具有低功率、低成本、高安全性、稳固等特性，易于使用、即时连接。

Android 平台支持蓝牙网络协议栈，实现蓝牙设备之间数据的无线传输。

蓝牙设备之间的通信主要包括以下 4 个步骤。

（1）设置蓝牙设备。

（2）寻找局域网内可能或者匹配的设备。

（3）连接设备。

（4）设备之间的数据传输。

Android 所有关于蓝牙开发的类都在 android.bluetooth 包下，一共有 8 个类，以下是建立蓝牙连接需要的一些基本类。

BluetoothAdapter 类：代表了一个本地的蓝牙适配器，是所有蓝牙交互的入口点。

BluetoothDevice 类：代表了一个远端的蓝牙设备，使用它请求远端蓝牙设备连接或者获取远端蓝牙设备的名称、地址、种类和绑定状态。

Bluetoothsocket 类：代表了一个蓝牙套接字的接口（类似于 TCP 中的套接字），是应用程序通过输入流、输出流与其他蓝牙设备通信的连接点。

Bluetoothserversocket 类：打开服务连接来监听可能到来的连接请求（属于 Server 端），为了连接两个蓝牙设备，必须有一个设备作为服务器以打开一个服务套接字。

9.1.4 Android 中的 Wi-Fi 基础

Wi-Fi 即无线保真，又称 802.11b 标准。它是一种可以将个人电脑、手持设备（如 PDA、手机）等终端以无线方式互相连接的技术，即一种无线联网技术。它是通过无线电波来联网的。Wi-Fi 为用户提供了无线的宽带互联网访问，能够访问 Wi-Fi 网络的地方被称为热点。Wi-Fi 或 802.11b 在 2.4GHz 频段工作，所支持的速度最高达 54Mb/s。

蓝牙和 WiFi 相比，WiFi 是一个更加快速的协议，覆盖范围更大。虽然两者使用相同的频率范围，但是 WiFi 需要更加昂贵的硬件。蓝牙设计被用来在不同的设备之间创建无线连接，而 WiFi 是无线局域网协议。

Android 开发 WiFi 时主要包括以下几个类。

（1）ScanResult 类：该类主要是通过 WiFi 硬件的扫描来获取一些周边 WiFi 热点的信息。

（2）WifiConfiguration：该类主要用来进行 WiFi 网络的配置，包括安全配置等。

（3）WifiInfo：该类主要进行 WiFi 无线连接的描述。

（4）WifiManager：该类提供了管理 WiFi 连接的大部分 API。

9.2 WebView 组件

9.2.1 WebView 组件基础知识

Android 手机中内置了一款高性能 webkit 浏览器，在 SDK 中封装为 WebView 组件。用户可以直接使用 WebView 组件显示网页的内容，或者将一些指定的 HTML 文件嵌入进来，除了支持各个浏览器的“前进”“后退”等功能之外，最为强大的是 WebView 还支持 JavaScript 的操作，能与 JavaScript 相互通信。

android.webkit.WebView 类的层次关系如下：

```
java.lang.Object
      android.view.View
            android.view.ViewGroup
                  android.widget.AbsoluteLayout
                        android.webkit.WebView
```

要在 Android 程序中使用 WebView 组件，必须在程序中使用下面的语句。

```
import android.webkit.WebView;                    //导入 webkit.WebView 类
```

Android 中 WebView 组件提供了如表 9-2 所示的常用方法。

表 9-2　WebView 组件的常用方法

方法	描述
public WebView(Context context)	取得 WebView 类的实例化对象
public void addJavascriptInterface(Object obj, String interfaceName)	绑定一个 JavaScript 的对象
public boolean canGoBack()	判断能否实现后退操作
public boolean canGoBackOrForward(int steps)	判断是否可以后退或前进指定步数
public boolean canGoForward()	判断是否可以前进
public boolean canZoomIn()	判断是否可以缩小
public boolean canZoomOut()	判断是否可以放大
public void clearCache(boolean includeDiskFiles)	清空缓存，如果是 false，则只清空 RAM
public void clearFormData()	清空表单的填写记录
public void clearHistory()	清空历史信息
public int getProgress()	得到访问进度
public String getTitle()	取得当前访问页面的标题
public void goBack()	后退一步
public void goBackOrForward(int steps)	后退或前进指定的步数
public void goForward()	前进一步
public void loadData(String data, String mimeType, String encoding)	通过指定的字符串进行页面的加载
public void loadUrl(String url)	读取指定的 URL 地址数据
public void reload()	重新加载页面

续表

方法	描述
public void savePassword(String host, String username, String password)	保存密码
public void setDownloadListener(DownloadListener listener)	对下载文件进行监听
public void setWebChromeClient(WebChromeClient client)	使用 Google Chrome 作为客户端
public void setWebViewClient(WebViewClient client)	使用 WebView 作为客户端
public boolean zoomIn()	是否缩小
public boolean zoomOut()	是否放大
public WebSettings getSettings()	返回 WebSettings 对象

WebView 组件有下面几个比较重要的子类。

（1）WebChromeClient 类：会在一些影响浏览器 UI 交互动作发生时被调用，例如，WebView 关闭和隐藏、页面加载进展、JS 确认框和警告框、JS 加载前、JS 操作超时、WebView 获得焦点，等等。其常用方法如表 9-3 所示。

表 9-3　WebChromeClient 类常用方法

方法	描述
public void onCloseWindow(WebView window)	窗口关闭操作
public boolean onCreateWindow(WebView view, boolean dialog, boolean userGesture, Message resultMsg)	创建新的 WebView
public boolean onJsAlert(WebView view, String url, String message, JsResult result)	弹出警告框互操作
public boolean onJsBeforeUnload(WebView view, String url, String message, JsResult result)	页面关闭互操作
public boolean onJsConfirm(WebView view, String url, String message, JsResult result)	弹出确认框互操作
public boolean onJsPrompt(WebView view, String url, String message, String defaultValue, JsPromptResult result)	弹出提示框互操作
public boolean onJsTimeout()	计时器已到互操作
public void onProgressChanged(WebView view, int newProgress)	进度改变互操作
public void onReceivedTitle(WebView view, String title)	接收页面标题更改
public void onRequestFocus (WebView view, String title)	WebView 显示焦点

（2）WebViewClient 类：在一些影响内容变化的动作发生时被调用，例如，表单错误提交后需要重新提交、页面开始加载及加载完成、资源加载中、接收到 HTTP 认证需要处理、页面键盘响应、页面 URL 打开处理，等等。其常用方法如表 9-4 所示。

表 9-4　WebViewClient 常用方法

方法	描述
public void doUpdateVisitedHistory(WebView view, String url, boolean isReload)	更新历史记录
public void onFormResubmission (WebView view, Message dontResend, Message resend)	应用程序重新请求网页数据
public void onLoadResource(WebView view, String url)	加载指定地址提供的资源
public void onPageFinished(WebView view,String url)	网页加载完毕
public void onPageStarted(WebView view,String url,Bitmap favicon)	网页开始加载
public void onReceivedError(WebView view, int errorCode,String description,String failingUrl)	报告错误信息
public void onScaleChanged(WebView view,float oldScale, float newScale)	WebView 发生改变
public boolean shouldOverrideUrlLoading(WebView view,String url)	控制新的连接在当前 WebView 中打开

例如：

```
webView.setWebViewClient(webViewClient);          //加载 WebViewClient
```

（3）WebSettings 类：用来对 WebView 的配置进行配置和管理，例如，是否可以进行文件操作、缓存的设置、页面是否支持放大和缩小、字体及文字编码设置、是否允许 JS 运行、是否允许图片自动加载、是否允许数据及密码保存，等等。其常用方法如表 9-5 所示。

表 9-5 WebSettings 类的常用方法

方法	描述
public void setBuiltInZoomControls(boolean enabled)	设置是否支持缩放控制
public synchronized void setJavaScriptEnabled(boolean flag)	设置是否启动 JavaScript 支持
public void setSaveFormData(boolean save)	设置是否保存表单数据
public void setSavePassword(boolean save)	设置是否保存密码
public synchronized void setGeolocationEnabled(boolean flag)	设置是否可以获得地理位置
public void setAllowFileAccess (boolean allow)	启用或禁止 WebView 访问文件数据
public synchronized void setBlockNetworkImage (boolean flag)	是否显示网络图像
public void setCacheMode (int mode)	设置缓冲的模式
public synchronized void setDatabaseEnabled (boolean flag)	设置是否使用数据库
public synchronized void setDefaultFontSize (int size)	设置默认字体的大小
public synchronized void setDefaultTextEncodingName(String encoding)	设置在解码时使用的默认编码
public synchronized void setFixedFontFamily (String font)	设置固定使用的字体
public synchronized void setLayoutAlgorithm (WebSettings.LayoutAlgorithm l)	设置布局方式

WebView 在 Android 中的主要作用如下。

（1）加载网页，直接显示网页。

（2）加载 HTML 文件，可以利用 HTML 做界面布局，在手机上实现复杂的显示布局效果。

（3）加载 JSP 文件和 JavaScript 相互通信。

9.2.2 使用 WebView 加载网页

要实现 WebView 加载网页，按照下述的步骤进行即可。

（1）在要 Activity 中实例化 WebView 组件。

例如：

```
WebView webView = new WebView(this);            //实例化 WebView 组件
```

或者在布局文件中声明 WebView，然后在 Activity 中实例化 WebView。

例如：

```
this.webview = (WebView) super.findViewById(R.id.webview) ;   //取得组件
```

（2）调用 WebView 的 loadUrl()方法，设置 WebView 要显示的网页。

例如：

```
webView.loadUrl("http://www.hnist.cn");
```

（3）调用 Activity 的 setContentView( )方法来显示网页视图。

（4）为了让 WebView 支持回退功能，需要覆盖 Activity 类的 onKeyDown()方法，如果不做任何处理，可单击系统回退键，整个浏览器会调用 finish()方法结束，而不是回退到上一页面。

（5）需要在 AndroidManifest.xml 文件中添加权限，否则会出现错误。例如：

```
<uses-permission android:name="android.permission.INTERNET" />
```

实例 9-1：使用 WebView 加载网页实例

新建一个项目，项目命名为 Exam9_1，利用 WebView 组件加载网页，设置简单的后退与前进功能。

（1）修改 activity_main.xml 文件，代码如下：

```
    …
      <Button
          android:id="@+id/open"
          android:layout_width="wrap_content"
          android:layout_height="wrap_content"
          android:text="打开网页" />
      <Button
          …
          android:text="网页后退" />
      <Button
          …
          android:text="网页前进" />
    </LinearLayout>
    <WebView                                               //加一个 WebView 组件
        android:id="@+id/webview"                          //名为 webview
        android:layout_width="fill_parent"
        android:layout_height="fill_parent"/>
    </LinearLayout>
```

（2）修改 MainActivity.java 文件，代码如下：

```
…
import android.webkit.WebView;                            //导入 webkit.WebView 类
public class MainActivity extends AppCompatActivity {
private Button open = null ;
private Button back = null ;
private Button forward = null ;
private WebView webview = null ;                                  //定义 WebView 组件
private String urlData[] = new String[] { "http://www.hnist.cn",
       "http://law.hnist.cn/", "http://xw.hnist.cn",
       "http://art.hnist.cn/", "http://dyb.hnist.cn/" }; //定义选项数据
@Override
public void onCreate(Bundle savedInstanceState) {
 super.onCreate(savedInstanceState);
 super.setContentView(R.layout.activity_main);
 this.open = (Button) super.findViewById(R.id.open) ;
 this.back = (Button) super.findViewById(R.id.back) ;
 this.forward = (Button) super.findViewById(R.id.forward) ;
 this.webview = (WebView) super.findViewById(R.id.webview) ; //取得 WebView
 this.open.setOnClickListener(new OpenOnClickListenerImpl());//单击事件
 this.back.setOnClickListener(new BackOnClickListenerImpl());
 this.forward.setOnClickListener(new ForwardOnClickListenerImpl());}
private class OpenOnClickListenerImpl implements OnClickListener {
    public void onClick(View view) {
       MainActivity.this.showUrlDailog() ; }   }         //显示对话框
private class BackOnClickListenerImpl implements OnClickListener {
    public void onClick(View view) {
       if (MainActivity.this.webview.canGoBack()) {      //可以后退
          MainActivity.this.webview.goBack() ; }}}      //后退
private class ForwardOnClickListenerImpl implements OnClickListener {
    @Override
    public void onClick(View view) {
       if (MainActivity.this.webview.canGoForward()) {  //可以前进
```

```
                MainActivity.this.webview.goForward() ;  }}} //前进
    private void showUrlDailog(){                                //显示对话框
        Dialog dialog = new AlertDialog.Builder(this)            //实例化Dialog对象
            .setTitle("请选择要打开的网站")                        //设置显示标题
            .setNegativeButton("取消",                             //增加取消按钮
                new DialogInterface.OnClickListener() {           //设置操作监听
                    public void onClick(DialogInterface dialog,
                            int whichButton) {   }})              //单击事件
            .setItems(this.urlData,                               //设置列表选项
                new DialogInterface.OnClickListener() {
                    public void onClick(DialogInterface dialog,
                        int which) {                              //设置显示信息
                        MainActivity.this.webview.loadUrl(
                                urlData[which]);}                 //打开网址
            }).create();                                          //创建Dialog
        dialog.show();}}                                          //显示对话框
```

（3）修改 AndroidManifest.xml 文件的配置权限，在 AndroidManifest.xml 中添加如下代码：

```
<uses-permission android:name="android.permission.INTERNET"/>
```

运行该项目，进入界面如图 9.1（a）所示，单击“打开网页”按钮，进入界面如图 9.1（b）所示，选择第三个选项，进入界面如图 9.1（c）所示。

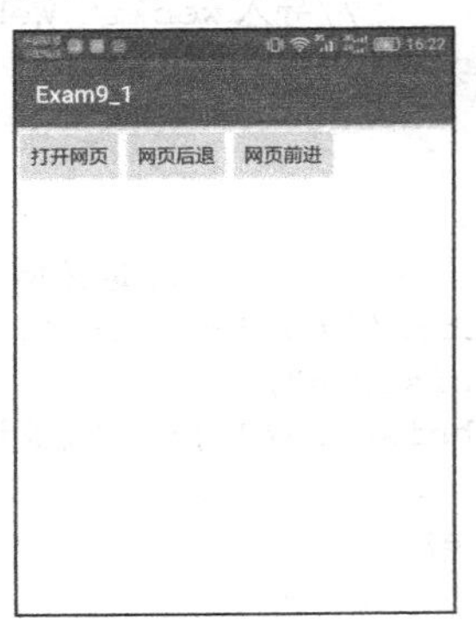

（a）运行界面

（b）打开网页

（c）网页界面

图 9.1　Exam9-1 运行结果

9.2.3　使用 WebView 加载 HTML 文件

使用 WebView 组件除了可以浏览 Web 站点外，它还可以加载项目中的 HTML 文件，将 HTML 运行的结果显示在手机上，因此可以利用 HTML 进行较为复杂的 Android 手机界面的编写，这主要是通过 android.webkit.WebSettings 类实现的。

按照如下步骤加载 HTML 文件。

（1）在 Activity 中实例化 WebView 组件。

例如：

```
WebView webView = new WebView(this);                    //实例化WebView组件
```

或者在布局文件中声明 WebView，然后在 Activity 中实例化 WebView 组件。

例如：

```
this.webview = (WebView) super.findViewById(R.id.webview);    //取得组件
```

（2）启用 WebView 的支持 JavaScript 的功能。

例如：

```
this.webview.getSettings().setJavaScriptEnabled(true);  //启用 JavaScript
```

（3）通过 WebView 的 loadUrl 方法载入相应 HTML 文件，该 HTML 文件中包含了 JavaScript 方法。例如：

```
this.webview.loadUrl("file:/android_asset/show.html") ; //读取网页文件
```

（4）需要在 AndroidManifest.xml 文件中添加权限，否则会出现错误。例如：

```
<uses-permission android:name="android.permission.INTERNET" />
```

实例 9-2：使用 WebView 加载 HTML 文件实例

新建一个项目，项目命名为 Exam9_2，包名称为 org.hnist.demo，利用 WebView 组件加载项目中的 HTML 文件。

（1）选择“New→Folder→Asseets Folder”选项，建立一个 assets 文件夹，右击 assets，选择“New→Directory”选项，在其中建立 html、images 文件夹，分别建立 show.html 文件和用到的图片文件，如图 9.2 所示。

其中，show.html 文件代码如下：

```
<meta http-equiv="Content-Type" content="text/html ; charset=utf-8">
<script language="javascript">
function openurl(url) {
    window.location = url ; }   //实现网页跳转
</script>
<center>
<img src="../images/xh.jpg" width="150" height="160">
<h3>请选择您要打开的网站：</h3>
<select name="url" onchange="openurl(this.value)">   //选择指定 URL 的值
    <option value="http://www.hnist.cn">湖南理工学院</option>
    <option value="http://xw.hnist.cn">新闻学院</option>
    <option value="http://law.hnist.cn/">政法学院</option>
    <option value="http://art.hnist.cn/">美术学院</option>
    <option value="http://61.187.92.238:8105/">信息学院</option>
</select>
</center>
```

（2）修改 activity_main.xml 文件，代码如下：

```
…
<WebView
    android:id="@+id/webview"
    android:layout_width="fill_parent"
    android:layout_height="fill_parent"/>
</LinearLayout>
```

（3）建立 Activity 文件 MainActivity.java，读取 HTML 文件，代码如下：

```
…
import android.webkit.WebView;
public class MainActivity extends AppCompatActivity {
private WebView webview = null ;
@Override
public void onCreate(Bundle savedInstanceState) {
 super.onCreate(savedInstanceState);
```

```
    super.setContentView(R.layout.activity_main);
    this.webview = (WebView) super.findViewById(R.id.webview) ; //取得组件
    this.webview.getSettings().setJavaScriptEnabled(true) ; //启用 JavaScript
    this.webview.getSettings().setBuiltInZoomControls(true);//控制页面缩放
    this.webview.loadUrl("file:/android_asset/html/show.html");}}//读取网页
```

（4）需要在 AndroidManifest.xml 文件中添加权限，否则会出现错误。代码如下：

```
  <uses-permission android:name="android.permission.INTERNET" />
```

注意：要将这个项目设置为 utf-8，步骤如下：选择“File→Settings→Editor→File Encodings”选项，否则可能出现乱码。运行该项目，进入如图 9.3 所示界面，单击下拉按钮，可以选择要打开的网站。

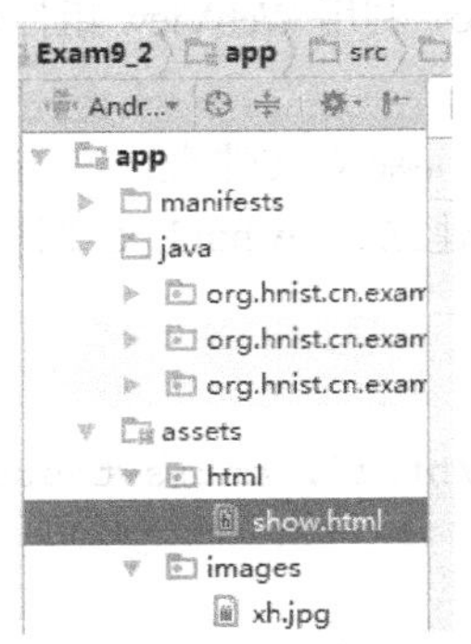

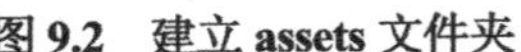

图 9.2 建立 assets 文件夹

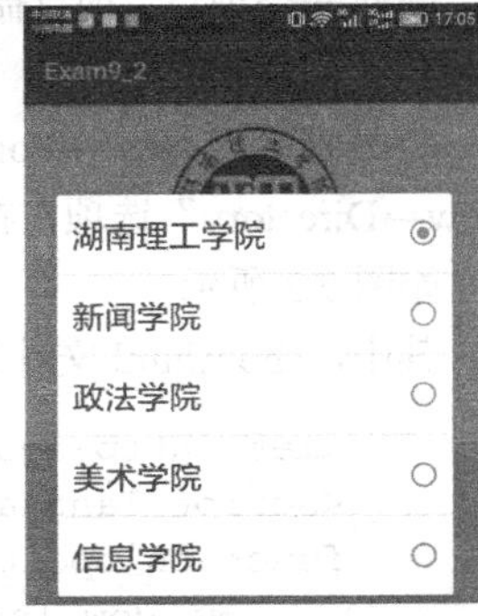

图 9.3 Exam9-2 运行结果

9.2.4 使用 WebView 加载 JSP 文件

类似的可以使用 WebView 组件加载 JSP 文件，将 JSP 程序的运行结果显示在手机上，但是前提是要先在浏览器中正常浏览该 JSP 文件。

按照如下步骤加载 JSP 文件。

（1）确保该 JSP 文件能在浏览器中正确浏览。

① 下载、安装 Tomcat，安装完成，在浏览器地址栏中输入 http://localhost:8080，如果进入如图 9.4 所示的界面，表示安装设置成功。

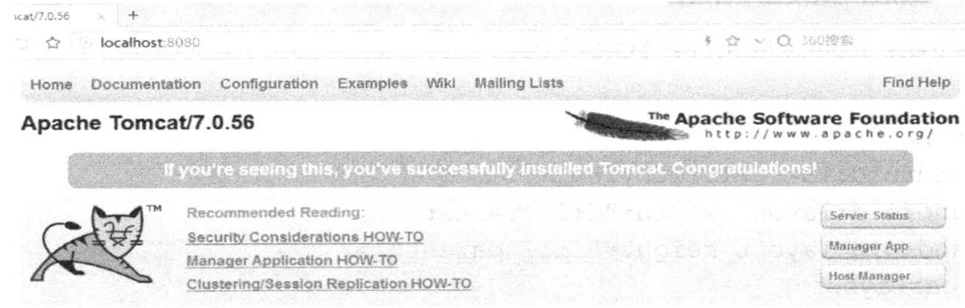

图 9.4 安装 Tomcat 成功

② Tomcat 的安装目录中有 webapps 文件夹，在其中建立一个文件夹（如 JSP），将 exam1.jsp 复制到该目录中，在浏览器地址栏中输入 http://localhost:8080/jsp/exam1.jsp，进入如图 9.5 所示的界面。

（2）进入手机编程环境，在 Activity 中实例化 WebView 组件。

例如：

```
WebView webView = new WebView(this);              //实例化 WebView 组件
```

或者在布局文件中声明 WebView，然后在 Activity 中实例化 WebView 组件。

例如：

```
this.webview = (WebView) super.findViewById(R.id.webview) ;//取得组件
```

（3）启用 WebView 的支持 JavaScript 的功能。

例如：

```
this.webview.getSettings().setJavaScriptEnabled(true) ;//启用 JavaScript
```

（4）通过 WebView 的 loadUrl 方法载入相应 JSP。

例如：

```
this.webview.loadUrl("http://10.0.2.2:8080/jsp/exam1.jsp"); //读取 JSP 文件
```

注意：这里的 10.0.2.2 指本地计算机的 IP 地址（仅仅限于虚拟机上运行），利用该计算机充当 Tomcat 服务器，也可以利用在 Internet 中实际存在的一个 Tomcat 服务器的 IP 地址。如果在真实机上运行，就不能使用 10.0.2.2 了，可以在计算机上开热点，手机连接到电脑的网络，再查看电脑的 IP 地址。

（5）需要在 AndroidManifest.xml 文件中添加权限，否则会出现错误。

```
<uses-permission android:name="android.permission.INTERNET" />
```

实例 9-3：使用 WebView 加载 JSP 文件实例

新建一个项目，项目命名为 Exam9_3，利用 WebView 组件加载 JSP 文件。

（1）建立一个 JSP 文件，如 exam1.jsp，将它保存在 webapps\jsp 文件夹中，exam1.jsp 代码如下：

```
<%@ page language="java" import = "java.util.*" pageEncoding = "gb2312" %>
<HTML>
    <HEAD>
     <TITLE>
      JSP 测试
     </TITLE>
    </HEAD>
    <BODY>
<center>
     <% out.println("<h1>www.hnist.cn<br>湖南理工学院</h1>"); %>
</center>
    </BODY>
    </HTML>
```

（2）建立布局管理文件 activity_main.xml，代码如下：

```
…
<WebView
     android:id="@+id/webview"
     android:layout_width="fill_parent"
     android:layout_height="fill_parent"/>
</LinearLayout>
```

（3）建立 Activity 文件 MainActivity.java，读取 JSP 文件，代码如下：

```
…
import android.webkit.WebView;
public class MainActivity extends AppCompatActivity {
private WebView webview = null ;
@Override
public void onCreate(Bundle savedInstanceState) {
```

```
        super.onCreate(savedInstanceState);
        super.setContentView(R.layout.activity_main); //默认布局管理器
        this.webview = (WebView) super.findViewById(R.id.webview) ; //取得组件
        this.webview.getSettings().setJavaScriptEnabled(true); //启用 JavaScript
        this.webview.getSettings().setBuiltInZoomControls(true); //控制页面缩放
        this.webview.loadUrl("http://10.0.2.2:8080/jsp/exam1.jsp");}}//读取 JSP 文件
```

（4）需要在 AndroidManifest.xml 文件中添加权限，否则会出现错误。

```
    <uses-permission android:name="android.permission.INTERNET" />
```

保存所有文件，运行该项目，进入如图 9.6 所示界面。

www. hnist. cn
湖南理工学院

图 9.5　JSP 在 Tomcat 中的运行界面

www.hnist.cn
湖南理工学院

图 9.6　Exam9-3 运行结果

9.3　利用 HttpURLConnection 开发 HTTP 程序

9.3.1　HttpURLConnection 基础

在 Android 中除了使用 WebView 组件访问网络以外，还可以用代码的方式访问网络，代码方式有时候会显得更加灵活。其中常用的类就是 Android SDK 中的 HttpURLConnection 类。

HttpURLConnection 是 Java 的标准类，HttpURLConnection 继承自 URLConnection，包含在包 java.net.*中，它的层次关系如下：

```
    java.lang.Object
       java.net.URLConnection
          java.net.HttpURLConnection
```

要在 Android 的 Java 程序中使用 HttpURLConnection 类，必须在程序中使用下面的语句。

```
    import java.net.HttpURLConnection;            //导入 HttpURLConnection 类
```

HttpURLConnection 类提供了如表 9-6 所示的常用方法。

表 9-6　HttpURLConnection 常用方法

方法	描述
public int getResponseCode()	获取服务器的响应代码
public String getResponseMessage()	获取服务器的响应消息
public String getResponseMethod()	获取发送请求的方法
public void getOutputStream()	接收数据流
public void getInputStream()	发送数据库
public void flush()	刷新对象输出流
public void close()	关闭流对象
public setUseCaches(boolean newValue)	设置是否使用缓存
public void setRequestMethod(String method)	设置发送请求的方法
public void setDoOutput(boolean newValue);	设置是否向 HttpURLConnection 输出

续表

方法	描述
public void setDoInput(boolean newValue);	设置是否向 HttpURLConnection 输入
public void setRequestProperty("Content-type", String)	设置请求报文头
public void setConnectTimeout(int timeout)	设置连接超时

HTTP 通信中使用最多的是 GET 请求和 POST 请求，GET 请求可以使用静态页面，也可以把参数放在 URL 字串的后面，传递给服务器；POST 的参数并不放在 URL 字串中，而是放在 HTTP 请求数据中。

一个重要的类是 URL，通常用来生成一个指向特定地址的 URL 实例。例如：

```
URL url = new URL("http://www.hnist.cn"); //HttpURLConnection 实例初始化
```

HttpURLConnection 和 URLConnection 都是抽象类，无法直接实例化对象，其对象主要通过 URL 的 openConnection 方法获得，创建一个 HttpURLConnection 连接的代码如下所示：

```
URL url = new URL("http://www.hnist.cn");
HttpURLConnection urlConn = (HttpURLConnection) url.openConnection();
```

openConnection 方法只创建 URLConnection 或者 HttpURLConnection 实例，并不进行真正的连接操作，在连接之前可以对其属性进行设置，例如：

```
connection.setDoOutput(true);                //设置输出流
connection.setDoInput(true);                 //设置输入流
connection.setRequestMethod("GET");          //设置请求的方式为 GET
connection.setRequestMethod("POST");         //设置请求的方式为 POST
connection.setUseCaches(false);              //设置 POST 请求方式不能够使用缓存
urlConn.disconnect();                        //关闭 HttpURLConnection 连接
```

在完成 HttpURLConnection 实例的初始化以后，即可通过 GET 方式或 POST 方式与服务器通信。

9.3.2 HttpURLConnection 通信：GET 方式

HttpURLConnection 对网络资源的请求在默认情况下会使用 GET 方式，GET 请求可以获取静态页面，也可以把参数放在 URL 字符串后面，传递给服务器。

此时会用到 InputStreamReader 读取字节并将其解码为字符，其 read()方法每次读取一个或多个字节。BufferedReader 流能够读取文本行，读取的量比 InputStreamReader 要多，通过向 BufferedReader 传递一个 Reader 对象，来创建一个 BufferedReader 对象，常用的方法如下。

（1）BufferedReader(Reader in)：创建一个使用默认大小输入缓冲区的缓冲字符输入流。参数 in 是一个 Reader。

（2）BufferedReader(Reader in, int sz) ：创建一个使用指定大小输入缓冲区的缓冲字符输入流。参数 in 是一个 Reader，sz 是输入缓冲区的大小。

实例 9-4：获取网络上的文件

新建一个项目，项目命名为 Exam9_4，利用 HttpURLConnection 的 GET 方式获取网络上的文件，编程直接将该文件的内容在手机上显示出来。

（1）在指定服务器指定位置建立一个文件 hnist.txt，内容如图 9.7 所示。

（2）定义布局文件 activity_main.xml，只有一个 TextView 组件，代码略。

（3）定义一个 Activity 文件 MainActivity.java，代码如下：

```
...
```

```
public class MainActivity extends AppCompatActivity {
    //定义路径
    private static final String PATH = "http://10.0.2.2:8080/jsp/hnist.txt";
    private TextView msg = null;                  // 定义文本显示组件
    private Handler handler = null;               //定义 Handler 对象
    private String ssr="";
    @Override
    public void onCreate(Bundle savedInstanceState) {
        super.onCreate(savedInstanceState);
        super.setContentView(R.layout.activity_main);       // 调用布局管理器
        this.msg = (TextView) super.findViewById(R.id.mytxt);  //取得文本组件
        new Thread() {
            @Override
            public void run() {//要执行的方法
                try { ssr=MainActivity.this.getData();
                } catch (Exception e) {  e.printStackTrace(); }
   handler.sendEmptyMessage(0); } }.start();//执行完毕后给 handler 发送一个空消息
   handler = new Handler() {
            @Override
   public void handleMessage(Message mg) { //有消息发送出时执行 Handler 方法
            super.handleMessage(mg);
            MainActivity.this.msg.setText(ssr);   } }; }//处理 UI
    public String getData() throws Exception {    // 定义 getData()方法
        URL url = new URL(PATH);                  // 定义 URL
        // 打开连接
        HttpURLConnection conn = (HttpURLConnection) url.openConnection();
        //读取文件到缓冲区，以 GBK 编码，否则可能出现乱码
  InputStreamReader inputReader = new InputStreamReader(conn.getInputStream(),
"GBK");
        //从缓冲区读取内容
        BufferedReader bufReader = new BufferedReader(inputReader);
        String line = "";
        String Data = "";
        while ((line = bufReader.readLine()) != null) {
            Data += line;          }                  //循环
        return Data; } }                              //返回获得的数据
```

（4）在 AndroidManifest.xml 文件中添加权限，否则会出现错误。代码如下：

```
<uses-permission android:name="android.permission.INTERNET" />
```

运行该项目，结果如图 9.8 所示。

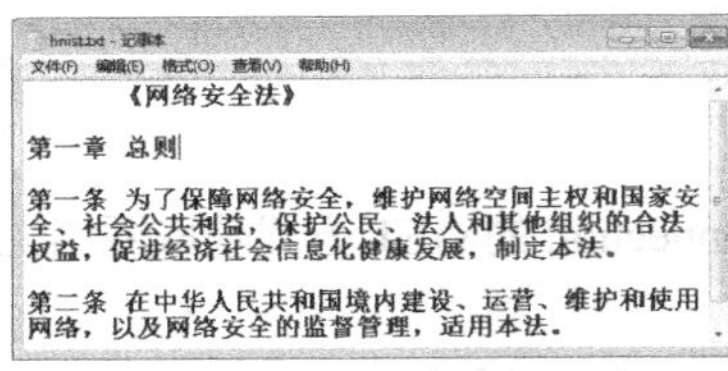

hnist.txt - 记事本

文件(F) 编辑(E) 格式(O) 查看(V) 帮助(H)

《网络安全法》

第一章 总则

第一条 为了保障网络安全，维护网络空间主权和国家安全、社会公共利益，保护公民、法人和其他组织的合法权益，促进经济社会信息化健康发展，制定本法。

第二条 在中华人民共和国境内建设、运营、维护和使用网络，以及网络安全的监督管理，适用本法。

图 9.7 Exam9-4 输入的内容

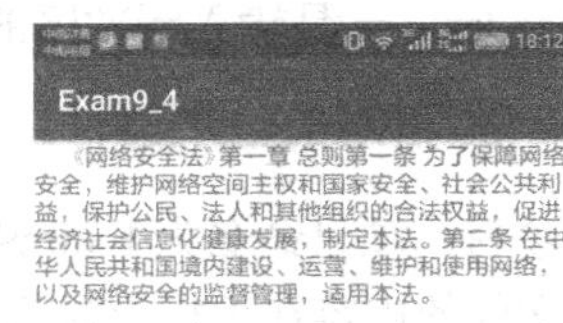

图 9.8 Exam9-4 运行结果

除了可以读取文本文件外，还可以读取图片、MP3 文件、视频文件，但读取数据的方式可能会有所不同，建议用字节数组流 ByteArrayOutputStream 来读取。

ByteArrayOutputStream 可以捕获内存缓冲区的数据，转换成字节数组。

ByteArrayOutputStream 类在创建它的实例时，程序内部创建了一个 byte 型数组的缓冲区，利用 ByteArrayOutputStream 和 ByteArrayInputStream 的实例向数组中写入或读出 byte 型数据。

项目 Exam9_5 利用了 HttpURLConnection 的 GET 方式来获取网络上的图片并在手机上显示出来，限于篇幅，本书中不再介绍，感兴趣的读者可自行下载学习。

利用 GET 请求可以把参数放在 URL 字符串后面，也可以把数据传递给服务器的 JSP 程序。对上面的代码做适当的修改，例如：

```
//定义一个字符串
String httpUrl = "http://10.0.2.2:8080/jsp/get.jsp?name=aa&password=123456";
URL url = new URL(httpUrl);  //实例化 URL
//打开连接
HttpURLConnection urlConn = (HttpURLConnection) url.openConnection();
```

代码表示变量 name 值为 aa、变量 password 值为 123456，发送到服务器端的 get.jsp 程序中，在 get.jsp 中应该由相应的语句接收，例如：

```
String name = request.getParameter("name ") ;
String password = request.getParameter("password") ;
```

实例 9-5：使用 GET 方式传递数据到 JSP 文件中

建立一个 Android 项目，项目名称为 Exam9_6，利用 HttpURLConnection 的 GET 方式传递数据到指定的 JSP 文件中。

（1）建立一个 get.jsp 文件，在 Tomcat 上发布后，在浏览器地址栏中输入地址，显示结果如图 9.9 所示，可以看到并没有参数值，get.jsp 代码如下：

```
<%   //接收发送来的请求参数
String name = request.getParameter("name") ;
String password = request.getParameter("password") ;
if("hnist".equals(name) && "8648870".equals(password)) {
out.println("Receive name is:"+name);
out.println("Receive password is:"+password); %>
Your message are right!   <%}
else {
out.println("Receive name is:"+name);
out.println("Receive password is:"+password);  %>
    Your message are Wrong!<%}%>
```

（2）定义布局文件 activity_main.xml，只有一个 TextView 组件，代码略。

（3）建立 Activity 文件 MainActivity.java 文件，代码如下：

```
…
import java.net.HttpURLConnection;
import java.net.URL;
import android.os.Handler;
import android.os.Message;
public class MainActivity extends AppCompatActivity {
private Handler handler = null;
private TextView msg=null;
private String temp="";
@Override
public void onCreate(Bundle savedInstanceState) {
```

```
        super.onCreate(savedInstanceState);
        super.setContentView(R.layout.activity_main);     //调用布局管理器
        msg = (TextView) super.findViewById(R.id.msg);    //取得文本组件
        new Thread() {
            @Override
            public void run() {//要执行的方法
                try {// 连接地址，同时传递参数
    URL url=
      new URL("http://10.0.2.2:8080/jsp/get.jsp?name=hnist&password=8648870");
    HttpURLConnection conn = (HttpURLConnection) url.openConnection();
    byte [] data = new byte[200]        ;                //开辟空间
    int len = conn.getInputStream().read(data) ;         //接收数据
        if(len > 0){
            temp = new String(data,0,len).trim() ;        }
            conn.getInputStream().close() ;              //关闭输入流
                } catch (Exception e) {
                    e.printStackTrace() ;
                    msg.setText("服务器连接失败。") ;    }
    handler.sendEmptyMessage(0);}}.start();//执行完毕后给 handler 发送一个空消息
    handler = new Handler() {
        @Override
    public void handleMessage(Message mg) {  //有消息发出时执行 Handler 方法
        super.handleMessage(mg);             //处理 UI
        msg.setText(temp) ;} }; } }
```

（4）在 AndroidManifest.xml 文件中添加权限，否则会出现错误，代码如下：

```
<uses-permission android:name="android.permission.INTERNET" />
```

运行该项目，结果如图 9.10 所示，由图可知已经将参数传递给了 JSP 文件。

图 9.9 get.jsp 文件浏览结果　　　　图 9.10 Exam9-6 运行结果

9.3.3 HttpURLConnection 通信：POST 方式

POST 与 GET 的不同在于，POST 的参数不能放在 URL 字符串后面，而是放在 HTTP 请求数据中，这些参数会通过 cookie 或者 session 等方式以键值对的形式（key=value）传送到服务器中。

按照下面的步骤来实现 POST 方式的通信。

（1）和 GET 方式一样，需要创建 HttpURLConnection 实例，例如：

```
URL url=new URL("http://10.0.2.2:8080/jsp/post.jsp");//定义一个 URL
HttpURLConnection urlConn =(HttpURLConnection) url.openConnection();
```

（2）在 POST 方式中，openConnection 方法只创建了 HttpURLConnection 实例，但是并不进行真正的连接操作，连接之前需要对其属性进行设置，如超过时间等。下面是对 HttpURLConnection 实例的属性设置。

```
urlConn.setDoOutput(true);//设置是否向 urlConn 输出，设为 true，默认为 false
```

```
    urlConn.setDoInput(true); //设置是否从urlConn读入，默认为true;
    /*设定传送的内容类型是可序列化的 Java 对象（如果不设此项，在传送序列化对象时，当 Web 服务默认的不是这种类型时，可能抛 java.io.EOFException 出错信息）*/
    urlConn.setRequestProperty("Content-type","application/x-java-serialized-object");
    urlConn.setRequestMethod("POST");       //设置方式为 POST，默认为 GET 方式
    urlConn.setUseCaches(false);            //请求不能使用缓存
    …                                       //注意，这些配置必须在连接之前完成
```

（3）建立连接。例如：

```
    urlConn.connect();          //建立 HttpURLConnection 连接
    //getOutputStream()方法会隐含地进行连接，开发中不调用 connect()方法也可建立连接
    OutputStream outStrm = urlConn.getOutputStream();
```

（4）HttpURLConnection 写数据与发送数据。

① 定义要上传的数据并传送到服务器中，接收返回数据并读出。

```
    //定义 DataOutputStream 对象，传送数据
    DataOutputStream out = new DataOutputStream(httpconn.getOutputStream());
    String content ="par="+URLEncoder.encode("hello", "gb2312");//定义要上传的参数
    out.writeBytes(content);            //将要上传的内容写入流
    out.flush();                        //刷新对象输出流
    out.close();                        //关闭输出流对象
    //将 HTTP 请求发送到服务端中，获取返回的内容
    BufferedReader reader = new BufferedReader(new InputStreamReader
                         (urlConn.getInputStream())) ;
    String inputLine = null;
    while(((inputLine = reader.readLine()) != null)){//使用循环来读取获得的数据
    resultData += inputLine + "\n"; }       //在每一行后面加上"\n"来换行
    reader.close();                         //关闭输入流对象
    urlConn.disconnect();                   //关闭 HttpURLConnection 连接
```

② 也可以采用下面的方式进行数据的读写：

```
    //构建一个对象输出流对象，以输出可序列化的对象
    ObjectOutputStream out = new ObjectOutputStream(outStrm);
    //向对象输出流写出数据，这些数据将存到内存缓冲区中
    out.writeObject(new String("This is a test!"));
    out.flush();            //刷新对象输出流，将字节都写入流
    //关闭流对象，不能再向对象输出流写入数据，之前写入的数据保存在缓冲区中
    out.close();
    /*调用 HttpURLConnection 连接对象的 getInputStream()函数，将缓冲区中封装好的完整的
      HTTP 请求发送到服务端*/
    InputStream inStrm = urlConn.getInputStream(); //将 HTTP 请求发送到服务端
```

一旦调用 getInputStream()就表示本次 HTTP 请求已结束，后面向对象输出流的输出已无意义，既使对象输出流没有调用 close()方法，后面的操作也不会向对象输出流写入任何数据。因此，要重新发送数据到服务器时，就需要重新创建连接、重新设置参数、重新创建流对象、重新写数据。

如果要将返回的数据读出来，可用上面的方式对 inStrm 进行操作。

（5）在 AndroidManifest.xml 文件中添加权限，否则会出现错误，代码如下：

```
    <uses-permission android:name="android.permission.INTERNET" />
```

实例 9-6：使用 POST 方式传递数据到 JSP 文件中

新建一个项目，项目命名为 Exam9_7，利用 HttpURLConnection 的 POST 方式传递数据到指定的 JSP 文件中。

（1）在服务器上建立一个 get1.jsp 文件，在 Tomcat 上发布后，显示结果如图 9.11 所示，代码如下：

```
<%  //接收发送来的请求参数
String name = request.getParameter("name") ;
if("hnist".equals(name) ) {
out.println("Receive name is:"+name); %>
    Your message are right!
<%}
else {   out.println("Receive name is:"+name); %>
    Your message are Wrong!
<%}%>
```

（2）定义布局文件 activity_main.xml，其中只有一个 TextView 组件，代码略。

（3）建立 Activity 文件 MainActivity.java 文件，代码如下：

```
…
public class MainActivity extends AppCompatActivity {
 private Handler handler = null;
 private URL url = null;
 private String resultData = "";        //获得的数据
 private TextView msg = null;
 public void onCreate(Bundle savedInstanceState){
     super.onCreate(savedInstanceState);
     setContentView(R.layout.activity_main);
     msg = (TextView)this.findViewById(R.id.msg);
     String httpUrl = "http: //10.0.2.2:8080/jsp/get1.jsp";
     try{
         url = new URL(httpUrl);        //构造一个 URL 对象
     }catch (MalformedURLException e){     }
     new Thread() {
         @Override
         public void run() {            //要执行的方法
             if (url != null){
                 try {
     // 使用 HttpURLConnection 打开连接
     HttpURLConnection urlConn = (HttpURLConnection) url.openConnection();
     urlConn.setDoOutput(true);                 //POST 请求需要设置为 true
         urlConn.setDoInput(true);
         urlConn.setRequestMethod("POST");      // 设置使用 POST 方式
         urlConn.setUseCaches(false);           // Post 请求不能使用缓存
         // 配置本次连接的 Content-type，配置为 application/x-www-form-urlencoded
         urlConn.setRequestProperty("Content-Type",
                     "application/x-www-form-urlencoded");
         urlConn.connect();
     //DataOutputStream 流
     DataOutputStream out = new DataOutputStream(urlConn.getOutputStream());
         //定义要上传的参数
```

```
            String content = "name=" + URLEncoder.encode("hnist", "gb2312");
            out.writeBytes(content); //将要上传的内容写入流
            out.flush();//刷新、关闭
            out.close();
            BufferedReader reader =
    new BufferedReader(new InputStreamReader(urlConn.getInputStream()));
    //获取数据
            String inputLine = null;
        while (((inputLine = reader.readLine()) != null)) {//使用循环来读取数据
                resultData += inputLine + "\n";}  //在每一行后面加上"\n"来换行
                    reader.close();
                    urlConn.disconnect();     //关闭 HTTP 连接
            }catch (Exception e){ }}
    handler.sendEmptyMessage(0);}}.start();  //执行完毕后给 handler 发送一个空消息
        handler = new Handler() {
            @Override
            //当有消息发送出来的时候就执行 Handler 方法
            public void handleMessage(Message mg) {
                super.handleMessage(mg);
                if ( resultData != null ){        //处理 UI
                    msg.setText(resultData);      //设置显示取得的内容
                }else{
                    msg.setText("读取的内容为 NULL"); } } }; } }
```

（4）在 AndroidManifest.xml 文件中添加权限，否则会出现错误，代码如下：

```
    <uses-permission android:name="android.permission.INTERNET" />
```

运行该项目，结果如图 9.12 所示，发现已经将参数传递给了 JSP 文件。

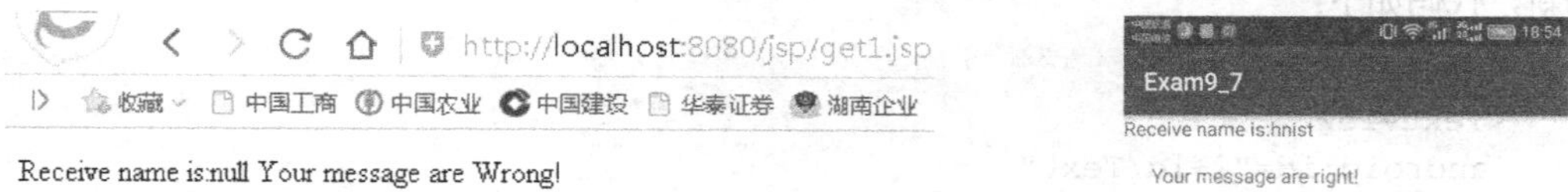

图 9.11　get1.jsp 文件浏览结果　　　　图 9.12　Exam9-7 运行结果

使用 HttpURLConnection 时要注意以下几点。

（1）HttpURLConnection 的 connect()函数，实际上只是建立了一个与服务器的 TCP 连接，并没有实际发送 HTTP 请求。HTTP 请求实际上直到 HttpURLConnection 的 getInputStream()调用这个函数才会正式发送出去。

（2）在用 POST 方式发送 URL 请求时，URL 请求参数的设定顺序很关键，对 connection 对象的配置都必须在 connect()函数执行之前完成。对于 outputStream 的写操作，又必须在 inputStream 的读操作之前，否则会抛出异常。

```
    java.net.ProtocolException: Cannot write output after reading input......
```

（3）HTTP 请求实际上由两部分组成：一部分是 HTTP 头，所有关于此次 HTTP 请求的配置都在 HTTP 头中定义；另一部分是正文。connect()函数会根据 HttpURLConnection 对象的配置值生成 HTTP 头部信息，因此在调用 connect 函数之前，就必须把所有的配置准备好。

（4）HTTP 头后面紧跟着的是 HTTP 请求的正文，正文的内容是通过 outputStream 写入的，实际上，

outputStream 不是一个网络流，向其中写入的东西不会立即发送到网络，而是存在于内存缓冲区中，当 outputStream 关闭时，根据输入的内容生成 HTTP 正文。此时，HTTP 请求的东西才全部准备就绪。在 getInputStream()函数调用的时候，就会把准备好的 HTTP 请求正式发送到服务器中，然后返回一个输入流，用于读取服务器对于这次 HTTP 请求的返回信息。因此，在 getInputStream()函数之后对 connection 对象进行设置或写入 outputStream 都没有意义，执行这些操作会导致异常的发生。

9.3.4 数据的实时更新

前面介绍的只是简单地一次性获取网页数据，而在实际开发中更多的是需要实时获取最新数据，例如，实时天气信息、实时交通信息等。可以通过一个线程来控制视图的更新，要实时地从网络中获取数据，简单说就是把获取网络数据的代码写到线程中，不停地进行更新。这里使用 Handler 来实现更新。

实例 9-7：数据实时更新实例

新建一个项目，项目命名为 Exam9_8，利用 HttpURLConnection 和 Handler 实现数据的实时更新。例如，创建一个网页来显示系统当前的时间，然后每隔 5 秒系统会自动刷新一次视图。

（1）创建一个显示当前系统时间的网页文件 date.jsp，在 Tomcat 上发布该文件，显示界面如图 9.13 所示。date.jsp 代码如下：

```
<%@ page language="java" import = "java.util.*" pageEncoding = "gb2312" %>
<%java.text.SimpleDateFormat simpleDateFormat=new java.text.
               SimpleDateFormat ("yyyy-MM-dd HH:mm:ss");
java.util.Date currentTime = new java.util.Date();
String time = simpleDateFormat.format(currentTime).toString();
out.println(time); %>
```

（2）定义布局文件 activity_main.xml，其中有一个文本显示框，用于显示当前系统时间；一个按钮，用于实时刷新，代码如下：

```
…
    <TextView
     android:id="@+id/Text"
     android:layout_width="fill_parent"
     android:layout_height="wrap_content"
     android:layout_gravity="center" />
    <Button
     android:id="@+id/But"
     android:layout_width="fill_parent"
     android:layout_height="wrap_content"
     android:text="刷新" />
  </LinearLayout>
```

（3）建立 Activity 文件 MainActivity.java 文件，代码如下：

```
…
public class MainActivity extends AppCompatActivity {
    private Button But;
    private TextView msg;
    private URL url = null;
    private String resultData = "";
    private Handler handler = null;
    public void onCreate(Bundle savedInstanceState) {
```

```
        super.onCreate(savedInstanceState);
        setContentView(R.layout.activity_main);
        msg = (TextView) this.findViewById(R.id.Text);
        But = (Button) this.findViewById(R.id.But);
        But.setOnClickListener(new Button.OnClickListener() {
            public void onClick(View arg0) {
                refresh();     } });           //刷新
        new Thread(mRunnable).start();}      //开启线程
    private void refresh() {                  //刷新网页显示
        String httpUrl = "http://10.0.2.2:9090/jsp/date.jsp";
        try {
            url = new URL(httpUrl);// 构造一个 URL 对象
        } catch (MalformedURLException e) { }
        new Thread() {
            @Override
            public void run() {
                if (url != null) {//要执行的方法
                    try {
                        // 使用 HttpURLConnection 打开连接
        HttpURLConnection urlConn = (HttpURLConnection) url.openConnection();
        InputStreamReader in = new InputStreamReader(urlConn.getInputStream(),
"GBK");// 得到读取的内容(流)
     BufferedReader buffer = new BufferedReader(in);// 为输出创建 BufferedReader
        String inputLine = null;
   while (((inputLine = buffer.readLine()) != null)) {// 使用循环来读取获得的数据
                        resultData += inputLine + "\n";
                } // 在每一行后面加上"\n"来换行
                in.close();// 关闭 InputStreamReader
                urlConn.disconnect();// 关闭 HTTP 连接
               } catch (IOException e) { } }
    handler.sendEmptyMessage(0);  } }.start();//执行完毕后给 handler 发送一个空消息
        handler = new Handler() {
            @Override
            //当有消息发出时就执行 Handler 方法
            public void handleMessage(Message mg) {
                super.handleMessage(mg);
                if (resultData != null) {//处理 UI
                    msg.setText(resultData);// 设置显示取得的内容
                } else {
                    msg.setText("读取的内容为 NULL"); } } }; }
    private Runnable mRunnable = new Runnable() {
        public void run() {
            while (true) {
                try {
                    Thread.sleep(5 * 1000);
                    mHandler.sendMessage(mHandler.obtainMessage());//发送消息
                } catch (InterruptedException e) { } } } };
     Handler mHandler = new Handler() {
        public void handleMessage(Message msg) {
            super.handleMessage(msg);  //接收消息
            refresh(); } }; }          //刷新
```

（4）在 AndroidManifest.xml 文件中添加权限，否则会出现错误，代码如下：

```
<uses-permission android:name="android.permission.INTERNET" />
```

保存所有文件，运行该项目，结果如图 9.14 所示，每 5 秒自动刷新一次，显示的时间来自于服务器。

图 9.13　文件浏览结果

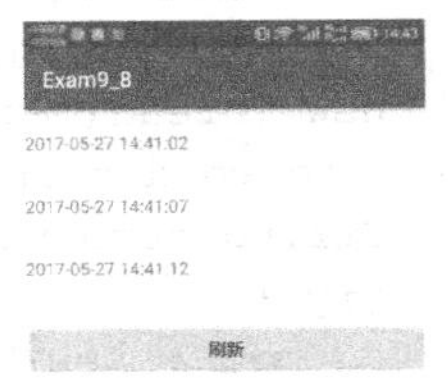

图 9.14　Exam9-8 运行结果

9.4　利用 Volley 框架进行数据交互

开发 Android 网络数据通信应用的时候一般会使用 HTTP 协议来发送和接收网络数据，Android 系统以前主要提供了两种方式来进行 HTTP 通信——HttpURLConnection 和 HttpClient。Android 6.0 明确表示不再支持 HttpClient，而 HttpURLConnection 使用比较麻烦，Android 开发团队在 2013 年推出了一个新的网络通信框架——Volley。Volley 能使网络通信更快、更简单、更健壮。

Volley 提供了如下功能：JSON，图像等的异步下载、网络请求的排序、网络请求的优先级处理、缓存、多级别取消请求、Activity 和生命周期的联动（Activity 结束时同时取消所有网络请求）。

Volley 适用于进行数据量不大，但通信频繁的网络操作，不适合进行大数据量的网络操作，如下载文件。因为 Volley 会保持在解析的过程中的所有响应。若要下载大量的数据，则可以考虑使用 DownloadManager。

9.4.1　Volley 框架的使用

1．添加 Volley.jar 包到项目中

（1）在使用 Volley 框架前先要添加 Volley 库，需要下载 Volley 的 JAR 包，下载完毕后保存在电脑中，如 D:\androidspace\AndroidVolleyJar，选择“File→New→New Module...”选项，如图 9.15 所示，在弹出的选项中选择“Import .JAR/AAR Package”选项，如图 9.16 所示，在弹出的对话框中输入刚才保存 Volley 的路径，如图 9.17 所示，单击“Finish”按钮。

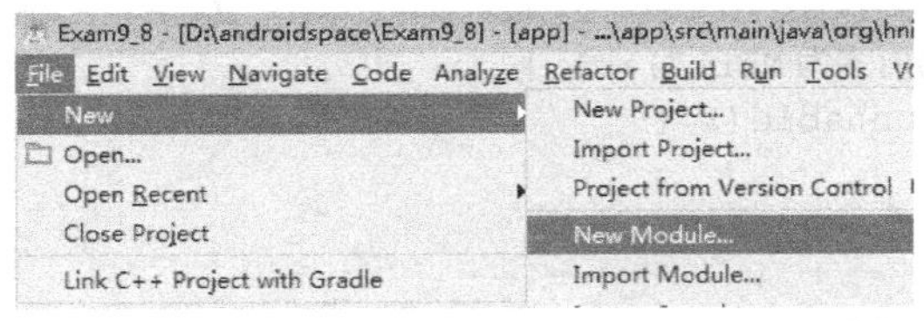

图 9.15　选择 New Module ...

图 9.16　选择 Import .JAR

（2）把包导入后再引用，要配置 bulid.gradle，选择“File→Project Structure”选项，如图 9.18 所示。

（3）选择左侧的 app 选项，然后选择“Dependencies”选项卡，单击“+”按钮，进入如图 9.19 所示界面。选择“3 Module dependency”选项，在弹出的窗口中选择导入的 Volley 模块，此后即可使用 Volley 框架。

图 9.17　输入路径

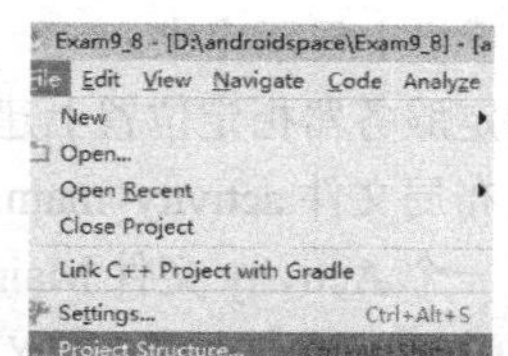

图 9.18　配置 bulid.gradle

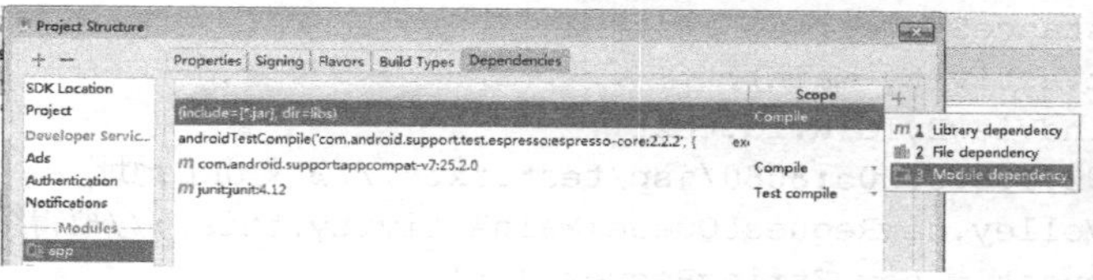

图 9.19　选择 Volley 模块

2. 使用 Volley 框架的基本步骤

使用 Volley 框架实现网络数据请求时主要有以下 3 个步骤。

（1）创建 RequestQueue 对象，定义网络请求队列，例如：

```
//实例化 Volley 全局请求队列
RequestQueue sRequestQueue = Volley.newRequestQueue(getApplicationContext());
```

RequestQueue 是一个请求队列对象，它可以缓存所有的 HTTP 请求，然后按照一定的算法并发地发出这些请求。不必为每一次 HTTP 请求都创建一个 RequestQueue 对象，建议用单例模式定义这个对象。RequestQueue 的成员方法主要有下面几个：

```
public void start();  //请求队列开始进行调度
public void stop();   //队列退出调度
public Request add(Request request);//添加一个请求，通过调用 start()来执行
void finish(Request request);//这个方法应该是释放请求资源的方法
public void cancelAll();//取消当前的请求
```

（2）创建 XXXRequest 对象(XXX 代表 String、JSON、Image)，定义网络数据请求的接收。

前面定义了请求对象，那么自然就有接收响应的对象，StringRequest 类可以用来从服务器获取 String，ImageRequest 类可以用来从服务器获取图片，JsonArrayRequest 类可以用来接收 JSON Array，不支持 JSON Object。它们都继承自 Request，然后根据不同的响应数据来进行特殊的处理。例如：

```
String url = "http://10.0.0.2:8080/jsp/date.jsp:8080"; //定义请求 URL
//Volley request，参数：请求方式，请求的 URL，请求成功的回调，请求失败的回调
StringRequest stringRequest = new StringRequest(Request.Method.GET, url, new
Response.Listener<String>(), new Response.ErrorListener());
```

WEB 请求方式有 GET 和 POST 两种，默认使用 GET 方式。

（3）把 XXXRequest 对象添加到 RequestQueue 中，开始执行网络请求。例如：

```
requestQueue.add(stringRequest); //将 stringRequest 加入队列
```

（4）在 AndroidManifest 中注册 Application，添加网络权限。例如：

```
<uses-permission android:name="android.permission.INTERNET"/>
```

9.4.2　Volley 框架使用实例

实例 9-8：Volley 框架使用实例 1

新建一个项目，项目命名为 Exam9_9，利用 Volley 框架的 GET 方式获取网络上的文件，直接将该

文件的内容在手机上显示出来。

（1）在指定服务器指定位置上建立一个文件 test.txt，内容如图 9.20 所示。

（2）定义布局文件 activity_main.xml，其中只有一个 TextView 组件，代码略。

（3）定义一个 Activity 文件 MainActivity.java，代码如下：

```
public class MainActivity extends AppCompatActivity {
   private TextView txt=null;
   protected void onCreate(Bundle savedInstanceState) {
     super.onCreate(savedInstanceState);
     setContentView(R.layout.activity_main);
     txt=(TextView) super.findViewById(R.id.text);
     String url = "http://192.168.0.205:8080/jsp/test.txt";//定义 URL 地址
     RequestQueue mQueue = Volley.newRequestQueue(MainActivity.this);//取得组件
     StringRequest stringRequest = new StringRequest(url,
             new Response.Listener<String>() {//请求成功的响应
                 @Override
                 public void onResponse(String response) {
                     String msg= response; //取得接收到的内容
                     txt.setText(msg); }}, //显示接收到的内容
             new Response.ErrorListener() {//请求失败的响应
                 @Override
                 public void onErrorResponse(VolleyError error) {
                     Log.e("TAG", error.getMessage(), error);  } });
    mQueue.add(stringRequest); }}// 将创建的请求添加到请求队列中
```

（4）在 AndroidManifest 中注册 Application，添加网络权限，代码如下：

```
<uses-permission android:name="android.permission.INTERNET"/>
```

保存所有文件，运行该项目，结果如图 9.21 所示。

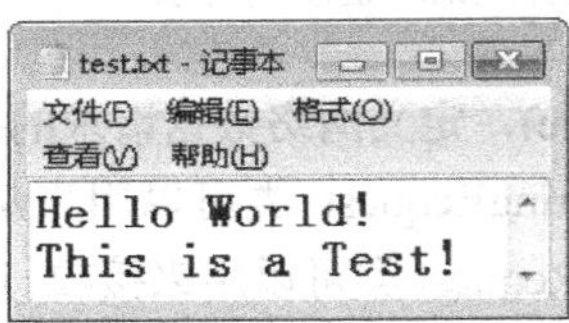

图 9.20 test.txt 文件内容

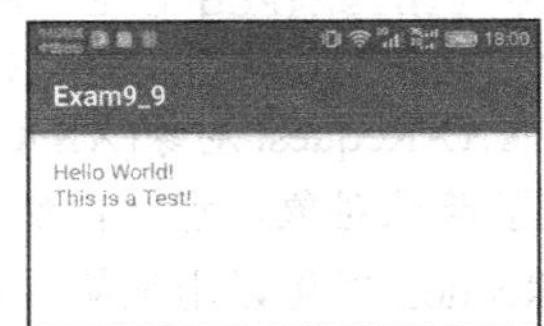

图 9.21 Exam9_9 运行结果

从这个例子可以看出，采用 Volley 框架的方式读取网络中的文件比使用 HttpURLConnection 要简单得多，类似的，也可以读出网络上的图片，详细代码参考 Exam9_10。

前面介绍了 GET 方式，也可使用 POST 方式，携带参数向服务器提交请求，服务器接收后，做出相应的响应，只需要将 Request.Method.GET 修改为 Request.Method.POST，再增加一个携带参数的集合即可。

实例 9-9：Volley 框架使用实例 2

新建一个项目，项目命名为 Exam9_11，利用 Volley 框架的 POST 方式向服务器传递参数，并将反馈结果在手机上显示出来。

（1）在服务器上建立一个 get.jsp 文件，在 Tomcat 上发布后，显示结果如图 9.9 所示。

（2）定义布局文件 activity_main.xml，其中只有一个 TextView 组件，代码略。

（3）定义一个 Activity 文件 MainActivity.java，代码如下：

```
public class MainActivity extends AppCompatActivity {
private TextView txt=null;
protected void onCreate(Bundle savedInstanceState) {
super.onCreate(savedInstanceState);
    setContentView(R.layout.activity_main);
    txt=(TextView) super.findViewById(R.id.txt);
    String url = "http://192.168.0.205:8080/jsp/get.jsp";
    RequestQueue mQueue = Volley.newRequestQueue(getApplicationContext());
    StringRequest stringRequest = new StringRequest(Request.Method.POST,url,
            new Response.Listener<String>() {//请求成功的响应
                @Override
           public void onResponse(String response) {
                      String msg= response;
                      txt.setText(msg);
                }}, new Response.ErrorListener() {//请求失败的响应
        @Override
        public void onErrorResponse(VolleyError error) {
            String msg= "没有接收到数据";
            txt.setText(msg);;}}) {
        @Override
        protected Map<String, String> getParams() {
        //在这里设置需要 post 的参数
       Map<String, String> map = new HashMap<String, String>();
            map.put("name", "hnist");
            map.put("password", "8648870");
            return map;       } };
   mQueue.add(stringRequest);}}
```

（4）在 AndroidManifest 中注册 Application，添加网络权限，代码如下：

```
<uses-permission android:name="android.permission.INTERNET"/>
```

保存所有文件，运行该项目，结果如图 9.10 所示，可以发现参数值已经传递给服务器了。

9.5 利用 Socket 交换数据

Socket 是 TCP/IP 协议上的一种通信，在通信的两端各建立一个 Socket，从而在通信的两端形成网络虚拟链路。一旦建立了虚拟的网络链路，两端的程序即可通过虚拟链路进行通信。

Socket 通信有两部分：一部分为监听的 Server 端，另一部分为主动请求连接的 Client 端。Server 端会一直监听 Socket 中的端口直到有请求为止，当 Client 端对该端口进行连接请求时，Server 端就给予应答并返回一个 Socket 对象，以后 Server 端与 Client 端的数据交换即可使用这个 Socket 来进行操作。

Socket 有两种主要的通信方式： 基于 TCP 协议和基于 UDP 协议的通信。

9.5.1 基于 TCP 协议的 Socket 通信

一个客户端若要发起一次通信，首先必须知道运行服务器端的主机 IP 地址，通过指定的端口和服务器建立连接，然后进行通信，通信方式如图 9.22 所示。

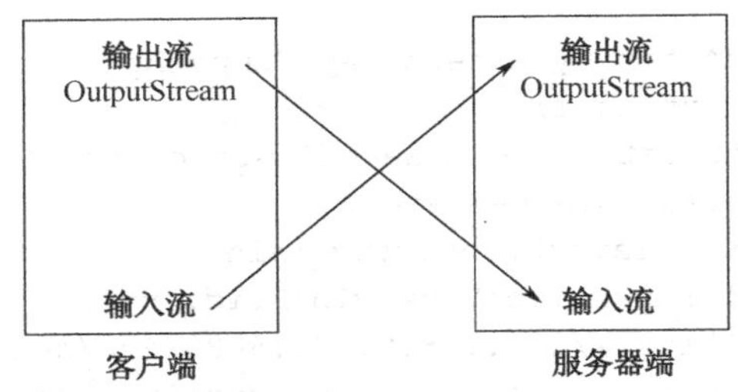

图 9.22　Socket 通信示意图

Socket 类的方法包含在 java.net.Socket 包中，下面通过实例给出常用的几个方法。

第一种：客户端 Socket 通过构造方法连接服务器，例如：

```
Socket socket = new Socket("211.69.1.17",8000); //新建 Socket，指定 IP 及端口号
```

第二种：通过 connect 方法连接服务器，例如：

```
Socket socket  = new Socket(); //新建 Socket，未指定 IP 及端口号
socket.connect(new InetSocketAddress("211.69.1.1",80)); //使用默认的连接超时
socket.connect(new InetSocketAddres s("211.69.1.1",80),2000); //连接超时 2s
socket.setSendBufferSize(8096);         //设置输出流的发送缓冲区大小，默认是 8KB
socket.setReceiveBufferSize(8096);      //设置输入流的接收缓冲区大小，默认是 8KB
socket.setKeepAlive(true);    //防止服务器端无效时，客户端长时间处于连接状态
OutputStream os = socket.getOutputStream();
                              //客户端向服务器端发送数据，获取客户端向服务器端输出流
//判断 Socket 是否处于连接状态
if((socket.isConnected() == true) && (socket.isClosed() == false)){ …}
InputStream is = socket.getInputStream();
                              //客户端接收服务器端的响应，读取服务器端向客户端的输入流
byte[] buffer = new byte[is.available()];     //设置缓冲区
is.read(buffer);                              //读取缓冲区
String responseInfo = new String(buffer);     //转换为字符串
Log.i("TEST", responseInfo);                  //日志中输出
socket.close();                               //关闭连接
```

Socket 通信的实现要分别设计服务器端和客户端。

（1）服务器端的操作如下。

① 指定端口实例化一个 ServerSocket，会自动对传入的端口号进行监听，例如：

```
ServerSocket server = new ServerSocket(9090);     //设置监听端口 9090
```

② 收到连接请求后调用 ServerSocket 的 accept()，返回一个连接的 Socket 对象，例如：

```
Socket client = server.accept();                  //接收客户端请求
```

③ 获取位于该层 Socket 的流以进行读写操作，例如：

```
PrintStream out = new PrintStream(client.getOutputStream()); //获得客户端的输出流
```

④ 将数据封装成流，例如：

```
BufferedReader msg = new BufferedReader(new InputStreamReader(client.
                    getInputStream())); //对收到的数据缓冲区进行读取
```

⑤ 对 Socket 进行读写，例如：

```
StringBuffer info = new StringBuffer();       //接收客户端发送回来的信息
info.append("I'm server!:");                  //回应给客户端的数据
info.append(msg.readLine());                  //接收客户端的数据
```

```
out.print(info);                                    //发送信息到客户端
```

⑥ 关闭打开的流，例如：

```
out.close();                                        //关闭输出流
msg.close();                                        //关闭输入流
client.close();                                     //关闭客户端连接
server.close();                                     //关闭服务器端连接
```

（2）客户端的操作如下。

① 通过 IP 地址和端口实例化 Socket，请求连接服务器，例如：

```
Socket client = new Socket("10.0.2.2", 9090);     //指定服务器及端口号
```

注意：10.0.2.2 是本地 PC 的 IP 地址，模拟机的 IP 地址是 127.0.0.1，在没有固定服务器的情况下，可以将 PC 当做服务器进行实验。

通过上面的步骤后，Server 端和 Client 端就可以连接起来了，要进行数据交换时，需要用到 I/O 流中的 OutputStream 类和 InputStream 类。

OutputStream：当应用程序需要对流进行数据写操作时，可以使用 Socket.getOutputStream()方法返回的数据流进行操作。

InputStream：当应用程序要从流中取出数据时，可以使用 Socket.getInputStream()方法返回的数据流进行操作。

② 获取 Socket 上的流以进行读写，例如：

```
//客户端向服务器端发送数据，获取客户端向服务器端的输出流
PrintStream out = new PrintStream(client.getOutputStream());
```

③ 把流包装到 BufferReader/PrintWriter 对象中，例如：

```
//对返回的数据流进行缓冲区读取
BufferedReader msg = new BufferedReader(new InputStreamReader
                    (client.getInputStream()));
```

④ 对 Socket 进行读写，例如：

```
out.println("已经连接上服务器");                     //发送数据到服务器中
```

⑤ 关闭打开的流，例如：

```
out.close();                                        //关闭输出流
msg.close() ;                                       //关闭输入流
client.close();                                     //关闭连接
```

实例 9-10：基于 TCP 协议的 Socket 实例

新建一个项目，项目命名为 Exam9_12，建立一个基于 TCP 协议的 Socket 通信实例。

（1）用 Eclipse 在服务器端建立一个项目——Server9_12，主程序 Server.java 的代码如下：

```
import java.io.BufferedReader;
import java.io.InputStreamReader;
import java.io.PrintStream;
import java.net.ServerSocket;
import java.net.Socket;
public class Server {
public static void main(String[] args) throws Exception { //所有异常抛出
    ServerSocket server = new ServerSocket(9090);          //设置监听端口 9090
```

```
    Socket client = server.accept();                        //接收客户端请求
    //获得客户端的数据流
    PrintStream out = new PrintStream(client.getOutputStream());
    BufferedReader msg = new BufferedReader(new InputStreamReader
    (client.getInputStream()));                  //对收到的数据在缓冲区进行读
取
    StringBuffer info = new StringBuffer();      //接收客户端发送回来的信息
    info.append("I'm server!:");                 //回应给客户端的数据
    info.append(msg.readLine());                 //接收客户端的数据
    out.print(info);                             //发送信息到客户端
    out.close();                                 //关闭输出流
    msg.close();                                 //关闭输入流
    client.close();                              //关闭客户端连接
    server.close(); }}                           //关闭服务器端连接
```

（2）建立客户端程序，建立一个布局管理文件 activity_main./xml，代码如下：

```
…
<Button
    android:id="@+id/send"
    android:layout_width="fill_parent"
    android:layout_height="wrap_content"
    android:text="连接到服务器" />
<TextView
    android:id="@+id/info"
    android:layout_width="fill_parent"
    android:layout_height="wrap_content"
    android:text="正在连接到服务器..." />
</LinearLayout>
```

（3）建立客户端程序，建立一个 Activity 文件 MainActivity.java，代码如下：

```
…
public class MainActivity extends AppCompatActivity {
 private Button send = null;            // 定义按钮组件
 private TextView info = null;          // 定义文本组件
 private Handler handler = null;
 private String str="";
 @Override
 public void onCreate(Bundle savedInstanceState) {
     super.onCreate(savedInstanceState);
     super.setContentView(R.layout.activity_main);              // 调用布局文件
     this.send = (Button) super.findViewById(R.id.send);  // 取得按钮组件
     this.info = (TextView) super.findViewById(R.id.info);// 取得文本显示组件
     this.send.setOnClickListener(new SendOnClickListenerImpl());}//设置事件
 private class SendOnClickListenerImpl implements OnClickListener{
     @Override
     public void onClick(View view) {
         new Thread() {
             @Override
             public void run() {                              //要执行的方法
                 try {
         Socket client = new Socket("10.0.2.2", 9090);//指定服务器及端口号
         //客户端向服务器端发送数据，获取客户端到服务器端的输出流
```

```
            PrintStream out = new PrintStream(client.getOutputStream());
            BufferedReader  msg  =  new  BufferedReader(new  InputStreamReader
(client.getInputStream())); // 对返回的数据流进行缓冲区读取
            out.println("已经连接上服务器");  // 发送数据
            str = msg.readLine();
            out.close();                    // 关闭输出流
            msg.close() ;                      // 关闭输入流
            client.close(); }catch (Exception e) {// 关闭连接
            e.printStackTrace();}
    handler.sendEmptyMessage(0); } }.start();//执行完毕后给 handler 发送一个空消息
            handler = new Handler() {
                @Override
                //当有消息发送出来的时候就执行 Handler 方法
                public void handleMessage(Message mg) {
                    super.handleMessage(mg);
                    info.setText(str);   } }; }}}// 设置文本内容
```

（4）在 AndroidManifest.xml 中配置相应的权限，在其中添加下面的代码：

```
    <uses-permission android:name="android.permission.INTERNET"/>
```

保存所有文件，先运行服务器端的 Server.java，再运行客户端项目的 Exam9_12，客户端运行结果如图 9.23 所示。

图 9.23　Exam9_12 运行结果

9.5.2　基于 UDP 协议的 Socket 通信

使用 TCP 协议时要先和接收方建立连接，然后发送数据，保证数据成功到达目的地，但是速度慢。而使用 UDP 协议方式时要先把数据打包成数据包，然后直接发送给接收方，不需要建立连接，速度快，但是有可能丢失数据。

使用基于 UDP 的 Socket 通信也包含服务器端的设计和客户端的设计。

服务端的设计如下：

```
    //创建服务端 socket，并使之监听 9999 端口
    DatagramSocket socket = new DatagramSocket (9999);
    byte data[]=new byte[1024];
    //准备接收数据
    DatagramPacket packet = new DatagramPacket (data,data.length);
    //接收到数据报文，并将报文中的数据复制到指定的 DatagramPacket 实例中
    socket.receive(packet);
    String s=packet.getData();  //接收 DatagramPacket 实例中的数据，转换成字符串
```

特别的，当数据小于 1024 时，由于 data 指定的是 1024，所以会出现乱码，解决的办法如下

```
        String s=new String(packet.getData(),packet.getOffset(),packet.getLength());
```

客户端的设计如下：

```
    DatagramSocket socket = new DatagramSocket (9999);    //创建客户端 socket
    InetAddress serverAddress=InetAddress.getByName("211.699.1.1");//服务端地址
```

```
DatagramPacket packet = new DatagramPacket (data,data.length,
 serverAddress, 9999);                                   //打包要发送的数据
socket.send(packet );                                    //发送 DatagramPacket 对象
```

这里给出一个简单的实例：使用 UDP 协议完成数据的传送。

（1）服务端主要代码如下：

```
//创建一个 DatagramSocket 对象，并指定监听的端口号
DatagramSocket socket = new DatagramSocket(9999);
byte data [] = new byte[1024];
//创建一个空的 DatagramPacket 对象
DatagramPacket packet =new DatagramPacket(data,data.length);
//使用 receive 方法接收客户端发来的数据，如客户端未发来数据，就停滞等待
socket.receive(packet);
String result = new String(packet.getData(),packet.getOffset(),packet.getLength());
System.out.println("result--->" + result); //输出收到的数据
```

（2）客户端主要代码如下：

```
public static void main(String[] args) {
  try {
      //创建一个 DatagramSocket 对象
    DatagramSocket socket = new DatagramSocket(9999);
     //创建一个 InetAddree
         InetAddress serverAddress = InetAddress.getByName("211.69.1.1");
    String str = "This is a test! ";                //这是要传输的数据
    byte data [] = str.getBytes();                  //把传输内容分解成字节
    DatagramPacket packet= new DatagramPacket(data,data.length,
        serverAddress,9999);
                  //创建一个 DatagramPacket 对象，并指定发送的目的 IP 地址及端口号
     socket.send(packet);           //调用 socket 对象的 send 方法，发送数据
       } catch (Exception e) {
      e.printStackTrace();}}
```

（3）配置 AndroidManifest.xml 文件，添加相应的权限，添加如下语句：

```
<uses-permission android:name="android.permission.INTERNET"/>
```

保存所有文件，先运行服务器端程序，再运行客户端程序，可以在服务器端发现已经接收到了客户端发来的数据，如图 9.24 所示。

```
Console
<terminated> MyServer (1) [Java Applicatio
result--->This is a test!
```

图 9.24　服务器端的显示界面

9.5.3　利用 Socket 实现简易的聊天室

前面的例子中实现了一个客户端和一个服务器的单独通信，并且只能通信一次，在实际中，往往需要在服务器上运行一个程序，它用来接收来自多个客户端的请求，并提供相应服务，这就需要使用多线程来实现。服务器总是在指定的端口上监听是否有客户请求，一旦监听到客户请求，服务器就会启动一个专门的服务线程来响应该客户的请求，而服务器本身在启动完线程后马上又进入监听状态，等待下一个客户。

使用 Socket 通信编写一个简单的聊天室程序，Client1 发信息给 Client2，Client1 的信息先发送到服务器端，服务器端接收到信息后再把 Client1 的信息广播发送给所有的客户端。

1．服务器端的操作

（1）在服务器上建立一个 ServerSocket，ServerSocket 对象用于监听来自客户端的 Socket 连接，如果没有连接，它将一直处于等待状态。

```
ServerSocket ss = new ServerSocket(10000);
                          //创建一个 ServerSocket，监听客户端 Socket 的连接请求
while (true){ …           //采用不断循环的方式接收来自客户端的请求
```

（2）Socket accept()：如果接收到一个客户端 Socket 的连接请求，则该方法将返回一个与客户端 Socket 对应的 Socket。

```
Socket s = ss.accept();//接收到客户端 Socket 的请求，服务器端也对应产生一个 Socket
```

（3）使用 Socket 和客户端进行通信。

2. 客户端的操作

（1）创建连接到服务器、指定端口为 30000 的 Socket。

```
Socket s = new Socket("192.168.2.214" , 30000);//和服务器建立连接
```

（2）使用 Socket 和服务器进行通信。

项目 Exam9_13 就是一个使用 Socket 通信实现的简单的聊天室程序，包括服务器端程序 Server9_13 和 Android 客户端程序 Client9_13，感兴趣的读者可下载学习。注意：运行时要先运行服务器端，客户端的 IP 地址要根据实际情况修改。

9.6 蓝牙通信

9.6.1 蓝牙通信基础

蓝牙是一种支持设备短距离通信（一般在 10m 以内，且无阻隔媒介）的无线电技术。它能在包括移动电话、PDA、无线耳机、笔记本电脑、蓝牙打印机等众多设备之间进行无线信息交换。在开发时，蓝牙需要硬件支持，模拟器上不能模拟蓝牙，需要在真正的手机上进行蓝牙功能的测试，如果要实现数据的传输，那么至少需要两台带蓝牙功能的设备。Android 系统的手机一般都带有蓝牙功能，可以很方便地进行程序的调试。

Android 所有关于蓝牙开发的类都在 android.bluetooth 包下，一共有 8 个类，如图 9.25 所示。以下是建立蓝牙连接时所需要的一些基本类。

android.bluetooth 类

BluetoothAdapter
BluetoothClass
BluetoothClass.Device
BluetoothClass.Device.Major
BluetoothClass.Service
BluetoothDevice
BluetoothServerSocket
BluetoothSocket

图 9.25 bluetooth 的 8 个子类

（1）BluetoothAdapter 类：代表一个本地的蓝牙适配器，是所有蓝牙交互的入口点。利用它可以发现其他蓝牙设备，查询绑定的设备，使用已知的 MAC 地址创建 BluetoothDevice，建立一个 BluetoothServerSocket（作为服务器端）来监听来自其他设备的连接。其常用量和方法如表 9-7 和表 9-8 所示。

表 9-7 BluetoothAdapter 类的常用常量

方法	描述
int STATE_OFF	蓝牙已经关闭
int STATE_ON	蓝牙已经打开
int STATE_TURNING_OFF	蓝牙处于关闭过程中
int STATE_TURNING_ON	蓝牙处于打开过程中
int SCAN_MODE_CONNECTABLE	表明该蓝牙可以扫描其他蓝牙设备
int SCAN_MODE_CONNECTABLE_DISCOVERABLE	表明该蓝牙设备可以扫描其他蓝牙设备，也可以被其他蓝牙设备扫描到

续表

方法	描述
int SCAN_MODE_NONE	该蓝牙不能扫描以及被扫描
ACTION_STATE_CHANGED	蓝牙状态值发生改变
ACTION_SCAN_MODE_CHANGED	蓝牙扫描状态(SCAN_MODE)发生改变
ACTION_DISCOVERY_STARTED	蓝牙扫描过程开始
ACTION_DISCOVERY_FINISHED	蓝牙扫描过程结束
ACTION_LOCAL_NAME_CHANGED	蓝牙设备 Name 发生改变
ACTION_REQUEST_DISCOVERABLE	请求用户选择是否使该蓝牙被扫描
ACTION_REQUEST_ENABLE	请求用户选择是否打开蓝牙
ACTION_FOUND	蓝牙扫描时，扫描到任一远程蓝牙设备时，会发送此广播

表 9-8　BluetoothAdapter 类的常用方法

方法	描述
public static synchronized BluetoothAdapter getDefaultAdapter()	获得本设备的蓝牙适配器实例，若设备具备蓝牙功能，则返回 BluetoothAdapter 实例，否则返回 null 对象
public boolean enable()	打开蓝牙设备
public boolean disable()	关闭蓝牙设备
public boolean startDiscovery()	扫描蓝牙设备
public boolean cancelDiscovery()	取消扫描过程
public boolean isDiscovering()	是否正在处于扫描过程中
public boolean isEnabled ()	是否已经打开蓝牙
public String getName()	获取蓝牙设备 Name
public String getAddress()	获取蓝牙设备的硬件地址
public boolean setName (String name)	设置蓝牙设备的 Name
public Set<BluetoothDevice> getBondedDevices()	获取与本机蓝牙所有绑定的远程蓝牙信息
public static boolean checkBluetoothAddress(String address)	验证蓝牙设备 MAC 地址是否有效
public BluetoothDevice getRemoteDevice (String address)	根据蓝牙地址获取蓝牙设备
public String getState()	获取本地蓝牙适配器当前状态
public BluetoothServerSocket listenUsingRfcommWithServiceRecord (String name, UUID uuid)	创建一个正在监听的 RFCOMM 蓝牙端口

关于 UUID：如果正试图连接蓝牙串口，那么可以使用众所周知的 SPP UUID 00001101-0000-1000-8000-00805F9B34FB；如果正试图连接 Android 设备，那么要生成自己的专有 UUID，查询 RFCOMM 通道的服务记录的 UUID。

（2）BluetoothDevice 类：代表一个远端的蓝牙设备，使用它可请求远端蓝牙设备连接或者获取远端蓝牙设备的名称、地址、种类和绑定状态（其信息封装在 bluetoothsocket 中）。其常用常量与方法如表 9-9 和表 9-10 所示。

表 9-9　BluetoothDevice 类的常用常量

方法	描述
String ACTION_BOND_STATE_CHANGED	指明一个远程设备的连接状态的改变
String ACTION_FOUND	发现远程设备
String ACTION_NAME_CHANGED	指明一个远程设备的名称第一次找到，或者自从最后一次找到该名称已经开始改变

续表

方法	描述
int BOND_BONDED	表明蓝牙已经绑定
int BOND_BONDING	表明蓝牙正在绑定过程中
int BOND_NONE	表明没有绑定

表 9-10 BluetoothDevice 类的常用方法

方法	描述
public BluetoothSocketcreateRfcommSocketToServiceRecord (UUID uuid)	根据 UUID 创建并返回一个 BluetoothSocket
public String getAddress()	返回该蓝牙设备的硬件地址
public BluetoothClass getBluetoothClass()	获取远程设备的蓝牙类
public int getBondState()	获取远程设备的连接状态
public String getName()	获取远程设备的蓝牙名称
public String toString()	返回该蓝牙设备的字符串表达式

（3）BluetoothSocket 类：蓝牙通信分为服务器端和客户端，它们使用 BluetoothSocket 类的不同方法来获取数据，该类代表了一个蓝牙套接字的接口（类似于 TCP 中的套接字），是应用程序通过输入、输出流与其他蓝牙设备通信的连接点。其常用方法如表 9-11 所示。

表 9-11 BluetoothSocket 类的常用方法

方法	描述
public void close()	与服务器断开关闭
public void connect()	与服务器建立连接
public InputStream getInputStream()	获取输入流
public OutputStream getOutputStream()	获取输出流
public BluetoothDevice getRemoteDevice()	获取 BluetoothSocket 指定连接的那个远程蓝牙设备
public boolean isConnected()	是否与远程蓝牙设备建立连接

（4）BlutoothServerSocket 类：打开服务连接来监听可能到来的连接请求（属于 Server 端），为了连接两个蓝牙设备，必须有一个设备作为服务器打开一个服务套接字。当远端设备发起连接请求且已经连接到的时候，BluetoothServerSocket 类将会返回一个 BluetoothSocket。其常用方法如表 9-12 所示。

表 9-12 BluetoothServerSocket 类的常用方法

方法	描述
public void close()	马上关闭端口，并释放所有相关的资源
public BluetoothSocket accept()	返回一个已连接的 BluetoothSocket 类
public BluetoothSocket accept(int timeout)	返回一个指定了过时时间已连接的 BluetoothSocket 类

在服务器端，使用 BluetoothServerSocket 类来创建一个监听服务端口。当一个连接被 BluetoothServerSocket 所接收时，它会返回一个新的 BluetoothSocket 来管理该连接。在客户端，使用一个单独的 BluetoothSocket 类来初始化一个外接连接和管理该连接。

最常使用的蓝牙端口是 RFCOMM，它是被 Android API 支持的类型。RFCOMM 是一种面向连接、通过蓝牙模块进行数据流传输的方式，也被称为串行端口规范（Serial Port Profile，SPP）。

为了创建一个对准备好的新来的连接进行监听的 BluetoothServerSocket 类，可使用 BluetoothAdapter. listenUsingRfcommWithServiceRecord()方法。调用 accept()方法监听该连接的请求。使用 BluetoothSocket 类管理该连接，如果不再需要接收连接，则调用在 BluetoothServerSocket 类下的 close()

方法，会马上放弃外界操作并关闭服务器端口。更多的功能可以参考 Android API 进行学习。

9.6.2 蓝牙通信实现

要实现蓝牙操作，一般可按照下面几个流程进行。

1. 获取本地蓝牙

```
BluetoothAdapter mAdapter= BluetoothAdapter.getDefaultAdapter();
```

2. 打开、关闭蓝牙

可以在系统设置中开启蓝牙，也可以在应用程序中启动蓝牙功能，有以下两种方法。

（1）直接调用函数 enable()打开蓝牙设备。

例如：

```
boolean result = mBluetoothAdapter.enable();
```

（2）以系统 API 的方式打开蓝牙设备，该方式会弹出一个对话框样式的 Activity 供用户选择是否打开蓝牙设备。例如：

```
if (!mBluetoothAdapter.isEnabled()) //如果未打开蓝牙功能，则用下面的语句打开
{ Intent intent = new Intent(BluetoothAdapter.ACTION_REQUEST_ENABLE);
    startActivityForResult(intent, REQUEST_OPEN_BT_CODE); } /*以 Dialog 样式
    显示一个 Activity，可以在 onActivityResult()方法下处理返回值*/
```

注意：如果蓝牙已经开启，则不会弹出该 Activity 对话框。在目前的 Android 手机中，是不支持在飞行模式下开启蓝牙的。如果蓝牙已经开启，那么蓝牙的开关状态会随着飞行模式的状态而发生改变。

```
//打开本机的蓝牙功能
Intent discoveryIntent=new Intent(BluetoothAdapter .ACTION_REQUEST_DISCOVERABLE);
discoverableIntent.putExtra(BluetoothAdapter.EXTRA_DISCOVERABLE_DURATION, 300);
                                        //设置持续时间（最多 300 秒）
mAdapter.disable();                     //关闭蓝牙
```

蓝牙功能开启后，即可查找周边存在的蓝牙设备。

3. 查找设备

使用 BluetoothAdapter 的 startDiscovery()方法搜索蓝牙设备，startDiscovery()方法是一个异步方法，调用后会立即返回。该方法会进行对其他蓝牙设备的搜索，该方法调用后，搜索过程实际上是在 System Service 中进行的，所以可以调用 cancelDiscovery()方法来停止搜索（该方法可以在未执行 discovery 请求时调用）。

请求 Discovery 后，系统开始搜索蓝牙设备，在这个过程中，系统会发送以下三个广播。

ACTION_DISCOVERY_START：开始搜索。

ACTION_DISCOVERY_FINISHED：搜索结束。

ACTION_FOUND：找到设备，这个 Intent 中包含两个 extra fields——EXTRA_DEVICE 和 EXTRA_CLASS，分别包含 BluetoothDevice 和 BluetoothClass。

可以注册相应的 BroadcastReceiver 来接收响应的广播，以便实现某些功能，例如：

```
//创建一个接收 ACTION_FOUND 广播的 BroadcastReceiver
private final BroadcastReceiver mReceiver = new BroadcastReceiver() {
   public void onReceive(Context context, Intent intent) {
       String action = intent.getAction();
       //发现设备
```

```
            if (BluetoothDevice.ACTION_FOUND.equals(action)) {
                //从 Intent 中获取设备对象
                BluetoothDevice device = intent.getParcelableExtra
                    (BluetoothDevice.EXTRA_DEVICE);
                //将设备名称和地址放入 array adapter，以便在 ListView 中显示
                mArrayAdapter.add(device.getName() + "\n" + device.getAddress());
            }       }  };
    //注册 BroadcastReceiver
    IntentFilter filter = new IntentFilter(BluetoothDevice.ACTION_FOUND);
    registerReceiver(mReceiver, filter); //不要忘记解除绑定
```

4．建立连接

两个蓝牙设备之间进行连接时，要有一个服务器端和客户端，这两个设备要在同一个 RFCOMM channel 下且拥有一个连接的 BluetoothSocket，才可能建立连接。

服务器设备与客户端设备获取 BluetoothSocket 的途径是不同的。服务器设备是通过 accepted 的一个 incoming connection 来获取的，而客户端设备则是通过打开一个到服务器的 RFCOMM channel 来获取的。

（1）服务器端的实现。

首先，调用 listenUsingRfcommWithServiceRecord(String, UUID)方法来获取 BluetoothServerSocket 对象。

```
    BluetoothServerSocket serverSocket = mAdapter.listenUsingRfcommWith-
                                ServiceRecord(serverSocketName,UUID);
```

其次，调用 accept()方法来监听可能到来的连接请求，当监听到后，返回一个连接上的蓝牙套接字 BluetoothSocket。

```
    serverSocket.accept();
```

最后，在监听到一个连接以后，调用 close()方法来关闭监听程序。

```
    serverSocket. close ();
```

例如：

```
    private class AcceptThread extends Thread {
        private final BluetoothServerSocket mmServerSocket;
        public AcceptThread() {
            BluetoothServerSocket tmp = null;
            try {
                tmp = mBluetoothAdapter.listenUsingRfcommWithServiceRecord(NAME, MY_UUID);
            } catch (IOException e) { }
            mmServerSocket = tmp;       }
        public void run() {
            BluetoothSocket socket = null;
            while (true) {
                try {
                    socket = mmServerSocket.accept();
                } catch (IOException e) {
                    break;              }
                //连接被接收
                if (socket != null) {
                    manageConnectedSocket(socket);
                    mmServerSocket.close();
```

```
            break;     }        }     }
    public void cancel() {
        try {
            mmServerSocket.close();
        } catch (IOException e) { }      } }
```

（2）客户端的实现。

首先，通过 BluetoothDevice 对象来获取 BluetoothSocket 并初始化连接，使用 BluetoothDevice 的方法 createRfcommSocketToServiceRecord(UUID)来获取 BluetoothSocket。

```
BluetoothSocket clienSocket=dcvice. createRfcommSocketToServiceRecord(UUID);
```

其次，调用 connect()方法。如果远端设备接收了该连接，则在通信过程中将共享 RFFCOMM 信道，并使用 connect()方法返回。

```
clienSocket.connect();
```

最后，数据传输完成后调用 close()方法关闭连接。

例如：

```
private class ConnectThread extends Thread { private final BluetoothSocket mmSocket;
    private final BluetoothDevice mmDevice;
    public ConnectThread(BluetoothDevice device) {
        BluetoothSocket tmp = null;
        mmDevice = device;
        try {
            tmp = device.createRfcommSocketToServiceRecord(MY_UUID);
        } catch (IOException e) { }
        mmSocket = tmp;      }
    public void run() {
        mBluetoothAdapter.cancelDiscovery();
        try {
            mmSocket.connect();
        } catch (IOException connectException) {
            try {
                mmSocket.close();
            } catch (IOException closeException) { }
            return;          }
        manageConnectedSocket(mmSocket);      }
    public void cancel() {
        try {
            mmSocket.close();
        } catch (IOException e) { }      } }
```

5. 数据传递

通过以上操作，已经建立了 BluetoothSocket 连接，数据传递使用了流的形式。

当设备连接上以后，每个设备都拥有各自的 BluetoothSocket，即可实现设备之间数据的共享。

首先，调用 getInputStream()和 getOutputStream()方法获取输入输出流，再调用 read(byte[]) 和 write(byte[])方法读取或者写数据。

例如：

```
private class ConnectedThread extends Thread {
    private final BluetoothSocket mmSocket;
```

```
    private final InputStream mmInStream;
    private final OutputStream mmOutStream;
    public ConnectedThread(BluetoothSocket socket) {
        mmSocket = socket;
        InputStream tmpIn = null;
        OutputStream tmpOut = null;
        try {
            tmpIn = socket.getInputStream();
            tmpOut = socket.getOutputStream();
        } catch (IOException e) { }
        mmInStream = tmpIn;
        mmOutStream = tmpOut;     }
    public void run() {
        byte[] buffer = new byte[1024];
        int bytes;
        while (true) {
            try {
                //从 InputStream 读数据
                bytes = mmInStream.read(buffer);
                mHandler.obtainMessage(MESSAGE_READ, bytes, -1, buffer)
                        .sendToTarget();
            } catch (IOException e) {
                break;           }         }     }
    public void write(byte[] bytes) {
        try {
            mmOutStream.write(bytes);
        } catch (IOException e) { }     }
    public void cancel() {
        try {
            mmSocket.close();
        } catch (IOException e) { }     }  }
```

6. 修改 AndroidManifest.xml 文件配置权限

注意：在使用这些类时要修改 AndroidManifest.xml 文件的配置权限，添加如下语句：

```
<uses-permission android:name="android.permission.BLUETOOTH_ADMIN" />
<uses-permission android:name="android.permission.BLUETOOTH" />
```

9.6.3 蓝牙通信实例

实例 9-11：蓝牙通信实例

新建一个项目，项目命名为 Exam9_13，编程实现蓝牙设备的开启、关闭，并发现配对设备列表。
（1）定义布局文件 activity_main.xml，代码如下：

```
…
<LinearLayout
      android:layout_width="match_parent"
      android:layout_height="match_parent"
      android:orientation="vertical" >
      <Button
         android:id="@+id/buton"
         android:layout_width="wrap_content"
```

```
            android:layout_height="wrap_content"
            android:onClick="on"I      //触发单击事件
            android:text="打开蓝牙" />
        <Button
            …
            android:onClick="visible"
            android:text="寻找配对" />
        <Button
            …
            android:onClick="list"
            android:text="设备列表" />
        <Button
            …
            android:onClick="off"
            android:text="关闭蓝牙" />
        <ListView
            android:id="@+id/listView1"
            android:layout_width="match_parent"
            android:layout_height="wrap_content"
            android:visibility="visible" >
        </ListView>
    </LinearLayout>
  </ScrollView>
</RelativeLayout>
```

（2）定义 Activity 文件 MainActivity.java，代码如下：

```
 public class MainActivity extends AppCompatActivity {
 private Button On,Off,Visible,list;
private BluetoothAdapter Badapter;
private Set<BluetoothDevice> pairedDevices;
private ListView listView;
@Override
protected void onCreate(Bundle savedInstanceState) {
    super.onCreate(savedInstanceState);
    setContentView(R.layout.activity_main);
    Button  On = (Button)findViewById(R.id.buton);
    Button  Off = (Button)findViewById(R.id.butoff);
    Button  Visible = (Button)findViewById(R.id.butvisible);
    Button  list = (Button)findViewById(R.id.butlist);
    listView= (ListView)findViewById(R.id.listView1);
    Badapter = BluetoothAdapter.getDefaultAdapter();       }
public void on(View view){//单击事件
    if (!Badapter.isEnabled()) {
        Intent turnOn = new Intent(BluetoothAdapter.ACTION_REQUEST_ENABLE);
        startActivityForResult(turnOn, 0);
        Toast.makeText(getApplicationContext(),"打开蓝牙",
                Toast.LENGTH_LONG).show();  }
    else{
        Toast.makeText(getApplicationContext(),"设备已打开",
                Toast.LENGTH_LONG).show();          }    }
public void list(View view){
    pairedDevices = Badapter.getBondedDevices();
    ArrayList list = new ArrayList();
```

```
        for(BluetoothDevice bt : pairedDevices)
           list.add(bt.getName());
        Toast.makeText(getApplicationContext(),"显示配对设备",
              Toast.LENGTH_LONG).show();
        final ArrayAdapter adapter = new ArrayAdapter
              (this,android.R.layout.simple_list_item_1, list);
        listView.setAdapter(adapter);    }
     public void off(View view){
        Badapter.disable();
        Toast.makeText(getApplicationContext(),"关闭蓝牙" ,
              Toast.LENGTH_LONG).show();        }
     public void visible(View view){
        Intent getVisible =new Intent(BluetoothAdapter.ACTION
     REQUEST_DISCOVERABLE);
        startActivityForResult(getVisible, 0);    }}
```

（3）修改 AndroidManifest.xml 文件配置权限，添加如下语句：

```
<uses-permission android:name="android.permission.BLUETOOTH_ADMIN" />
<uses-permission android:name="android.permission.BLUETOOTH" />
```

保存所有文件，运行结果如图 9.26 所示，注意最上面一行蓝牙标识的变化。

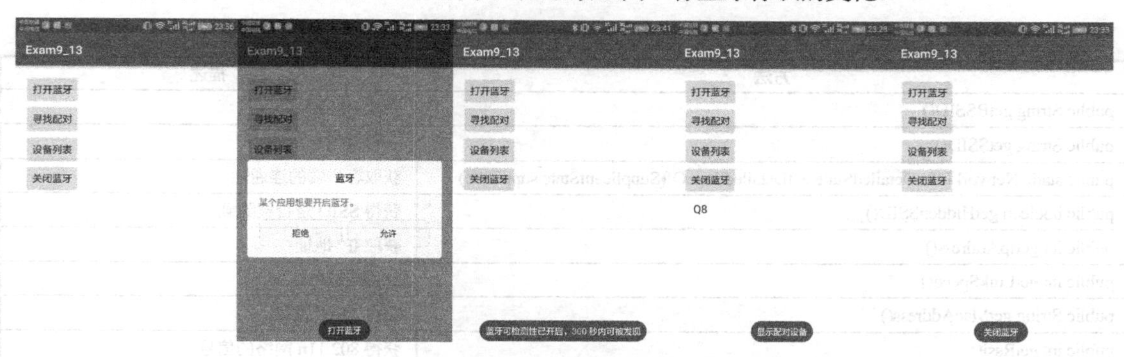

图 9.26　Exam9_13 运行结果

9.7　Wi-Fi 通信

1．WiFi 通信基础

Android 中提供了 android.net.wifi 包对 Wi-Fi 进行操作，主要包括以下几个类。

（1）ScanResult 类：该类主要通过 Wi-Fi 硬件的扫描来获取一些周边的 Wi-Fi 热点的信息，包括接入点的地址、接入点的名称、身份认证、频率、信号强度等信息。其具体的变量如表 9-13 所示。

表 9-13　ScanResult 类常用的变量

方法	描述
public String BSSID	Wi-Fi 热点的地址
public String SSID	网络名称
public String capabilities	描述 Wi-Fi 热点的认证、密钥管理以及加密方案等相关信息
public int frequency	客户端与 Wi-Fi 热点通信信道的频率
public int level	主要用来判断网络连接的优先数

这里只提供了一个方法将获得信息变成字符串，即 public String toString()。

（2）WifiConfiguration 类：该类主要用来进行 Wi-Fi 网络的配置。其常用的变量如表 9-14 所示。

表 9-14　WifiConfiguration 类常用的变量

方法	描述
public String BSSID	当设置好后，这个网络配置入口只能在指定 BSSID 的 AP 时才调用
public String SSID	设置该网络的 SSID
public boolean hiddenSSID	隐藏的 SSID，即该网络不对 SSID 进行广播
public int networkId	客户端与 Wi-Fi 热点通信信道的频率
public int level	这个网络配置入口的 ID
public int priority	配置的优先级
public int status	当前配置状态

这里提供的方法也是 public String toString()。

（3）WifiInfo 类：Wi-Fi 已经连接成功以后，可以通过这个类获得一些已经连通的 Wi-Fi 连接的信息，获取当前连接的信息，包括接入点、网络连接状态、隐藏的接入点、IP 地址、连接速度、MAC 地址、网络 ID、信号强度等信息。其具体方法如表 9-15 所示。

表 9-15　WifiInfo 类常用方法

方法	描述
public String getBSSID()	获取 BSSID
public String getSSID()	获得 SSID
public static NetworkInfo.DetailedState getDetailedStateOf(SupplicantState suppState)	获取客户端的连通性
public boolean getHiddenSSID()	获得 SSID 是否被隐藏
public int getIpAddress()	获取 IP 地址
public int getLinkSpeed()	获得连接的速度
public String getMacAddress()	获得 MAC 地址
public int getRssi()	获得 802.11n 网络的信号
public SupplicantState getSupplicanState()	返回具体客户端状态的信息
public String toString()	转换为字符串
public int getnetworkId()	获得通信信道的频率

（4）WifiManager 类：该类用来管理 Wi-Fi 连接，这是最重要的一个类。其常量如表 9-16 所示，方法如表 9-17 所示。

表 9-16　WifiManager 类常用常量

方法	描述
public static final int WIFI_STATE_DISABLING	0 表示网卡正在关闭
public static final int WIFI_STATE_DISABLED	1 表示网卡不可用
public static final int WIFI_STATE_ENABLING	2 表示网卡正在打开
public static final int WIFI_STATE_ENABLED	3 表示网卡可用
public static final int WIFI_STATE_UNKNOWN	4 表示未知网卡状态

表 9-17 WifiManager 类常用方法

方法	描述
public int addNetwork(WifiConfiguration config)	通过获取的网络连接状态信息来添加网络
public static int compareSignalLevel(int rssiA, int rssiB)	对比连接 A 和连接 B
public static int calculateSignalLevel (int rssi , int numLevels)	计算信号的等级
public boolean disableNetwork(int netId)	使一个网络连接失效
public boolean disconnect()	断开连接
public boolean enableNetwork(int netId, Boolean disableOthers)	连接一个连接
public WifiManager.WifiLock createWifiLock (String tag)	创建一个新的 Wi-Fi 锁，锁定当前的 Wi-Fi 连接
public WifiManager.WifiLock createWifiLock (int lockType, String tag)	创建一个指定类型的 Wi-Fi 锁，锁定当前的 Wi-Fi 连接
public List<WifiConfiguration> getConfiguredNetworks()	获取网络连接的状态
public WifiInfo getConnectionInfo()	获取当前连接的信息
public List<ScanResult> getScanResulats()	获取扫描测试的结果
public DhcpInfo getDhcpInfo()	获取 DHCP 的信息
public int getWifiState()	获取一个 Wi-Fi 接入点是否有效
public boolean isWifiEnabled()	判断一个 Wi-Fi 连接是否有效
public boolean pingSupplicant()	ping 一个连接，判断是否能连通
public boolean ressociate()	连接没有准备好，是否还要连通
public boolean reconnect()	如果连接准备好了，是否连通
public boolean removeNetwork()	是否移除某一个网络
public boolean saveConfiguration()	是否保留一个配置信息
public boolean setWifiEnabled()	是否使一个连接有效
public boolean startScan()	是否开始扫描
public int updateNetwork(WifiConfiguration config)	更新一个网络连接的信息

更多常量和方法的使用可以参考 android.net.wifi 包的 API，参考网址为 http://developer.android.com/reference/android/net/wifi/package-summary.html。

2．对 Wi-Fi 网卡的基本操作

对 Wi-Fi 网卡进行操作时，需要通过 WifiManger 对象来进行，获取该对象，方法如下：

```
WifiManger wifiManger = (WifiManger)Context.getSystemService(Service. WIFI_SERVICE);
wifiManger.setWifiEnabled(true);          //打开 Wi-Fi 网卡
wifiManger.setWifiEnablee(false);         //关闭 Wi-Fi 网卡
wifiManger.getWifiState();                //获取网卡 Wi-Fi 的当前状态
wifiManger.addNetwork();                  //添加一个配置好的网络连接
wifiManger.calculateSignalLevel();        //计算信号的强度
wifiManger.compareSignalLevel();          //比较两个信号的强度
wifiManger.createWifiLock();              //创建一个 Wi-Fi 锁
wifiManger.disconnect();                  //从接入点断开
wifiManger.updateNetwork();               //更新已经配置好的网络
```

要想对手机的 Wi-Fi 网卡进行操作，需要在 Manifest.xml 中配置相应的权限，例如：

```
<uses-permission                          //修改网络状态的权限
android:name="android.permission.CHANGE_NETWORK_STATE"></uses-permission>
<uses-permission                          //修改 Wi-Fi 状态的权限
```

```
android:name="android.permission.CHANGE_WIFI_STATE"></uses-permission>
<uses-permission                          //访问网络权限
android:name="android.permission.ACCESS_NETWORK_STATE"></uses-permission>
<uses-permission                          //访问 Wi-Fi 权限
android:name="android.permission.ACCESS_WIFI_STATE"></uses-permission>
```

3. Android Wi-Fi 开发实例

WiFi 与硬件设备有关，在模拟机上看不到真正的效果，需要在真机上才能正常运行。

项目 Exam9_14 是一个关于 Wi-Fi 通信的实例，实现设备 Wi-Fi 的开启、断开、扫描网络等功能，限于篇幅，这里代码没有给出，感兴趣的读者可以下载学习。

本章小结

本章着重介绍了几种常见网络通信技术：WebView 组件、HTTP 通信技术、Socket 通信技术、蓝牙和 Wi-Fi 通信技术。其中，最简单的就是 WebView 组件，蓝牙和 WiFi 通信技术要借助硬件设备运行，在模拟机上不能正常运行。

习题

（1）利用 WebView 组件，打开一个指定的网页，给出具体步骤和关键代码。

（2）利用 WebView 组件加载一个 HTML 文件和一个 JSP 文件，操作有区别吗？区别在哪里？

（3）HttpURLConnection 利用 GET 方式传递数据给 JSP 文件的步骤有哪些？写出关键代码。

（4）利用 Volley 框架的 POST 方式传递数据给 JSP 文件的步骤有哪些？写出关键代码。

（5）Socket 通信时要建立服务器端和客户端，关于端口的部分是如何设计的？分别写出服务器端和客户端关于端口的语句。

（6）编程实现两部带有蓝牙功能的手机的简单数据传递。

（7）编程实现如下功能：登录页面，输入的用户名和密码信息来自 MySQL 数据库对应的数据表，如果是合法的用户名和密码，则跳转到投票列表页面，否则弹出用户名和密码不正确的提示信息，如图 9.27 所示。

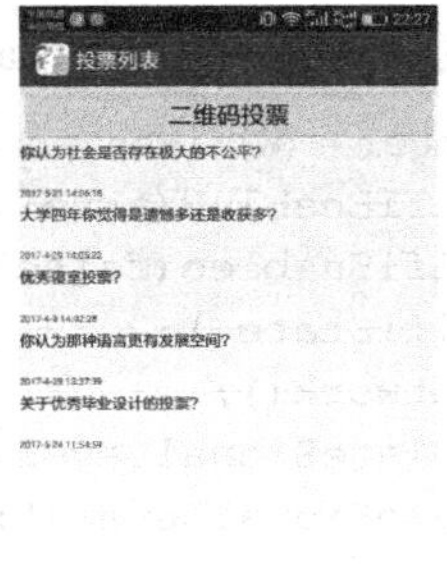

图 9.27　实现的页面

第 10 章　投票系统 APP 端设计

学习目标：

- 了解 Android　APP 开发流程。
- 掌握 Android　UI 组件的使用。
- 掌握 Android 事件处理的运用。
- 掌握 Android 与 Web 程序的通信。

前面介绍了 Android 的基本组件和事件处理、Intent 组件、数据通信等知识，本章就在这些知识的基础上开发一个简单的手机投票 APP。

10.1　需求分析

手机投票的 APP 是投票管理系统的一部分，投票管理系统包含两部分：Android 服务器端和客户端。其中，服务器端采用 JavaWeb 技术，能进行票数统计、添加投票、删除投票、添加投票用户、获得指定投票的二维码，以方便用户投票；Android 客户端能够浏览投票信息并进行投票。

10.1.1　系统基本需求

（1）系统用户分为管理员用户（服务器端）与普通用户（客户端）。管理员用户对投票内容进行管理，普通用户对投票信息进行浏览与投票。

（2）管理员用户使用网页端对投票内容进行浏览、删除，以及生成二维码。管理员对过期无效的投票进行清除。为了方便用户投票，可生成二维码进行公示，方便用户进行扫码投票。

（3）管理员可以对投票信息进行管理，增加、删除投票信息。

（4）管理员可以添加投票的普通用户，这样可方便管理员控制哪些用户能够进行投票。

（5）普通用户在管理员用户添加到系统后方能进行安卓端登录与投票；普通用户通过登录验证后，可通过手动查找进行投票。同时，为了便捷快速地找到指定的投票内容，也可通过二维码进行投票。

（6）服务器端录入投票信息，客户端的投票信息都保存于 MySQL 数据库中，便于统计。

服务器端和 Android 客户端功能需求分析如图 10.1 和图 10.2 所示。

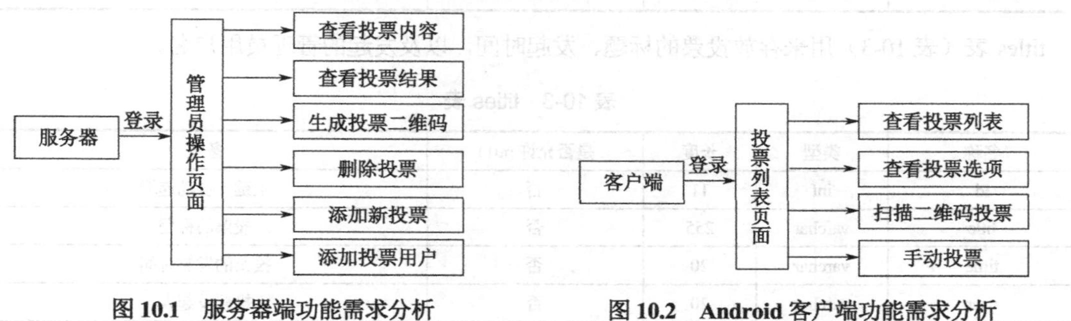

图 10.1　服务器端功能需求分析　　图 10.2　Android 客户端功能需求分析

10.1.2 系统开发参数

（1）系统使用技术：JavaSE、JavaEE、JavaScript、JQuery、JSON（Gson）、MySQL、Android 组件以及网络通信技术等。

（2）硬件环境：酷睿 i3-2310、2.1GHz 双核 CPU 计算机，内存 4GB、小米红米 1s 手机(Android 4.4.2)。

（3）软件环境：Microsoft Windows 7、MyEclipse、Tomcat 8.5、MySQL 5.5.36、Navicat for MySQL、JDK 1.8。

10.2 系统设计

10.2.1 数据库的设计与实现

1．创建数据库

启动 MySQL 服务，打开 Navicat for MySQL 并连接数据库服务端，新建数据库，数据库名称为“vote”，将字符集设置为“CHARSET=utf-8”，防止后续存取中文时出现乱码情况。

2．数据字表

adminuser 表（表 10-1）用来存放管理员的用户名和密码。

表 10-1 adminuser 表

名称	类型	长度	是否允许 null	备注
id	int	11	否	主键；自动递增
name	varchar	20	否	管理员用户名
password	varchar	20	否	管理员密码

user 表（表 10-2）用来存放普通用户的用户名和密码。

表 10-2 user 表

名称	类型	长度	是否允许 null	备注
id	int	11	否	主键；自动递增
username	varchar	20	否	用户名
password	varchar	20	否	密码

titles 表（表 10-3）用来存放投票的标题、发起时间，以及发起的管理员用户名。

表 10-3 titles 表

名称	类型	长度	是否允许 null	备注
id	int	11	否	主键；自动递增
title	varchar	255	否	投票的标题
time	varchar	20	否	投票的发起时间
user	varchar	20	否	投票发起人

options 表（表 10-4）用来存放投票的选项、投票 title 的 ID，以及该选项的内容和该选项的票数。title_id 来自 titles 表中的外键，并且设置为级联更新与级联删除，方便数据的更新和删除操作。

表 10-4 options 表

名称	类型	长度	是否允许null	备注
id	int	11	否	主键；自动递增
title_id	int	10	否	投票的标题ID（外键）
content	varchar	255	否	选项的内容
count	int	5	否	选项的票数

数据库的设计是非常重要的，数据库的设计关乎系统的性能和系统的可拓展性，在附带的代码中有 vote.sql 文件，在 Navicat for MySQL 中导入并执行此文件可以自动生成数据库及表。

10.2.2 服务器端设计与实现

后台管理通过 Web 站点的方式来实现，包含登录验证、查看投票内容及投票结果、生成投票二维码、删除投票、添加新投票、添加投票用户的功能。在 MyEclipse 中新建“Web Project”，项目名称是 VoteServer，服务器采用 Tomcat 7.0。项目结构如图 10.3 所示。

com.hnist.voteserver.dao 包中存放数据库操作的类。

com.hnist.voteserver.DBUtils 包中存放数据库的工具类。

com.hnist.voteserver.entity 包中存放实体类，对应数据库中的表。

com.hnist.voteserver.servlet 包中存放接收请求的 servlet。

Web.xml 文件中存放对 servlet 的配置。

c3p0-config.xml 文件存放配置的数据源参数，以方便连接参数的修改和数据库的移植，内容如下：

```
<?xml version="1.0" encoding="UTF-8"?>
<c3p0-config>
 <named-config name="voteapp">
 <property name="user">root</property>    \\注意，这里的用户名要修改为自己的用户名
 <property name="password">mysql</property>\\注意，这里的密码要修改为自己的密码
 ...
</c3p0-config>
```

由于本书的重点不是讨论服务器端的建设，所以这里不再详细介绍，读者可以把附带代码中的 VoteServer 文件夹复制到 C:\Tomcat 7.0\webapps 中，启动 Tomcat，这样服务器端就安装完成了。打开浏览器，输入 http://localhost:8080/VoteServer/，如图 10.4 所示的登录界面，输入测试的用户名“admin”和密码“123456”，单击“登录”按钮，进入如图 10.5 所示的后台主界面。

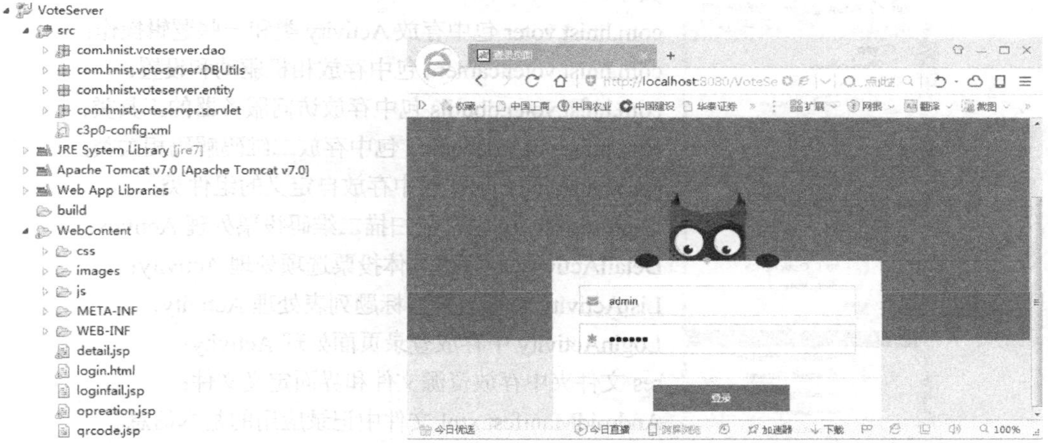

图 10.3 服务器端项目结构　　图 10.4 服务器端登录界面

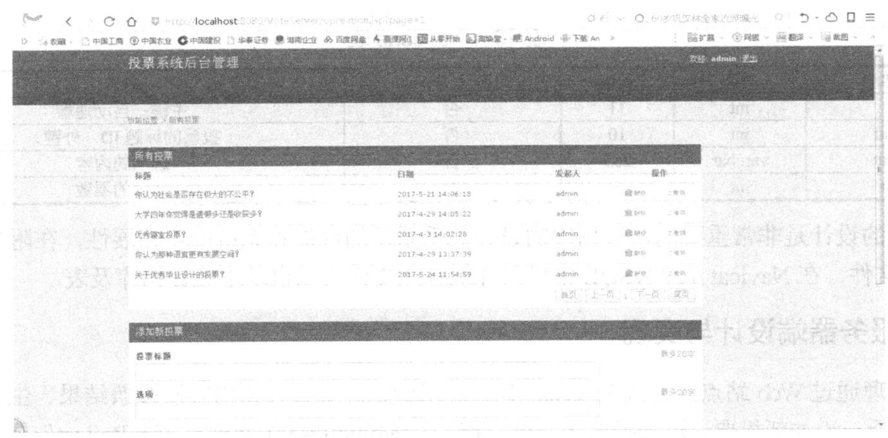

图 10.5　服务器端后台主界面

在后台主界面中可以添加、删除投票信息，增加投票用户，单击投票标题以查看投票结果，读者可自行测试，这里不再赘述。

10.2.3　Android 客户端设计与实现

用户通过在安卓设备上安装此 APP，登录后能够对后台发起的投票进行浏览；可通过单击相应的标题进入投票界面；还可以通过扫描二维码进入投票界面，以方便快速投票。

1．界面设计

根据需求分析设计如下几个页面：登录页面、投票列表页面、投票页面、二维码投票页面、扫码成功对话框页面，代码略。

2．功能设计

在 Android 客户端开发过程中，接收 Android 客户端请求的 servlet 放在后台服务器的项目中，只要服务器启动，后台和 Android 客户端都能进行连接通信，以减少不必要的工作量。Android 客户端的项目结构如图 10.6 所示。

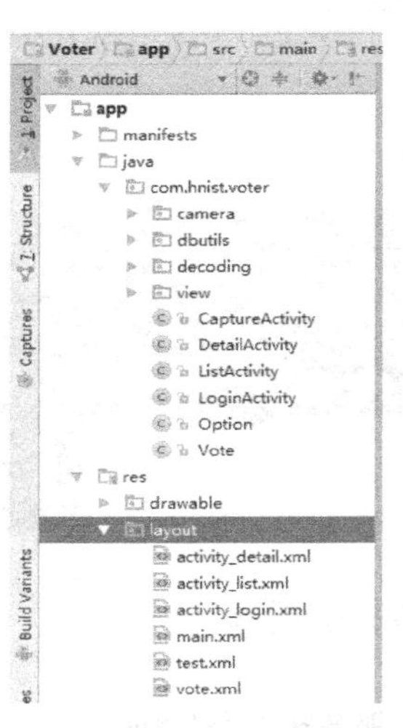

其中：

com.hnist.voter 包中存放 Activity 类和一些逻辑操作；
com.hnist.voter.camera 包中存放相机驱动和设置；
com.hnist.voter.dbutils 包中存放访问服务器的工具类；
com.hnist.voter.decoding 包中存放二维码解码相关类；
com.hnist.voter.view 包中存放自定义的组件类；
CaptureActivity 中存放扫描二维码投票处理 Activity；
DetailActivity 中存放具体投票选项处理 Activity；
ListActivity 中存放投票标题列表处理 Activity；
LoginActivity 中存放登录页面处理 Activity；
res 文件夹中存放资源文件和界面定义文件；
AndroidManifest.xml 文件中记录应用的基本信息。

图 10.6　Android 客户端项目结构

为了快速便捷地测试，所有操作都在真机上测试。由于涉及客户端服务器通信，还需搭建本地网络。主机在获得IP地址的情况下，开启Wi-Fi，再将手机通过Wi-Fi连接到主机上，客户端访问服务器使用的IP即为主机地址。

由于基于网络通信，项目中多处与服务器进行了通信，为了避免代码冗余，提高程序的可移植性，将网络通信编写为工具类HttpUtil.java，详细代码见本书附带的代码，要注意的是，代码中第11行中的IP地址要根据实际情况来填写。

```
public static final String BASE_URL = "http://192.168.0.205:8080/VoteServer/";
```

在AndroidManifest.xml文件中配置相应的权限，例如：

```
<uses-permission android:name="android.permission.INTERNET"></uses-permission>
<uses-permission android:name="android.permission.CAMERA"></uses-permission>
<uses-permission android:name="android.permission.WRITE_EXTERNAL_STORAGE" />
<uses-feature android:name="android.hardware.camera" />
<uses-feature android:name="android.hardware.camera.autofocus" />
<uses-permission android:name="android.permission.VIBRATE" />
<uses-permission android:name="android.permission.FLASHLIGHT" />
<application
…
     <activity
          android:name=".LoginActivity"        //程序入口，主LoginActivity
          …
       </activity>
       <activity
          android:name=".ListActivity"          //定义ListActivity
          android:label="@string/title_activity_list" >
       </activity>
       <activity
          android:name=".DetailActivity"        //定义DetailActivity
          android:label="@string/title_activity_detail" >
       </activity>
       <activity
          android:name=".CaptureActivity"       //定义CaptureActivity
          …
```

3. Android客户端登录功能的实现

Android用户启动应用之后，执行登录操作，对用户的身份进行验证，执行流程如图10.7所示，运行界面如图10.8所示。

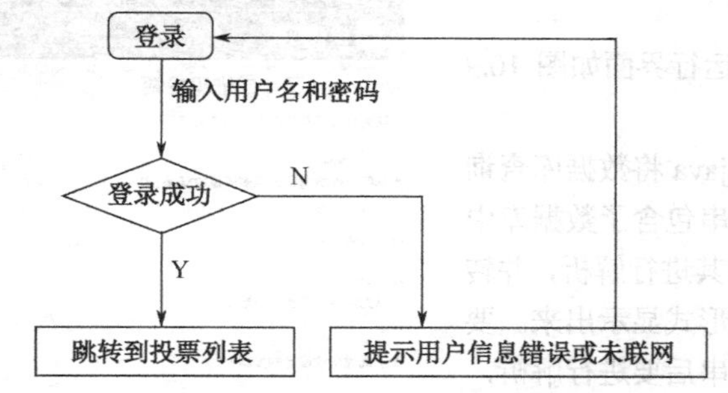

图10.7 Android客户端登录流程

图10.8 Android客户端登录界面

用户启动 Android 客户端之后会启动 LoginActivity，输入用户名和密码之后单击“登录”按钮，通过验证后会跳转到 ListActivity。这里新创建了一个线程来访问服务器，所以需要在单击“登录”按钮以后新建一个线程。

LoginActivity 的关键代码如下：

```
loginButton.setOnClickListener(new OnClickListener() {  // “登录”按钮监听
public void onClick(View v) {
    nameStr = loginName.getText().toString().trim();      // 取得编辑框用户名
    passStr = loginPass.getText().toString().trim();      // 取得编辑框密码
    if (nameStr.equals("") || passStr.equals("")) {     // 判断是否为空
Toast.makeText(LoginActivity.this, "用户名或密码不能为空", 0).show();}
        new Thread(new Runnable() {// 开启一个 HTTP 线程
            @Override
            public void run() {
            result = LoginActivity.this.query(nameStr, passStr);}//调用 query()
        }).start();
        try {                   // 等待数据传回
            Thread.sleep(2000);
        } catch (InterruptedException e) {
            e.printStackTrace();           }
       if ("".equals(result)) {// 判断用户名和密码是否正确，或者网络是否正常
            isLogin = false;
Toast.makeText(LoginActivity.this, "登录失败，请检查网络或者用户名密码是否正确！",
0).show();} else {
            isLogin = true;
            // 定义 Intent，登录成功，跳转到投票列表 ListActivity
            Intent it = new Intent(LoginActivity.this,  ListActivity.class);
            it.putExtra("votes", result); // 携带参数跳转到 ListActivity
            LoginActivity.this.startActivity(it);}}});
private String query(String nameStr, String passStr) {//调用 query 方法
// 设置查询参数
String queryString = "name=" + nameStr + "&pass=" + passStr + "&date="
        + new Date().getTime();
String url = HttpUtil.BASE_URL + "UserLoginservlet?" + queryString;
// 返回查询结果
return HttpUtil.queryStringForGet(url);}
```

4. Android 客户端投票列表显示功能的实现

用户通过验证之后，列表中会显示所有投票，运行界面如图 10.9 所示。

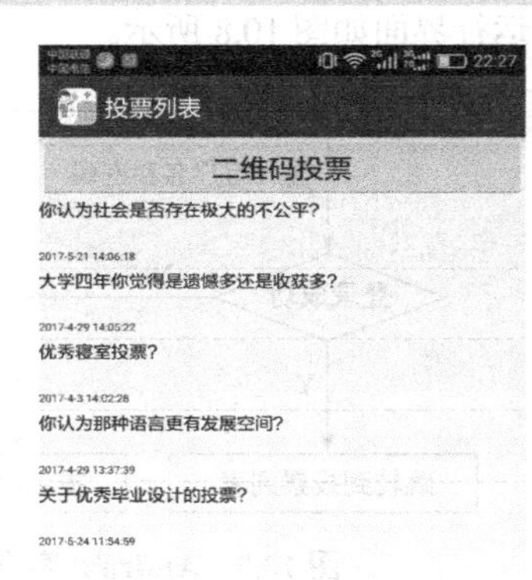

图 10.9　列表显示所有投票

用户通过验证之后，服务器的 UserLoginservlet.java 将数据库查询获得的结果集转换为 JSON 字符串作为响应，字符串包含了数据库中的投票信息，Android 客户端收到响应字符串后，对其进行解析，并转化为一个 List 集合，以便将数据以 ListView 组件的形式显示出来。要在项目中导入相应的 JAR 包，Android 端收到字符串后要进行解析，Android 中解析 JSON 时会使用 Google 的 Gson 来解析，需要在 Android 项目中导入“gson-2.1.jar”。

UserLoginservlet 的关键代码如下:

```
// 将数据转换成 JSON 格式，发送到 Android 客户端
JSONArray jsonArr = JSONArray.fromObject(titlesList);
```

ListActivty 的关键代码如下：

```
//将 JSON 数据转换成 List，再转换成适配器中的 List
public ArrayList<HashMap<String, String>> getData() {
ArrayList<HashMap<String,  String>>  list  =  new  ArrayList<HashMap<String,
String>>();
HashMap<String, String> map = null;
Intent it = super.getIntent();
voteStr = it.getStringExtra("votes");    // 接收数据
// 获得 titlesList
Gson g = new Gson();
voteList = (List<Vote>) g.fromJson(voteStr,new TypeToken<List<Vote>>() {
        }.getType());
for (Vote vote : voteList) {
    map = new HashMap<String, String>();
    map.put("id", vote.getId() + "");
    map.put("title", vote.getTitle());
    map.put("time", vote.getTime());
    list.add(map);   }
return list;}......
```

将 ArrayList<HashMap<String, String>>适配到 ListView 组件中，关键代码如下：

```
// 定义 ListView 适配器
SimpleAdapter simpleAdapter = new SimpleAdapter(this, this.getData(),
R.layout.vote, new String[] { "id", "title", "time" },
new int[] { R.id.id, R.id.title, R.id.time });
titleListView.setAdapter(simpleAdapter);......
```

5. Android 客户端手动投票的实现

用户单击投票标题，进入投票界面进行投票，执行流程如图 10.10 所示，运行界面如图 10.11 所示。

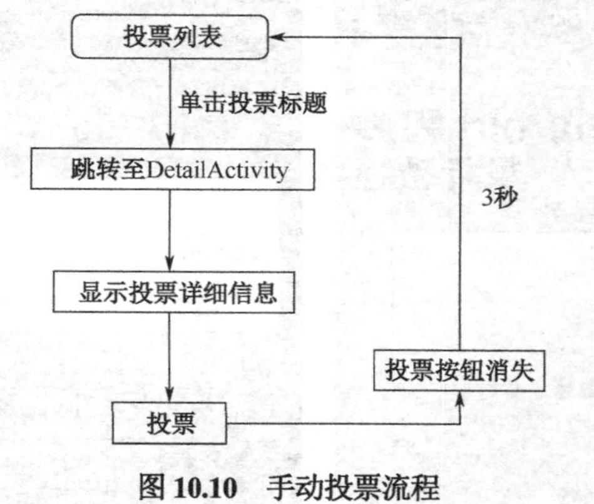

图 10.10 手动投票流程

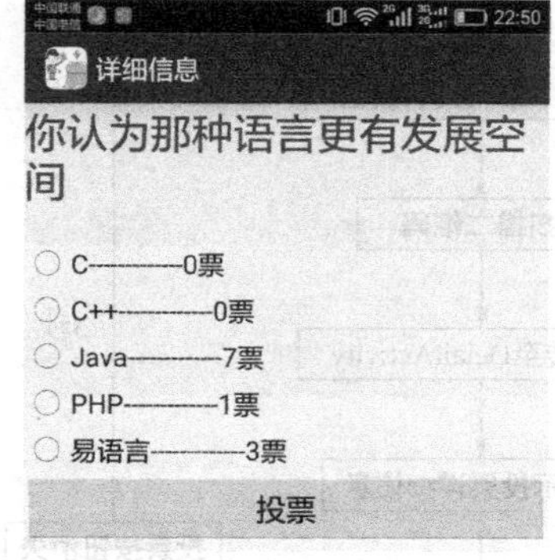

图 10.11 手动投票运行界面

用户通过选择列表中的投票进入 DetailActivity 对应的投票界面，ListActivity 会将投票的数据传递到 DetailActivity 中。选择投票的选项，单击投票按钮之后完成投票。后台会创建一个通信线程，将投票数据发送给服务器，后台再修改数据库中的票数。完成投票后投票按钮会消失，防止用户多次单击，

如图 10.10 所示，3 秒过后系统会自动跳转到投票列表界面。

DetailActivity 的关键代码如下：

```
button.setOnClickListener(new OnClickListener() {
  public void onClick(View v) {
  if (!("".equals(changedOptionId))) {  // 判断是否为空
    new Thread(new Runnable() {// 启动通信线程
        public void run() {
            try {
                DetailActivity.this.query(title_id, content);
                } catch (UnsupportedEncodingException e) {
                    e.printStackTrace();
                }}  }).start();
button.setVisibility(Button.INVISIBLE);
Toast.makeText(DetailActivity.this, "投票成功", 0).show();
} else {
    Toast.makeText(DetailActivity.this, "请选择投票项！", 0).show();  }
    // 定时跳转到 List
    new Handler().postDelayed(new Runnable() {
        public void run() {
            DetailActivity.this.finish();}
        }, 3000);        }    });
```

6．Android 客户端二维码投票的实现

用户单击投票列表上方的二维码后，进入二维码扫描界面，将摄像头对准二维码，高亮区域是二维码扫描区，将二维码置于高亮区域。扫描成功后，会跳转至相应的 DetailActivity 投票界面。其执行流程如图 10.12 所示，运行界面如图 10.13 所示。

二维码扫描时会启动相机扫描并对内容进行解析，将包含在二维码中的投票 ID 和标题内容解析出来，根据 ID 获得该投票的选项，在 DetailActivity 中显示出来，后续操作与手动投票一致。最后，要在 AndroidManifest.xml 文件中配置使用摄像头的权限，二维码扫描部分需要导入 zxing.jar 包。

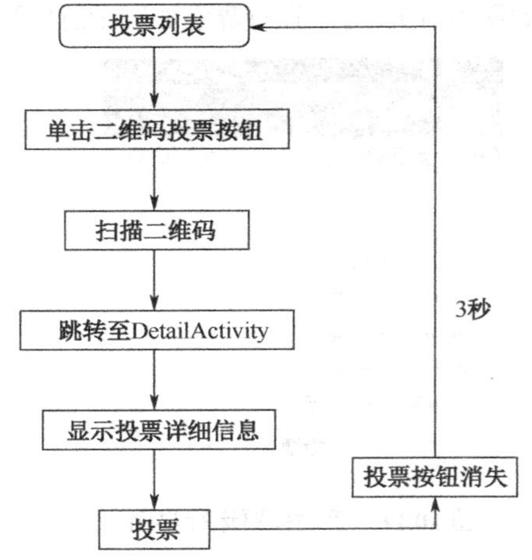

图 10.12　扫描二维码投票流程

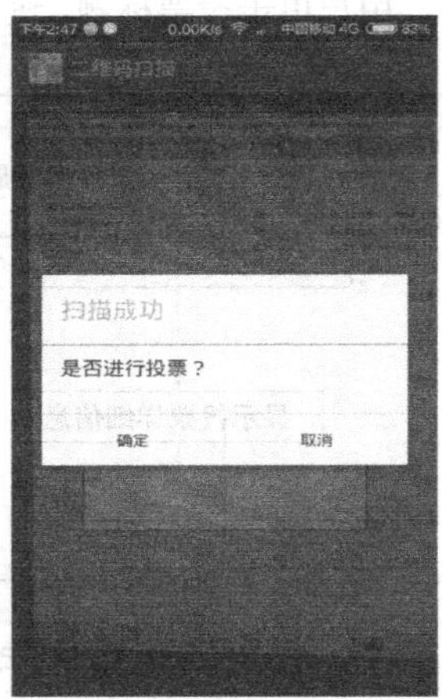

图 10.13　扫描二维码投票运行界面

CaptureActivity 的关键代码如下：

```
public void handleDecode(final Result obj, Bitmap barcode) {
```

```
    inactivityTimer.onActivity();
    playBeepSoundAndVibrate();
    AlertDialog.Builder dialog = new AlertDialog.Builder(this);
    if (barcode == null) {      }
    else {
       Drawable drawable = new BitmapDrawable(barcode);   }
    dialog.setTitle("扫描成功");
    try {
    String info = new String(obj.getText().getBytes(), "UTF-8");//扫描得到字符串
    dialog.setMessage("是否进行投票? ");  //弹出对话框
    id = info.split(",")[0];
    title = info.split(",")[1];
    } catch (UnsupportedEncodingException e) {
        e.printStackTrace(); }
dialog.setNegativeButton("确定", new DialogInterface.OnClickListener() {
@Override
public void onClick(DialogInterface dialog, int which) {
// 新建查询线程，获得该标题有哪些选项
new Thread(new Runnable() {
    @Override
    public void run() {
        result = CaptureActivity.this.query(id); }}).start();
    try {// 等待数据传回
    Thread.sleep(1000);
} catch (InterruptedException e) {
    e.printStackTrace(); }
// 启动 DetailActivity
Intent intent = new Intent(CaptureActivity.this, DetailActivity.class);
intent.putExtra("options", result);
intent.putExtra("title", title);
intent.putExtra("title_id", id);
CaptureActivity.this.startActivity(intent);
finish();        }    });
dialog.setPositiveButton("取消", new DialogInterface.OnClickListener() {
@Override
public void onClick(DialogInterface dialog, int which) {
    finish();        }    });
dialog.create().show();}
```

10.3 测试

测试的目的主要是测试界面是否与设计效果吻合、运行效果是否良好、功能是否完善、性能是否稳定。测试内容主要包括功能测试、效果测试、兼容性测试和安全性测试，并为开发人员提供优化方案。

1．测试环境

测试环境如表10-5所示。

表 10-5　测试环境

软件环境（相关软件、操作系统等）	硬件环境（网络、设备等）
操作平台：Windows 7 32 位	手机：红米 1s
JDK：1.7	内存：4GB
Android SDK：4.4	系统：MIUI6（基于 Android 4.4）
软件：MyEclipse + Tomcat 8.5 + MySQL 5.5.36	驱动：小米手机驱动

2. 项目部署

1）服务器端部署

（1）启动 MySQL 服务。

（2）将包含项目“VoteServer”的 Tomcat Web 服务器启动；在浏览器地址栏中输入 http://localhost:8080/ VoteServer/，测试的用户名是“admin”，密码是“123456”。

2）Android 客户端部署

（1）将服务器与测试 Android 设备连接在同一局域网内（仅测试），注意服务器 IP 地址的设置。

（2）将 Android 应用安装到 Android 设备上，启动之后，测试的用户名是“abc”、密码是“123”。

服务器端的功能测试略。

3. Android 客户端功能测试

Android 客户端功能测试结果如表 10-6 所示。

表 10-6　Android 客户端功能测试结果

功能名称	测试操作及记录	测试结果	备注
登录	启动 Android 应用，进入登录界面，用户名输入“abc”，密码输入“123”，正常跳转到投票列表界面。输入其他数据时会显示“登录失败，请检查网络或者用户名密码是否正确！”	正常	可添加验证码登录，防止暴力破解
列表显示	登录成功后，将数据库中的投票列表显示出来，在后台删除“你认为哪种语言更有发展空间”之后，刷新 Android 客户端，发现该投票已经消失	正常	无
手动投票	在投票列表中单击“优秀寝室投票”，进入投票界面，对选项“131 寝室”进行投票，3 秒后自动跳转到投票列表，再进入“131 寝室”，发现票数为 4 票（之前是 3 票）	正常	无
二维码投票	单击“二维码投票”按钮，正常进入二维码扫描界面，对网页端生成的“优秀寝室投票”进行扫描，扫描成功后，自动跳转至投票界面	正常	无

4. Android 客户端其他性能测试

Android 客户端兼容性测试通过云测试平台 Testin（Testin 是全球最大的移动游戏、应用真机和用户云测试平台）进行，通过对市场占有率较高的 105 款 Android 设备进行测试，结果如图 10.14 所示。

测试基本信息				
测试程序	投票器	测试通过率	99.05%	
测试平台	Android			
测试机数量	115			
未执行机数量	10			
测试结果	测试终端数	测试结果百分比	可能损失用户数/万	用户损失百分比
安装失败	0	0	0	0
启动失败	0	0	0	0
运行失败	1	0.87%	22	0.03%
测试结果	测试终端数	测试结果百分比	覆盖智能活跃用户数/万	用户覆盖率百分比
待改进	0	0	0	0
通过	104	90.43%	7991	11.70%
未执行	10	8.70%	1000	1.46%

图 10.14　Android 客户端兼容性测试结果

通过对以上数据进行分析可知，兼容机型达到 99%，说明兼容性很好。

Android 客户端的安全性测试在腾讯安全实验室进行，测试结果为安全，测试结果如图 10.15 所示。

图 10.15　Android 客户端安全测试结果

测试结论：从程序功能、运行兼容性、安全性 3 个方面进行测试，测试结果反映出此应用运行良好、兼容性好，有较好的安全性。

需要注意的是不同版本的 Android 系统运行结果可能会有差异，本系统是在 Android4.4.2 运行正常，在 Android6.0 版本下运行有些异常，要对程序稍作修改。

参 考 文 献

[1] 方欣，赵红岩．Android 程序设计教程[M]．北京：电子工业出版社，2014．

[2] 李刚．疯狂 Android 讲义[M]．2 版．北京：电子工业出版社，2013．

[3] 明日科技．Android 从入门到精通[M]．北京：清华大学出版社，2012．

[4] 陈承欢，赵志茹．Android 移动应用开发任务驱动教程（Android Studio+Genymotion）[M]．北京：电子工业出版社，2016．

[5] 李宁宁．基于 Android Studio 的应用程序开发教程[M]．北京：电子工业出版社，2016．

[6] 罗文龙．Android 应用程序开发教程（Android Studio 版）[M]．北京：电子工业出版社，2016．

举报电话：（010）88254396；（010）88258888
传　　真：（010）88254397
E-mail:　dbqq@phei.com.cn
通信地址：北京市万寿路173信箱
　　　　　电子工业出版社总编办公室
邮　　编：100036